PHYSICO-THEOLOGY

MEDICINE, SCIENCE, AND RELIGION
IN HISTORICAL CONTEXT
Ronald L. Numbers, Consulting Editor

PHYSICO-THEOLOGY

Religion and Science in Europe, 1650–1750

Edited by
ANN BLAIR AND KASPAR VON GREYERZ

Johns Hopkins University Press
Baltimore

Johns Hopkins University Press
2715 North Charles Street
Baltimore, Maryland 21218-4363
www.press.jhu.edu

Library of Congress Cataloging-in-Publication Data

Names: Blair, Ann, 1961– editor. | von Greyerz, Kaspar, editor.
Title: Physico-theology : religion and science in Europe, 1650–1750 /
    edited by Ann Blair and Kaspar von Greyerz.
Description: Baltimore : Johns Hopkins University Press, 2020. |
    Series: Medicine, science, and religion in historical context |
    Includes bibliographical references and index.
Identifiers: LCCN 2019047525 | ISBN 9781421438467 (hardcover : acid-free
    paper) | ISBN 9781421438474 (ebook)
Subjects: LCSH: Religion and science—Europe—History—17th century. |
    Religion and science—Europe—History—18th century.
Classification: LCC BL240.3 .P4975 2020 | DDC 261.5/509409032—dc23
LC record available at https://lccn.loc.gov/2019047525

A catalog record for this book is available from the British Library.

*Special discounts are available for bulk purchases of this book. For more
information, please contact Special Sales at specialsales@press.jhu.edu.*

Johns Hopkins University Press uses environmentally friendly book
materials, including recycled text paper that is composed of at least
30 percent post-consumer waste, whenever possible.

CONTENTS

Contributors vii   *Acknowledgments ix*

Introduction 1
Ann Blair and Kaspar von Greyerz

PART I. TERMS AND PURVIEW OF PHYSICO-THEOLOGY

1. Was Physico-theology Bad Theology and Bad Science? 23
John Hedley Brooke

2. What's in a Name? "Physico-theology" in Seventeenth-Century England 39
Peter Harrison

3. The Form of a Flower 52
Jonathan Sheehan

PART II. NATIONAL TRADITIONS

4. What Was Physico-theology For? 67
Scott Mandelbrote

5. Physico-theology in the Seventeenth-Century Dutch Republic: The Case of Willem Goeree (1635–1711) 78
Eric Jorink

6. Back to the Roots? J. A. Fabricius's "Register of Ancient and Modern Writers" of 1728 90
Kaspar von Greyerz

PART III. STYLES OF RELIGIOSITY

7. Miracles, Secrets, and Wonders: Jakob Horst and Christian Natural Philosophy in German Protestantism before 1650 103
Kathleen Crowther

8. "Rather Theological than Philosophical": John Ray's Seminal *Wisdom of God Manifested in the Works of Creation* 115
Katherine Calloway

9. Matters of Belief and Belief That Matters: German Physico-theology, Protestantism, and the Materialized Word of God in Nature 127
Anne-Charlott Trepp

10. Pascal's Rejection of Natural Theology: The Case of the Port-Royal Edition of the *Pensées* 141
Martine Pécharman

PART IV. ENGAGEMENT WITH THE NEW SCIENCE

11. Physico-theology or Biblical Physics? The Biblical Focus of the Early Physico-theologians 157
Rienk Vermij

12. *Maxima in minimis animalibus*: Insects in Natural Theology and Physico-theology 171
Brian W. Ogilvie

13. What Abbé Pluche Owed to Early Modern Physico-theologians 183
Nicolas Brucker

14. Antonio Vallisneri between Faith and Flood 194
Brendan Dooley

PART V. AESTHETIC SENSIBILITIES

15. A Language for the Eye: Evidence within the Text and Evidence as Text in German Physico-theological Literature 209
Barbara Hunfeld

16. A Hybrid Physico-theology: The Case of the Swiss Confederation 222
Simona Boscani Leoni

Bibliography 235   *Index 267*

Editors

Ann Blair is Carl H. Pforzheimer University Professor, Department of History, Harvard University.

Kaspar von Greyerz is Professor Emeritus of History, University of Basel.

Authors

Simona Boscani Leoni is Assistant Professor of History, University of Bern.

John Hedley Brooke is Professor Emeritus of Science and Religion, University of Oxford.

Nicolas Brucker is Professor of French Language and Literature, Université de Lorraine, Centre Écritures (EA 3943).

Katherine Calloway is Assistant Professor of English, Baylor University.

Kathleen Crowther is Associate Professor of the History of Science, University of Oklahoma.

Brendan Dooley is Professor of Renaissance Studies, College of Arts, University College Cork.

Peter Harrison is Professor of History and Philosophy and Australian Laureate Fellow at the University of Queensland.

Barbara Hunfeld is Researcher in Modern German Literature, at the Bavarian Academy of Sciences, University of Würzburg.

Eric Jorink is Teylers Professor of Enlightenment and Religion in Historical Perspective, University of Leiden.

Scott Mandelbrote is Fellow and Director of Studies in History at Peterhouse College, University of Cambridge.

Brian W. Ogilvie is Professor of History, University of Massachusetts Amherst.

Martine Pécharman is Director of Research in Philosophy at the French National Centre for Scientific Research (CNRS).

Jonathan Sheehan is Professor of History and Director of the Berkeley Center for the Study of Religion, University of California Berkeley.

Anne-Charlott Trepp is Professor of History at the University of Kassel.

Rienk Vermij is Professor of the History of Science, University of Oklahoma.

ACKNOWLEDGMENTS

We, the editors, wish to thank all the contributors to this volume for their inspiring research on physico-theology and particularly John H. Brooke and Rienk Vermij for helpful comments on our introduction. Many of the contributions assembled here were originally presented at an international conference held at the Herzog August Bibliothek (HAB) in Wolfenbüttel, Germany's outstanding research library, on June 14–16, 2017. We wish to address warm thanks to the director of the Wolfenbüttel library, Professor Peter Burschel, and to the head of the HAB section responsible for academic events, Dr. Volker Bauer, for hosting the meeting and for their active encouragement of the entire project, especially in securing financial support for our undertaking over and above the HAB's own unsparing assistance. Likewise, we wish to thank the Fritz Thyssen Stiftung for its generous sponsorship of the conference.

Since June 2017, the contributions to this volume have undergone substantial revisions; we are grateful to all the contributors for sharing our interest in creating a coherent and innovative publication. Warm thanks to Ronald L. Numbers and our editors Matthew McAdam and William Krause of Johns Hopkins University Press for their crucial support and guidance and to the anonymous reviewer for many helpful and substantial suggestions. Janis Bolster created the bibliography with careful attention to the endless variations of spelling and capitalization in the primary sources. Brian MacDonald copyedited the manuscript with great acuity, and Lin Maria Riotto composed a comprehensive index.

PHYSICO-THEOLOGY

Introduction

ANN BLAIR AND KASPAR VON GREYERZ

"We doubt that there is any occupation for a free man more worthy and delightful than to contemplate the exquisite works of nature and honor the infinite wisdom and goodness of the divine artificer," exclaimed the botanist John Ray (1627–1705) in the Latin preface of his *Catalogus plantarum circa Cantabrigiam nascentium* (Catalog of plants existing around Cambridge).[1] In terms that became the key words of physico-theology in the ensuing decades, Ray invited educated contemporaries in his first book-length publication to join him in an examination of nature conjoined with the praise of the Creator's wisdom and benevolence. John Ray was a physico-theologian of the first hour. Although he did not use the term as early as 1660, his *Three physico-theological Discourses* of 1693 played a crucial role in associating this term "physico-theologian" with a distinctive intellectual project that flourished across national and religious barriers starting in the late seventeenth century.

To this day, much anglophone research uses the terms "physico-theology" and "natural theology" synonymously.[2] A clear categorical differentiation is also missing in older German research.[3] Natural theology designates knowledge about God attained by human reason alone without the aid of biblical revelation. Although the term was coined in English only in the seventeenth century, natural theological arguments appealed to defenders of Christianity over many centuries because these arguments were expected to convince all skeptics, even those who rejected the authority of Christian texts and institutions. Despite its long history and many forms, natural theology is frequently equated with just one kind of argument that held special sway in the eighteenth and nineteenth centuries: the argument from design, in which the existence of the "divine artificer"—to use John Ray's words—is deduced from the apparent design visible in the interactions of natural objects and creatures. But the reduction of natural theology to the argument from design ignores other ways in which natural theologians, and among them physico-theologians, sought to persuade through reason. For example, a study of the wonders of nature that defied human understanding could also form a reasoned argument for the existence and greatness of God, as deployed by Ralph Cudworth and Henry More, among others.[4] In other cases early modern natural theologians were concerned less with rational arguments and sought instead "merely to offer exemplifications of Divine design in illustration or confirmation of revealed theology."[5] John Ray's *Three Physico-theological Discourses* is a case in point because it demonstrates, along with several other contemporary treatises, that another aim of physico-theologians was to establish the compatibility of the new science with the biblical narrative.[6] In fact, demonstrating this compatibility was the main component of the second largest physico-theological genre (after the argument from design) that dealt with the shape of the earth following the biblical Flood and especially with fossils and their relation to this cataclysm.

What, then, is unsatisfactory in treating natural theology and physico-theology as synonyms? First, because most current definitions restrict natural theology to the argument from design, equating natural theology with physico-theology has the effect of also limiting the latter to the argument from design, whereas both encompassed a broader array of arguments. Second, treating the two terms as synonymous eliminates the chronological specificity of physico-theology. While natural theology forms a tradition reaching back to antiquity, we use physico-theology to designate new developments after 1650. Physico-theology featured specific contents and linguistic codes, embraced the values of personal, empirical observation and of the "new science," and soon acquired transnational and transcultural features.[7] As a distinctive intellectual movement it flourished between the late seventeenth century and mid-eighteenth century across much of Europe as a way of reconciling Christianity (across many denominations) with the numerous scientific positions that began to prevail after 1650, including Copernicanism, mechanical philosophy, and new observational instruments and techniques.

This volume offers the first occasion for experts across multiple national and disciplinary perspectives to address together the complex interactions of science and religion through physico-theology in England and the Continent. We aim to highlight the transnational aspects of the movement, including the role of cities that served as hubs at various times, such as London, Amsterdam, Zurich, and Hamburg, while also noting the differences between national contexts. Physico-theology played a crucial role in diffusing new scientific ideas and assumptions and interest in the study of nature to a broad eighteenth-century public. All too often today religion and science are seen as antithetical. This view has been cultivated since the late nineteenth century but was quite alien to the period in which the "new science" became widely admired and diffused, transforming in the process the parameters of scientific thought from an earlier emphasis on interpreting ancient authorities and developing many of the principles that are still central to scientific thinking today. Contemporary with and crucial to the successful spread of the new science was physico-theology, which aimed at establishing the compatibility of the new science with the biblical tradition.

Historiographical Background

The field of science and religion is vast and constantly growing. The late nineteenth-century claims by John William Draper and Andrew Dickson White that science and religion are in conflict have prompted many refutations, although as one recent volume notes the conflict thesis has proved to be an "Idea That Wouldn't Die."[8] We now have overviews as well as case studies that offer instead a nuanced picture of a wide range of historical interactions between science and religion, including examples of conflict rooted in specific historical circumstances, of mutual indifference, and of mutual support.[9] Within this tripartite scheme, English natural theology has often been interpreted as an example of mutually beneficial cooperation between science and religion, at least in the pre-Darwinian periods. On the other hand, the predominance of natural theology among Anglican preachers in the first half of the nineteenth century may have unexpectedly contributed to the crisis in the English religious landscape triggered by Darwin's theory of evolution by natural selection. To the extent that the special creation of species adapted to their environment had been presented as chief grounds for religious belief, Darwin's explanation of adaptation without special providence was a threat. Nevertheless, the argument from design could also be adapted to this new scientific development, with the argument that God provided laws of biological evolution in order to create the admirable adaptations that natural theology had taught the faithful to admire. Indeed, William Paley's *Natural Theology* continued to be published, with adjustments, after the publication of Darwin's *Origin of Species*, and new variations on the argument from design continue to be deployed today. Thus, whether natural theology was a help or a hindrance to religion can probably be answered only within specific historical contexts.

Similarly, there have been multiple interpretations of the impact of physico-theology within this larger framework. Did this strand of natural theology have a secularizing impact on its readers or on the contrary act as an apologetics for religion? The interpretations have been segmented by language and national context within and beyond the English case. In the Netherlands a physico-theology based on Newtonianism appealed to clerics who feared the materialism latent in Cartesianism; the combination of physico-theology with religious edification ensured its popularity with Dutch readers even beyond the mid-eighteenth century.[10] In France, though physico-theology had fewer exponents, one was the widely published abbé Noël-Antoine Pluche (1688–1761); although he has been portrayed as a Christian apologist, this interpretation neglects the richness of Pluche's natural history and of his utilitarianism, which harnessed scientific knowledge for the solution of everyday problems.[11] The German historiography on the topic includes a succession of studies by theologians—Otto Zöckler (1833–1906), Wolfgang Philipp (1915–69), and Manfred Büttner (1923–2016)—who lumped natural and physico-theology together as agents of secularization.[12] By contrast a literary approach, notably to the poetry of Barthold Heinrich Brockes, has expanded the range of sources and emphasized the role of art as expressions of physico-theological sensibilities.[13]

Our goal in bringing together leading scholars in this field who span multiple national foci (including the Italian) is to facilitate a supranational perspective on the wide spread of physico-theology across many areas of Europe. We seek not to promote any grand narratives but to argue more modestly that physico-theology represents a symbiosis of science and religion characteristic of early modernity, before the invention of "science" in its modern sense.[14]

The Term and the Defining Features of Physico-theology

Describing the main features of physico-theology is not easy, particularly since, as we have seen, it has often been subsumed under the broader category of natural theology. The term itself was in use from the seventeenth century onward but is not a reliable indicator of what historians today regard as physico-theology. Rather than focusing on the term, we identify the main characteristics of the movement from a few widely circulated works that formed its backbone, and we distill from them some specific conventions of the field, which we categorize in five points later in this section.

Our research suggests that the term physico-theology did not exist in the sixteenth century. Peter Harrison reaches the same conclusion about British titles in his contribution to this volume. In the German lands a collective catalog covering 260 libraries dates to 1610 the first appearance of physico-theology in a title.[15] But this *Disputatio . . . Physico-Theologica de pane & vino* (A physico-theological disputation about bread and wine), focused on the Last Supper, had little to do with the kind of physico-theology we are addressing in this volume.[16] Similarly the *Meditationes Physico-Theologicae* published by Theodor Müller in 1642 is a book of spiritual edification and prayer, also unrelated to our object of study. The term also appears in a "Physico-theological disputation" held in Bern in 1660 on the *hexaemeron*, the six days of Creation described in Genesis.[17] But this work offers only a commentary on Genesis 1 and 2 with no reference to empirical perspectives of the kind that are central to a majority of physico-theological works. To be sure, hexaemeral language played an important role in the physico-theological genre focused on fossils and the history of the earth.[18] There were also connections between accounts of the six days of Creation and natural theology—for example, in early modern nature-poetry, from Salluste du Bartas's *La Sepmaine* (1578) to Richard Blackmore's *Creation* (1712).[19] On the whole, however, the *hexaemeron* was an independent, chiefly homiletic genre in use from Basil of Caesarea (ca. 330–79) to Josef Ratzinger (born 1927, later to become Pope Benedict XVI) and beyond.

We identify *The Wisdom of God Manifested in the Works of Creation* (London, 1691) by John Ray (1627–1705) as the earliest, very widely disseminated treatise of physico-theology. This work was based on a series of sermons of a much earlier date, for they were preached at Trinity College in Cambridge before 1662 when Ray resigned his fellowship on account of his refusal to sign the Act of Uniformity.[20] The first English treatise using the term "physico-theology" in its title was *The Darknes of Atheism Dispelled by the Light of Nature. A Physico-Theologicall-Treatise* (London, 1652) by Walter Charleton (1619–1707). However, there is no agreement in current scholarship whether this work can be considered a truly physico-theological treatise.[21] As Ray's *Wisdom of God* was a vernacular treatise, it reached a broad non-Latinate public. Between 1691 and 1714 it appeared in six English editions, and this was followed by translations into French (Utrecht, 1714), German (Goslar, 1717), and Dutch (Amsterdam, 1732).[22] *Physico-Theology* (London, 1713) by Ray's executor William Derham (1657–1735) was another best seller that spread the term itself through translations into Italian (Florence, 1719), French (Rotterdam, 1726), Dutch (Leiden, 1728), German (Hamburg, 1730), and Swedish (Stockholm, 1736). One likely driver of its popularity was its ambition to offer a physico-theological appreciation of *all* natural philosophy, from the place of the earth in the universe to an examination of human physiology.

At the same time, the Dutch physician and mathematician Bernard Nieu-wentijt (1654–1718) published a similarly wide-ranging treatise on "the right usage of observations of nature" (*Het regt gebruik der werelt beschouwingen*) in 1715. This weighty tome was primarily directed against the followers of Baruch Spinoza (1632–77) and radical Cartesianism more generally. The phi-losophy of René Descartes, by claiming to explain all natural phenomena as matter in motion, seemed to leave no place for God's active presence in the natural world—hence the alarm of physico-theologians who sought, and found, physical manifestations of God's power and wisdom in the fabric of nature. Nieuwentijt's work appeared in multiple editions in English and in French and German translations within the space of just fourteen years.[23] He lived in the small town of Purmerend, where he practiced as a medical doc-tor and was a regent of the town.[24] However, small-town life did not prevent him from participating in a larger Dutch network. In Amsterdam he contrib-uted his physico-theological perspective to the discussions of early followers of Newton centered around the Amsterdam Mennonite merchant Adriaan Verwer (1655–1717).[25]

New forms of communication, including the periodicals and learned societies formed in the second half of the seventeenth century, helped to spread physico-theology. Following a significant historiographical consensus, we focus our study on the mid-seventeenth century to about 1750, when physico-theology was most lively and homogeneous on a European scale. The tradition proved especially resilient in Britain, lasting well into the nineteenth century—witness William Paley's *Natural Theology*,[26] first published in 1802, which reached the twelfth edition within seven years, or the success of the *Bridgewater Treatises*, published between 1833 and 1840.[27] Although Paley's work was translated into German in 1823, physico-theology did not enjoy the same kind of continuous appeal outside Britain. However, during the second half of the eighteenth century and well beyond that time, physico-theological treatises in Dutch and German frequently, unlike during the preceding de-cades, focused on religious edification rather than new scientific questions.[28] Compared to the Enlightenment in France, the German *Aufklärung* and its British sister were much less anticlerical, and few of its exponents questioned the Christian tradition—hence the broad appeal of physico-theological argu-ments.[29] Nonetheless, the difference in the trajectories of physico-theology in the British and German contexts is striking. Historians frequently attribute the relative decline of interest in natural theological arguments among the German readership of the last decades of the eighteenth century to the influence of Immanuel Kant's philosophy. But then why did David Hume's criticism of natural theology not have a similar effect in Britain? These puzzling differences in the longer-term reception certainly invite further study.

Because neither an exact classification nor an analytically precise definition of physico-theology seems attainable, we offer instead five criteria that we consider characteristic of the core works of physico-theology:

- The physico-theological perspective assumes that God is a rational being. This notion accords well with the principle that nature does nothing in vain; although that maxim was Aristotelian in origin, English mechanical philosophers including Robert Hooke (1635–1703) and Robert Boyle (1627–91) also embraced it. Following their lead, physico-theologians reaffirmed as a basic axiom of natural philosophy that God has a plan for everything. The God of physico-theology was not a vengeful God, an understanding widespread during the first half of the seventeenth century, but was recognizable by his goodness and benevolence.
- Physico-theology included a utilitarian approach to nature. Not only did nature always follow a divinely ordained purpose, but that purpose was to ensure the well-being of humankind and the subservience of nature to that end. For this reason, physico-theologians felt they could argue from final causes even when the efficient cause was unknown or unknowable.
- Physico-theological works adopted a specific vocabulary, which functioned as a code of recognition among insiders, highlighting, for example, the wisdom, omnipotence, and goodness of God.[30]
- Physico-theological texts often contained polemics against the threat of atheism and deism. Whether this threat was real or only perceived depended largely on the specific cultural context as well as on the time of publication. Early texts addressed older polemics, such as those voiced, for example, by Johann Heinrich Alsted (1588–1638) or Marin Mersenne (1588–1648), and argued from a position of strength against abstract opponents.[31] As deism and atheism gained actual adherents, this polemic could assume the form of Christian apologetics against a threat seen as imminent. This was the tone of some of the British Boyle Lectures (especially *The Folly and Unreasonableness of Atheism* [1692] by Richard

Bentley [1662–1742]) and, even more so, of the Dutchman Bernard Nieu-wentijt's physico-theological classic of 1715, *Het regt gebruik der werelt beschouwingen*. By contrast, in Lutheran Germany Radical Pietism, rather than atheism, was perceived as the most urgent threat.[32]

- Finally, most physico-theological authors (even a majority of them) shared an evident interest in basing their texts on personal witnessing and experience. However, it would be wrong to overlook another group of authors, albeit much smaller in number, who followed a tendency initiated by William Derham's *Physico-Theology* (1713) to celebrate the wonderful order of Creation. In their cosmological considerations, such works were naturally unable to rely on empiricism. Linnaeus's *Oeconomia naturae* of 1749 and Johann Gottlieb Walpurger's *Cosmotheologische Betrachtungen* of 1748–54 are cases in point.[33]

This list of criteria does not exhaust the potential categorizations within physico-theology. In particular, the movement comprised several different genres. In quantitative terms, texts based on the argument from design clearly dominated. Most of them concentrated on one creature or object, hence the proliferation in the German lands of Testacea- (seashells), Melitto- (bees), Akrido- (locusts), Hydro- (water), and Pyro- (fire) theologies. Other treatises tried to cover the whole gamut of contemporary scientific knowledge. Central examples of this encyclopedic genre include Bernard Nieuwentijt's *Het regt gebruik der werelt beschouwingen* (1715); William Derham's *Physico-Theology*, first published in 1713; and Johann Jakob Scheuchzer's *Physica sacra* of 1731–35. The readership of these works has yet to be studied carefully, but we agree with John Brooke (chapter 1) that, even if these texts often did not operate at the forefront of contemporary research, they laid the ground for a social and cultural broadening of the audience for science.

Works on fossils, in quantitative terms the second physico-theological genre next to the argument from design, pose a special challenge to the project of identifying physico-theology. Many dozens of authors composed hundreds of treatises or contributions to learned journals concerning fossils during the first half of the eighteenth century, but only a few can be considered proponents of physico-theology. John Woodward (1665–1728) and Johann Jakob Scheuchzer (1672–1733) both located the origin of fossils in the biblical Flood following a diluvialist explanation, but their work gained only a few followers among authors on paleontology. Many German authors, for example, accepted diluvialism but maintained at the same time that fossils that could not be explained as resulting from a flood should be considered jokes of nature (*lusus naturae*).[34] However, the very idea of a "joke of nature" was at odds with the physico-theological focus on God's earnest purposefulness in nature.

Only a handful of physico-theological texts dealt with the possibility of the resurrection and parthenogenesis (virgin birth), yet these form an understudied part of the movement and an especially interesting one in that they spanned the Protestant-Catholic divide.[35] We have not addressed this understudied topic in this volume but offer it as an area that would warrant further investigation. The curé Jean Pierquin (1672–1742), who published a short treatise on the virgin birth of Christ in the womb of Mary in 1742, was a right-thinking Catholic.[36] While he did not question the miracle of the conception of Jesus Christ in Mary's body wrought by the Holy Ghost, he explained everything that followed from there in terms of *ovism*. Contemporary ovism highlighted the central role of the female egg in reproduction and was opposed to traditional Aristotelian notions of procreation. Likewise, when Robert Boyle approached the theme of the resurrection in an equally short treatise published in 1675, he admitted that the resurrection "is not to be brought to pass according to the common course of Nature, I presume; after the universal experience of so many Ages, which have afforded us no instances of it."[37] In some respects, Boyle's treatise expanded on the discussion by Thomas Aquinas (1225–74) whether a body "devoured by cannibals" could be resurrected.[38] This debate continued during the seventeenth century until, at least in the British context, John Locke argued in favor of a spiritual resurrection and thus questioned the hitherto widespread belief that humans would be resurrected "with the same bodies possessed during normal life."[39] This was a theme likewise treated by Bernard Nieuwentijt in his *Het regt gebruik der werelt beschouwingen* of 1715, in which he discussed this particular question in the context of his corpuscularian theory of matter.[40] Boyle's treatise of 1675 was more original in that he went on to explain a number of "chymical" processes that reenacted transformations analogous to the resurrection, in some cases owing to the "plastick power" inherent in matter.[41]

The commentaries on virgin birth and the resurrection introduced here, however briefly, stray from the usual physico-theological argument from design to prove the existence of God as a wise Creator and yet also display hallmarks of physico-theology in applying a rationalist explanation of a religious phenomenon informed by the latest natural philosophical concepts.[42]

Physico-theology and Religious Affiliation

Because of their emphasis on studying nature, physico-theological authors rarely discussed their personal religious positions. As a result, historians have proposed various interpretations of the religious positions characteristic of physico-theologians—from deism to various strands of Protestantism.

Deists, as the *Encyclopedia Britannica* of 1771 tells us, "are those persons in Christian countries, who, acknowledging all the obligation and duties of natural religion, disbelieve the Christian scheme, or revealed religion."[43] This article identified four variations of deism and concluded that the "only true deists" were those who "believe[d] the existence of a supreme Being, together with his providence in the government of the world, as also the obligations of natural religion; but so far only as these things are discoverable by the light of nature alone, without believing any divine revelation." In other words, for deists the divine providence subtending the "government of the world" was understood as "general" providence, operating only through the unchanging laws of nature. By contrast, the "special" providence of God, which deists rejected, comprised all the ways in which God intervened directly in nature or in the life of individual persons or groups (e.g., in response to prayer or through miracles). In opposition to deism, traditional theism affirmed that God exercises both general and special providence. But, of course, assigning even these actors' categories to specific authors is delicate, because many identified themselves publicly with more traditional positions than ones they espoused in private; and many changed positions over time as the spectrum of thought around them evolved (usually toward greater deism).

Barthold Heinrich Brockes (1680–1747), Hamburg senator and celebrated poet in his time, has for example been the object of multiple interpretations. While Hans-Georg Kemper points to Brockes's hermeticism,[44] others are certain that this poet was underneath it all a deist or even an unbeliever.[45] Most recently, Marc Chraplak has reinstated Brockes in the camp of theists.[46] But we should not overlook an unpublished poem, written during the last year of Brockes's life. Here, he unmistakably revealed his deism.[47] This does not necessarily mean that Chraplak's argument is entirely wrong, because Brockes's views may have shifted over time. William Derham is a case in point. In German works on physico-theology he is frequently labeled a deist, following an assessment of 1957 by Wolfgang Philipp, which made its way without verification into an encyclopedia entry of 1981.[48] On that basis the theologian Johann Anselm Steiger has suggested that all physico-theologians were ultimately deists in disguise.[49] We advocate for more nuance and close textual analysis in making these delicate assessments of religious positions.

Context must also play a role in interpreting religious statements. While the English in this period were haunted by "the specter of deism" (Jonathan Sheehan), and the Dutch were anxious about the spread of Spinozism (Eric Jorink and Rienk Vermij), eighteenth-century Germans were not similarly concerned with deism. Instead, Germans were trying to come to terms with the legacy of the destructive religious violence of the first half of the seventeenth century. Pietism can be considered "a product of these conflicts" (J. Sheehan), and it was above all this movement of religious and ecclesiastical renewal focused on personal religious experience that elicited fierce criticism from the mainstream Protestant churches and their orthodox representatives.[50] The connections of physico-theology to Pietism are as yet understudied. In this volume Anne-Charlott Trepp offers a welcome contribution to this discussion with special reference to the Lutheran theology of the Lord's Supper. What is more, the German reception of Newtonianism was complicated by the influence of Leibnizian thought. Leibniz (1646–1716), who styled himself as an opponent of Newton,[51] is not usually identified with physico-theology, although his most prominent student, the philosopher Christian Wolff (1679–1754), has occasionally been placed in that movement.[52]

Organization

We have assembled the sixteen contributions in this volume in five thematic parts that define the terms and purview of physico-theology, consider various national contexts, and then assess its confessional range, engagement with scientific developments, and its aesthetic sensibilities.

In part I, "Terms and Purview of Physico-theology," John Hedley Brooke places physico-theology in a broad chronological perspective. He shows how, despite weaknesses in their arguments, physico-theologians helped spread acceptance of the various new sciences. At the same time, in downplaying revelation, physico-theological authors unwittingly facilitated the contention of their adversaries that the experience of God in nature was possible without paying heed to revelation. As a result, it was not particularly difficult for Voltaire and Laplace to replace, in the course of the eighteenth century, Newtonian theism with a new, deist Newtonianism. Conversely, it was comparatively easy for Hume and Kant to question the physico-theological argument that claimed to prove God's transcendental existence from observations of his immanence in the created world. While physico-theology failed as a philosophy, it thrived as a popular epistemic and moral model for understanding the relationship between religion and science. This observation applies especially well to Britain, where physico-theology lasted to the mid-nineteenth century and beyond.

Peter Harrison detects an important incentive for the rise of physico-theology in the "hybridization" of scientific disciplines from the early seventeenth century onward. He first discusses different currents within physico-theology in order to counter the widespread misunderstanding, which dates back to Kant's critique, that physico-theology relied exclusively on the argument from design in an attempt to prove God's existence from final causes. He concedes that the argument from design was the prevalent mode during the eighteenth century but shows that we can find composite terms comparable to the meaning of the hyphenation in "physico-theology" earlier on. The common motive behind these *composita* was the wish to transcend the traditional disciplinary boundaries of mathematics and enter territories traditionally reserved to natural philosophy, a practice already under way in the work of figures like Kepler and Galileo. Harrison argues that physico-theology could have pursued similar objectives before the approach had coalesced around the observational sciences. He examines this hypothesis in looking at English authors from Samuel Parker (1640–88) to William Whiston (1667–1752) who used "-theology" hyphenation in a variety of meanings. Harrison establishes that the publication of William Derham's influential *Physico-Theology* in 1713 prompted for the first time a significant consensus on the meaning of physico-theology.

When the young Immanuel Kant considered the physico-theological proof of the existence of God in 1762, he viewed that proof as a *formal* argument and thus revived the Aristotelian category of form. Jonathan Sheehan examines this revival and shows how it grew from terrain prepared by physico-theology. At root physico-theology was a search for explanations about how nature and the universe are ordered and how the individual relates to that overarching order. Indeed, in 1728 the philosopher Christian Wolff offered this interpretation of physico-theology, which inspired an expansion by Kant in 1762. But Kant later rejected the philosophical validity of the physico-theological proof of the existence of God in arguing in the *Critique of Pure Reason* (1781) that it was not possible to resolve questions related to transcendence on the basis of merely immanent observations. This is why, as Sheehan argues, the flower demonstrated to the older Königsberg philosopher the contingent autonomy of the natural object, freed from its connection to final causes.

Part II, "National Traditions," considers examples from the English, Dutch, and German cases to highlight how physico-theology fit in each case into contemporary debates specific to these distinct geographic and linguistic contexts.

In England Scott Mandelbrote takes seriously Walter Charleton's physico-theological orientation. Whereas Harrison and Calloway (in parts I and III respectively) do not consider him a physico-theologian, Mandelbrote observes that Charleton emphasized the providential blessings inherent in the physiology of animals or the human body and drew for these points on the works of ancient authors. Mandelbrote also looks at other works of Charleton that lead him to reaffirm the author's commitment to physico-theology, beginning with his treatise of 1652. As Harrison does, Mandelbrote maintains that the work of William Derham had a formative influence on eighteenth-century physico-theological writing but adds that we should not underestimate a similar role played by the work of Derham's older contemporary John Ray in establishing physico-theology as a distinctive genre from the early eighteenth century onward.

Eric Jorink argues that the physico-theology that arrived in Holland from England in the period 1680–1710 treated Newtonianism and experimental natural philosophy as almost synonymous, thus supporting a point made by Rienk Vermij on earlier occasions.[53] In particular, Jorink investigates the work of the Amsterdam architect, draftsman, theologian, and printer Willem Goeree (1635–1711), whose vernacular publications are almost forgotten today. They culminated in his *Mosaize historie der Hebreeuwse kerke* (Mosaic history of the Hebrew church), which appeared in four substantial volumes in 1700, lavishly illustrated by the noted engraver Jan Luyken. Jorink's critical look at Goeree's attempt to prove the compatibility of the biblical tradition and the new science, especially concerning Noah's ark, reveals some surprising parallels with the work of Baruch Spinoza and his pantheistic notion of an absolute, infinite substance identical with nature and God.

A specific view of the forerunners of eighteenth-century physico-theology was presented by one of the German pioneers of the movement, Johann Albert Fabricius (1668–1736) in a lengthy foreword to his translation of William Derham's *Astro-Theology*, first published in Hamburg in 1728. In his study of Fabricius's "Register of Ancient and Modern Writers, Who Have Made It Their Concern to Lead Men to God by an Examination of Nature and [Its] Creatures" Kaspar von Greyerz shows that Fabricius referred to many physico-theological treatises that had appeared from the 1670s onward but also looked for the origins of physico-theology among the church fathers, to give the movement more authority. However, von Greyerz argues that the historian need not follow Fabricius in identifying such deep origins to a movement that originated rather in the mid-seventeenth century. But the "Register of Ancient and Modern Writers" does, however unwittingly, reflect on the preceding German tradition of physico-theological treatises published in Latin and its predominantly academic nature. From the 1720s, this tradition—initially encouraged by Fabricius's example–produced vernacular treatises at an astounding rate, thus reaching well beyond the boundaries of university life. But this clearly exceptional outreach to broad audiences occurred in German lands rather than in Britain and the Netherlands.

Part III, "Styles of Religiosity," focuses on the various religious persuasions represented by physico-theological authors, including a full range of Protestant affiliations—German Lutherans, Anglicans, Nonconformists, and Pietists, in particular their missionaries in South India—but also the case of Catholic Jansenism.

Kathleen Crowther examines one of the sixteenth-century sources for physico-theology.[54] *Occulta Naturae Miracula. Wunderbahrliche Geheimnisse der Natur* by the German Lutheran physician Jakob Horst (1537–1600) was first published in 1572 and reprinted many times in the course of the seventeenth century. Horst's work translated freely and expanded on the work of the Dutch Catholic physician Levinus Lemnius (1505–68). Considering its great popularity, Crowther takes Horst's expansive version of Lemnius's text as representative of the natural knowledge present in Lutheran culture of the late sixteenth and seventeenth centuries. She identifies three central aspects of that Lutheran outlook: first, the belief that natural knowledge leads to awe and amazement about the Creation; second, the assumption that it results in physically and morally correct conduct; and, third, the warning against a false use of natural knowledge. While some of the similarities to later physico-theological works are clear, Crowther notes that Horst did not offer a refutation of deism and atheism as later authors would.

Katherine Calloway offers a fine-grained textual analysis of John Ray (especially *The Wisdom of God Manifested in the Works of the Creation*, 1691), Henry More's *Antidote against Atheism* (1653), and Richard Bentley's *Folly and Unreasonableness of Atheism* (1693). She follows the path of English physico-theology from its inception during the 1650s to its first systematization in the annual Boyle Lectures endowed by Robert Boyle at his death in 1691, assessing the relationship of each author to the basic tenets of physico-theology. In her view, John Ray was the most genuine physico-theologian among them, whereas Bentley typically called his Boyle Lectures "sermons" rather than lectures. Bentley's main purpose was an apology directed against atheists and thus more limited in scope compared to the wider aims of physico-theology. In Calloway's view, Henry More, unlike his near contemporary John Ray, does not qualify as a physico-theologian owing to his pessimistic judgments regarding the contemporary world and his skeptical attitude toward empiricism.

The Protestant context of German physico-theology is treated by Anne-Charlott Trepp. According to her, German physico-theology owes specific aspects to the heritage of Martin Luther's Christology. Luther claimed that Christ could be simultaneously present bodily in the Eucharist and in heaven, as well as in all creation. Christ, for him, was ubiquitous as the all-pervasive presence of God's eternal word in nature and therefore could be understood, seen, touched, and believed in. Trepp argues that Johann Arndt (1555–1621) in his immensely popular *Vier Bücher, Vom wahrem Christhentumb* (Four Books of True Christianity, 1610) advocated a similar view and inspired the Pietists, especially those in Halle, to embrace physico-theology. The German Pietists showed more affinity toward physico-theology than conservative Lutherans, according to Trepp, and Pietist missionaries at Tranquebar in South India were still perpetuating the physico-theological understanding of nature around 1800. Thus, germanophone physico-theology continued to flourish despite the impact of Kant's critique of the argument from design.

Given the example of the abbé Pluche, one can wonder whether there were other French Jansenists inspired by physico-theology, in spite of Blaise Pascal's skepticism about natural theology. Martine Pécharman explores this question by examining the Port-Royal edition (1670) of Pascal's *Pensées*. She confirms Pascal's rejection of natural theology along with all the traditional proofs of God's existence. For him, there was no true experience of God without the mysteries of atonement, prophecy, and miracles. Nonetheless, other Jansenists were less hostile to natural theology, including Pierre Nicole (1625–95), who worried that Pascal's attitude removed an important argument in the fight against atheism. As a result, Nicole was as favorable to natural theology as the physico-theologians were. Pécharman concludes that in the late seventeenth century and at the beginning of the eighteenth, Port-Royal Jansenists shied away from Pascal's positions by omitting from publication his critique of natural theology.

In most cases, physico-theological writing consisted of a mixture of traditional apologetics (e.g., in rejecting ancient heathen or early modern deistic interpretations of Creation) and recently published scientific materials. Part IV, "Engagement with the New Science," discusses the role that physico-theology played in encouraging acceptance of the new science among a broad readership.

Rienk Vermij examines Bernard Nieuwentijt's *Regt gebruik* (1715). As we have seen, Nieuwentijt was the preeminent Dutch physico-theologian whose works sold widely in multiple translations. In comparing Nieuwentijt with John Ray and Johann Jakob Scheuchzer, Vermij argues that these three authors shared a desire to safeguard and strengthen the authority of the Bible, and thus he emphasizes the apologetic aspects of physico-theology. In the Netherlands, physico-theologians saw their main task as refuting Spinozism (which they equated with atheism), which they feared would spread among contemporaries. But after denouncing atheism, they declined to enter into the confessional disputes of their day. Instead the reasoned study of God's work in nature, informed by a knowledge of the Bible, offered an irenic alternative to dogmatic intolerance.

The motto in the title of Brian Ogilvie's article (*maxima in minimis animalibus*) stems from the title of Friedrich Christian Lesser's *Insecto-Theologia* of 1738. In the wake of the first experiments with microscopes in the 1660s, natural philosophers claimed this new instrument would permit them to get closer than ever before to the secrets of God's creation. However, Ogilvie makes clear that even before the microscope, sixteenth-century naturalists were fascinated with small creatures and with demonstrating their usefulness in nature, although he does not see a seamless continuity from the sixteenth to the eighteenth century. Instead he argues that a new perspective on the observation of small creatures developed relatively independently in Britain and, later, in Germany from the late seventeenth century onward. Both of these trends led to the foundation of modern entomology, which turned its back on natural history and that traditional form of insectology.

Nicolas Brucker's article on the Jansenist abbé Noël-Antoine Pluche shows that he owed a great deal to physico-theology and identified with it to a considerable extent. One of Pluche's favorite subjects was insectology, where he drew inspiration from René-Antoine Ferchault de Réaumur's six-volume history of insects of 1734–42. Pluche shared with Pascal the conviction that the book of nature does not yield any proof of God's existence. Despite this epistemological restraint, Pluche basically adhered to the belief of the physico-theologians, whose works he consulted in depth, that natural knowledge could lead to God morally, even though a deeper knowledge about God was not possible.

Only a handful of early eighteenth-century Italian naturalists—all brought up as Catholics—embraced basic tenets of physico-theology, mostly at the Academy of Bologna.[55] Other Italian naturalists developed different views about reading the book of nature, although they exchanged letters and specimens with physico-theologians in other European countries. Such was the case of Antonio Vallisneri (1661–1730), famous professor of medicine at the University of Padua. His work and theories are examined in this volume by Brendan Dooley. It is well known that Vallisneri's treatise *De' corpi marini que su monti si trovano* (Of the bodies of sea creatures found on mountains) of 1721 made clear to representatives of paleontological physico-theology (Louis Bourguet [1678–1742], Johann Jakob Scheuchzer, and others) that he parted ways with them. Less familiar are his publications on human and animal generation, where he positioned himself in the ongoing European debate on this question as an ovist attributing generation chiefly to the female egg, and as a preformist claiming that the entire human race at its outset was completely present in Eve's ovary. At first sight, these two main concerns of the Paduan professor look very different, but Dooley demonstrates convincingly that they had in common more than one might assume—as examples of natural phenomena that human understanding could not fully grasp.

Part V, "Aesthetic Sensibilities," concentrates on the aesthetic aspects of physico-theological reasoning, as expressed in the contemporary discourses about the various providential qualities of mountains and, especially, in poetry. As we have seen, there was plenty of nature-poetry in Italian, French, German, and English during the period considered here, but in view of our theme we must differentiate here, too, between natural theology and physico-theology. Barthold Heinrich Brockes, discussed by the literary historian Barbara Hunfeld, stands out as the archetypal physico-theological poet.

Poetry inspired by physico-theology was a widespread phenomenon in eighteenth-century Europe, extending all the way to Russia, for example.[56] Hunfeld's contribution studies the influence of English physico-theology on Germany's most prolific physico-theological lyricist. Between 1721 and 1748 (a year after Brockes's death), there appeared no fewer than eight volumes of poems composed by this indefatigable Hamburg poet and city councilor. His work enjoyed great popularity in the germanophone world of the time. Hunfeld offers a largely semiotic analysis of this vast source material. She shows how Brockes tried to fuse the actions of seeing and reading but also how this attempt was ultimately threatened with failure. To explain the poems' contemporary popularity, Hunfeld invokes instead Brockes's physico-theologically inspired, cognitive utopia, which responded to an obvious desire of readers to gain a certain knowledge of God from an inspiring perusal of the book of nature.

The contribution by Simona Boscani Leoni examines the place of mountains in the works of Thomas Burnet (ca. 1635–1715), John Ray, Johann Jakob Scheuchzer, and the less well-known Swiss physico-theologian of the later eighteenth century Elie Bertrand (1713–97). Ray, Scheuchzer, and Bertrand all reacted against Burnet's devaluation of mountains as "warts" on the surface of the earth. Typically for physico-theologians, they turned the story of the biblical Flood—for Burnet the reason for the defacement of the earth—into a story of human success, because the mountains, in their eyes, were witnesses to God's wisdom and providence, providing half of Europe with necessary water and ensuring the continuing good health of many Alpine peoples. They thus also contributed to the aesthetization of the Alps under way through the eighteenth century and celebrated the sublime in a way that diverged from the traditional path of physico-theology and signaled in the longer term the advent of romanticism.

NOTES

1. "[N]escimus an ulla ingenuo homine vel dignior vel delectabilior occupatio sit quam pulcherrima *naturae* opera contemplari, adeoque infinitam Divini opificis sapientiam simul & bonitatem celebrare." Ray, *Catalogus,* sig. **2r (Praefatio ad Lectorem).

2. For a recent example, see Peterfreund, *Turning Points in Natural Theology.*

3. Otto Zöckler, for example, makes no clear differentiation between physico-theology and natural theology in his foundational work: *Geschichte der Beziehungen,* vol. 2 (1879), e.g., part A, chap. 6 (74–92) and chap. 8 (104–22) entitled "Die physikotheologische Dichtung," which lists German, French, Italian, and English works presenting either natural-theological or physico-theological arguments. The same is true of Philipp, *Das Werden der Aufklärung* (1957), whose wide bibliographical knowledge is indebted to Zöckler.

4. Mandelbrote, "The Uses of Natural Theology," e.g., 468ff.

5. Topham, "Natural Theology and the Sciences," 60. See also Harrison, *The Territories,* 110: "As for physico-theology, it reflects the conviction that what natural philosophers study is God's activity, both his direct causal activity and his design of the creatures."

6. Ray, *Three physico-theological Discourses,* 1693. See in particular Discourse II (62–230): "Of the general Deluge in the Days of NOAH, its Causes and Effects."

7. See Harrison, "Physico-theology"; Mandelbrote, "The Uses of Natural Theology"; Krolzik, Physikotheologie"; and Sparn, "Natürliche Theologie," 89.

8. Hardin, Numbers, and Binzley, *The Warfare between Science and Religion.* See also Numbers, *Galileo Goes to Jail.*

9. Lindberg and Numbers, *God and Nature*; Brooke, *Science and Religion*; Harrison, *The Bible, Protestantism and the Rise of Natural Science* and *The Territories of Science and Religion*.

10. On the Dutch case, see Vermij, *Secularisering en natuurwetenschap*, with a short summary in English; Jorink, *Reading the Book of Nature*; H. Bots, *Tussen Descartes en Darwin*, particularly 60–81.

11. Gipper, *Wunderbare Wissenschaft*, 177 and 189.

12. On Zöckler, see note 3 in this chapter; Philipp followed Zöckler's lead. Büttner adopted from them in turn the claim that the Hamburg-based philologist and theologian Johann Albrecht Fabricius was the initiator of German physico-theology and called the philosopher Christian Wolff its most important representative. By contrast, we view Wolff as a rationalist who argued top-down, rather than from concrete experience as most physico-theologians did. We argue that physico-theology began in Germany several decades earlier; on Fabricius, see also Kaspar von Greyerz's contribution to this volume. See Büttner, "Kant und die Überwindung der physikotheologischen Betrachtung der geographisch-kosmologischen Fakten"; "Zum Übergang von der teleologischen zur kausalmechanischen Betrachtung der geographisch-kosmologischen Fakten"; and "Theologie und Klimatologie im 18. Jahrhundert."

13. Stebbins, *Maxima in minimis*.

14. Cunningham, "Getting the Game Right." See also Harrison, *The Territories*, 164–75.

15. Goclenius, *Disputatio Duplex Ordine VIII*, pt. II.

16. See Goclenius, *Disputatio Duplex Ordine VIII*, pt. II, 3v–10r.

17. Lüthard, *Disputatio physico-theologica de operibus του εξαμερου*.

18. Magruder, "Thomas Burnet."

19. See the useful overview by the Scottish literary historian Ida M. Kimber, "Barthold Heinrich Brockes' *Irdisches Vergnügen in Gott*," 66–67; also Blair, "Natural Philosophy and the 'New Science.'"

20. Calloway, *Natural Theology*, 96.

21. See the contributions of Peter Harrison and Scott Mandelbrote in this volume, as well as Steinmann, *Absehen—Wissen—Glauben*, 19–20, and Calloway, *Natural Theology*, 34.

22. See intervening English editions of 1692, 1701, 1704, 1709; Ray, *L'Existence et la sagesse de Dieu manifestées dans les oeuvres de la Création*; Ray, *Gloria Dei oder Spiegel der Weissheit und Allmacht Gottes*; and Ray, *Gods wysheid geopenbaard in de werken der schepping*.

23. Vermij, "Translating, Adapting, Mutilating."

24. Vermij, *The Calvinist Copernicans*, 352–53; Vermij, "Religion and Mathematics"; Vermij, "Nature in Defense of Scripture."

25. For the following, see Vermij, "Formation of the Newtonian Philosophy."

26. This work can be considered a physico-theological treatise because it champions the argument from design on the basis of the mechanical philosophy. It begins, for example, by invoking the analogy of God as the ingenious watchmaker typical of that argument.

27. See Topham, "Beyond the 'Common Context.'"

28. See Bots, *Tussen Descartes en Darwin*, 60–81, and Zöckler, *Geschichte der Beziehungen*, 2:92–104.

29. Von Greyerz, *Religion and Culture*, 189–95. Hans Bots also points to the longevity of Dutch physico-theology into the nineteenth century in *De Republiek der Letteren*.

30. See, for example, the changes applied to Fénelon's *Démonstration de l'existence de Dieu* in the German translation of this treatise by Johann Albert Fabricius, which were designed to broaden the audience for theology.

31. Alsted, *Theologia naturalis exhibens augustissimam naturae scholam*; Mersenne, *L'impieté*.

32. For the Boyle Lectures, see the comments in Scott Mandelbrote's and Katherine Calloway's contributions in this volume; for Germany, see Sheehan, *The Enlightenment Bible*. It is important to note that we mean *radical* Pietism rather than mainstream Pietism. Like Puritanism, the latter was a reform movement *within* the established church. Most radical Pietists, however, were separatists in that they placed themselves outside the established Lutheran and Reformed churches.

33. Linnaeus and Biberg, "Oeconomia Naturae." On Linnaeus's authorship of all the dissertations published in the *Amoenitates*, see Lindroth, "Linnaeus (or von Linné) Carl," 375; and Walpurger, *Cosmotheologische*. More generally on the two contrasting strands of natural theology, see Mandelbrote, "The Uses of Natural Theology."

34. Findlen, "Jokes."

35. See esp. Vidal, "Extraordinary Bodies"; Strickland, "The Doctrine of 'The Resurrection of the Same Body.'"

36. Pierquin, *Dissertation physico-théologique touchant la conception de Jésus-Christ dans le sein de la Vierge Marie sa Mère.*

37. Boyle, *Some Physico-theological Considerations*, 2.

38. Thomas Aquinas, *[Summa] contra gentiles*, bk. 4, trans. Charles J. O'Neil [Latin/English ed.] (New York: Hanover House 1955–57), chaps. 80 and 81 (online: dhspriory.org). Medieval debates about bodily fragmentation and immortality were famously discussed by Caroline Bynum. See, e.g., Caroline Walker Bynum, *Fragmentation and Redemption: Essays on Gender and the Human Body in Medieval Religion* (New York: Zone Books; Cambridge, MA; MIT Press, 1991).

39. Strickland, "The Doctrine of 'The Resurrection of the Same Body,'" 164.

40. Nieuwentijt, *Het regt gebruik*, 29th observation. See also Rienk Vermij's chapter in this volume.

41. Boyle, *Some Physico-theological Considerations*, 11.

42. See also Steinmann, *Absehen—Wissen—Glauben*, 70–71.

43. *Encyclopaedia Britannica* (1771), 2:413.

44. Kemper, "Brockes und das hermetische Schrifttum seiner Bibliothek."

45. J. Steiger, "Ist es denn ein Wunder?"

46. Chraplak, *B. H. Brockes' fröhliche Physikotheologie*, 112–19.

47. Grosse, "Abgründe der Physikotheologie."

48. Philipp, *Das Werden der Aufklärung*, 66–73.

49. Gestrich, "Deismus," 403; J. Steiger, *Bibel-Sprache*, 56; Steiger, "Ist es denn ein Wunder?" See also Michel, *Physikotheologie*, 12, who formulates a similar view.

50. Sheehan, *The Enlightenment Bible*, 55.

51. See Leibniz and Clarke, *Correspondence.*

52. On Wolff as physico-theologian, see the works of Büttner cited in note 12. For challenges to that position, see Schwaiger, "Philosophie und Glaube bei Wolff und Baumgarten," 225, n. 51; Stebbins, *Maxima in minimis*, 17; Toellner, "Die Bedeutung des Physico-theologischen Gottesbeweises," 78–79.

53. Vermij, "Formation of the Newtonian Philosophy."

54. For some other examples, see Blair, "Mosaic Physics"; and Crowther, "Sacred Philosophy, Secular Theology."

55. See, for example, the botanist Giuseppe Monti (1682–1760) and his son, Gaetano Lorenzo Monti (1712–97), discussed in Sarti, "Giuseppe Monti and Paleontology"; and Ferrone, *The Intellectual Roots of the Italian Enlightenment.*

56. Breitschuh, *Die Feoptija V. K. Trediakovskijs.*

PART ONE

TERMS AND PURVIEW
OF PHYSICO-THEOLOGY

Was Physico-theology Bad Theology and Bad Science?

JOHN HEDLEY BROOKE

It can come as a surprise to learn that the term "physico-theology" is not an invention of historians but actually featured in the titles of influential seventeenth- and eighteenth-century books. The English cleric William Derham gave the title *Physico-Theology* to lectures he delivered in London during 1711and 1712.[1] More than sixty years earlier the physician Walter Charleton described one of his books, *The Darknes of Atheism Dispelled by the Light of Nature*, as a "physico-theological treatise."[2] Dispelling darkness was Derham's aim, too. His subtitle was *A Demonstration of the Being and the Attributes of God, from His Works of Creation*.

What are we to make of this mixing of theology with study of the natural world? How could evidence of divine wisdom be found in the architecture of nature? There are, of course, modern perspectives from which any interpenetration of science and theology must be misconceived. Think of the judgment of Harvard biologist Stephen J. Gould that any overlap in the magisteria of science and religion is a recipe for disaster.[3] It is also difficult to examine the argument from design, on which physico-theology usually depended, without the specter of Darwin and his thesis that a perfecting process of natural selection could create the illusion of design in the parts of living things.[4] Physico-theology has had a bad press from theologians, too. In John Dillenberger's *Protestant Thought and Natural Science*, physico-theology was charged with having seriously damaged Christianity by promoting the doctrine of Creation at the expense of a theology of redemption.[5] Physico-theologians were eloquent on the power and wisdom of God but often reticent on God's Fatherhood and love. With Pascal in mind, the Jesuit historian Michael Buckley also sees in physico-theology one of the origins of modern atheism. In his view, science-based arguments employed by religious apologists were of a kind that positively invited refutation, leading inexorably to the atheism of Denis Diderot (1713–84).[6]

Was physico-theology such bad theology and bad science? My aim in this introductory chapter is to use this deceptively straightforward question as an aid in exploring the cultural phenomenon of physico-theology and to help clear the ground for the chapters that follow. I also wish to show why the question itself is problematic. After reflecting on the grounds and their insufficiencies for the "bad science" and "bad theology" verdicts, I indicate various respects (and their concomitant ironies) in which physico-theology nevertheless managed to dig its own grave. In conclusion I suggest that it is perhaps better described as doomed philosophy than as bad science or bad theology.

The Diversity of Physico-theology

Physico-theology was a cultural phenomenon that took too many forms to allow simple generalizations about its strengths and weaknesses. As a new genre of natural theology, it surfaced in England in the middle years of the seventeenth century, later spreading and diversifying in the Netherlands, France, and Germany. Its influence was probably longer-lasting in Britain than elsewhere, not least because, in the first half of the nineteenth century, it was a valuable resource when responding to dangerous naturalistic science emanating from revolutionary France.[7] Jean-Baptiste Lamarck with his evolutionary biology and Pierre-Simon Laplace with his nebular hypothesis for the emergence of the solar system were particular threats. But this persistence of physico-theology in Britain does not mean that it had failed to flourish elsewhere. It took root in the Netherlands, for example, where Bernard Nieuwentijt looked to Robert Boyle and Isaac Newton for an empiricist methodology that challenged the deductive mechanistic philosophy of Descartes.[8] Both Boyle and Newton found evidence of divine power and wisdom in nature, Boyle primarily from the exquisite craftsmanship displayed in even the minutest of creatures.[9] Newton was more impressed by the mathematical intelligence of a creator who had calculated the precise tangential component of each planet's velocity to ensure that it went into a stable orbit around the sun, without which life on earth would have been impossible. Such precision, in Newton's own words, "argued a deity very well skilled in mechanics and geometry."[10]

Imported into the Netherlands, physico-theology became something of a popular movement, where it was embraced as an antidote to the monism of the biblical critic Baruch Spinoza (1632–77). Some fifty physico-theology texts were published there in the eighteenth century. Although they were not so obviously dominated by an animus against Spinoza, comparable texts were produced in Germany with such curious titles as insect-theology, water-theology, and star-, thunder-, and snow-theology.[11] We might perhaps call them specifico-theologies. A typical argument was that only a providential God could have so determined that water, exceptionally, is less dense in its solid than in its liquid state, thus proving God's concern for the aquatic creatures that would otherwise be crushed by descending layers of ice.

Exceptional features of the natural world could certainly evoke a genuine sense of marvel that was easily translated into religious awe and gratitude. In John Ray's classic text, *The Wisdom of God Manifested in the Works of Creation* (1691), we are in the company of an acute observer of nature who could marvel at the migratory instincts of birds and their remarkable navigation skills. In Ray's writings, as in John Wilkins's (1614–72), we also find the ingenious argument that living things, when examined under the microscope, reveal a perfection completely lacking in human artifacts. A sharp needle looked like a botched job in comparison even with fish scales or the eye of a fly. This was an ingenious and pervasive argument because it presented evidence of divine transcendence in material form.[12] The Dutch microscopist Jan Swammerdam is often quoted for his boast: "I offer you the omnipotent finger of God in the anatomy of a louse: wherein you will find miracles heaped upon miracles."[13] Sermons were even preached on the proboscis of the flea, nature's own device for bloodletting![14] The power of the microscope was more than incidental. It was integral to, and facilitated, what became an important feature of the treatments of insects in works of physico-theology: the detailed investigation of their anatomy and the intricate workings and coordination of their parts. As Brian Ogilvie argues in chapter 12, physico-theology was not a mere byway in the history of natural history: it contributed to the formation of entomology as a discipline.

Not only the diversity of physico-theology precludes an absolute judgment. The fact that prevalent forms changed over time also makes it difficult to pin down an answer. Scott Mandelbrote has detected a significant shift in England from the mid-seventeenth century, when Platonist Christians such as Henry More were amassing evidence for a world of spirits and wonders, to the early eighteenth century, by which time the emphasis had shifted to the regularity of a natural world governed by divinely instituted laws.[15] A greater incredulity concerning supernatural apparitions and the agency of spirits accompanied this transition, though the story is never simple. Robert Boyle, in what one biographer describes as his "spirit-tinged alchemy," had continued to believe until his death that the philosopher's stone would attract angels and spirits in a suprasensual realm.[16] Judged retrospectively, a physico-theology of the Newtonian type, in which one could celebrate the mathematical elegance of divine laws and see in their regular operation evidence of God's faithfulness, would be more propitious for the natural sciences than one simply premised on nature's marvels. Accordingly, I return to my question.

Was Physico-theology Bad Science?

A pertinent question, posed by the organizers of the symposium on which this book is based, is whether physico-theology should be seen as a reaction *against* the new mechanical philosophy of the seventeenth century or as an attempt to legitimate mechanical science on religious grounds—or perhaps as both. It was, surely, both. It was certainly a reaction against the overmechanization of nature. We can see this in Boyle's critique of Descartes, the most ambitious of the mechanists, who had reduced animals to machines and called for the exclusion of final causes from the study of nature.[17] In Boyle's physico-theology, and that of John Ray, there were explicit reactions against Descartes's closed mechanistic universe.[18] On the other hand, Boyle was willing to speak of the excellence of the mechanical philosophy when, in the form of a corpuscular theory of matter, it helped to explain the properties of substances and the chemical reactions between them.[19] There undoubtedly was religious legitimation in that inferences to design could be drawn from mechanical models of nature. Machines are not the kinds of thing that create themselves. Moreover, the clockwork analogies of physico-theology were attractive to Boyle for an even more fundamental reason. They created space for the natural philosopher to investigate the inner workings of nature with a degree of autonomy, without prejudice to fundamental religious doctrines.

It is an obvious point, but particular forms of science must not be described as bad just because they were later superseded. That can be the fate of even the best science. Theories that appeared in physico-theology were, in many cases, up to date and the best available. When Boyle attacked Descartes, he illustrated the power of teleological reasoning by advertising one of the great scientific discoveries of the century, William Harvey's demonstration of the circulation of the blood. It was by asking questions about the *purpose* of valves in the veins that Harvey had been led to his conclusion. Or take the example of John Ray, building on advances in Copernican astronomy to argue that a sun-centered system was aesthetically far more elegant than the complex system of Ptolemy. In his *Wisdom of God* Ray reproached the notorious king

Alfonso the Wise of Aragon, who, struggling with the complexities of the Ptolemaic system, had allegedly boasted that, if only God had consulted him at Creation, he could have suggested something simpler. "Rash" and "profane" were the adjectives Ray used to describe such a boast.[20] But the point was that it could no longer be voiced. The superior elegance of the sun-centered system had eliminated the problem. When Newton's spokesman, Samuel Clarke, endeavored to quash the widely perceived skepticism of Hobbes and Spinoza, he similarly turned to the most recent developments in anatomy and astronomy. Features of living and inanimate things once considered defective or purposeless had been "discovered to serve the wisest and most exquisite ends imaginable." As Newton had shown, there was an "inexpressible nicety of the adjustment of the primary velocity and original direction of the annual motion of the planets, with their distance from the central body and their force of gravitation towards it."[21]

Was Newton's natural philosophy bad? There were respects in which Leibniz thought it was,[22] and it was for some time resisted in France; but it was also cutting edge in its explanation of the elliptical orbits of planets. By corresponding directly with Newton himself, the first Boyle Lecturer, Richard Bentley, took pains to show that, in his science, he was really up to speed.[23] It was a correspondence in which Bentley's theological concerns actually prompted further reflection from Newton on the distribution of stars.[24] It also gave Newton an opportunity to distance himself from materialism by insisting that gravity was not an innate property of matter.[25]

There is a further example of scientific awareness in William Derham's *Astro-Theology*, which was quickly translated into other languages, notably into German by Fabricius in 1728. This was a text in which Derham exulted in being abreast of recent science. He exploited a new cosmological system, which in crucial respects was an advance on the Copernican. Following Descartes and Christiaan Huyghens (1629–95), Derham expounded a science in which the stars were all suns, surrounded by their own planetary systems. Life could (and would) exist on these other worlds. In our universe, filled every day with more exoplanets, this can still sound prescient today. Yet Derham's design, in his own words, was "particularly for the conviction of infidels."[26] His argument was that the new cosmology, with its plurality of worlds, provided a "far more extensive, grand, and noble view of God's works" than the "old vulgar opinion, that all things were made for man."[27] Physico-theology is often criticized for being anthropomorphic in its description of God and anthropocentric in its focus. It was indeed incurably anthropocentric in the weight it attached to aesthetic criteria in the evaluation of scientific theories, in the subordination of natural objects to human use, and in its anthropomorphic presupposition of resemblance between the human and the divine mind.[28] But Derham could also rejoice in his scientific liberation from a naively anthropocentric cosmos. Earlier, Ray had challenged the same naive view that "all this visible world was created for man" alone. Wise men, he wrote, nowadays think otherwise.[29]

Even if physico-theology did not always contain the best available science, it could be good *for* science when it generated a passion to study the natural world. Anne-Charlotte Trepp has examined the case of Johann Rist, a German pastor writing in the middle years of the seventeenth century. Disillusioned by the "sick Lutheranism" of doctrinal dispute, Rist turned to the natural world for reassurance of God's providence and mercy.[30] He was a Copernican in astronomy and had studied botany and medicine. As Trepp has put it, "Knowing everything *about* creation was confirmation of God's love and care *for* his Creation."[31] Such motivation was not confined to Britain and Germany. Ann Blair has drawn attention to the influence of the French Jansenist Noël-Antoine Pluche and his multivolume *Spectacle de la nature* (1732–50). This was a text that helped to launch the career of the eminent Swiss naturalist and etymologist Charles Bonnet. Blair describes how Bonnet recalled stumbling on Pluche's description of the antlion, a group of insects whose larva conceal themselves in holes in the sand, from which they pounce on unsuspecting ants. Bonnet was sixteen when, enthralled by what he read, he suddenly found his passion. He says that he read the book many times and eventually knew it "almost by heart."[32] Reading works of natural theology certainly could be a point of entry into a world of scientific discourse. Bonnet's natural history was infused with natural theology, and he, in turn, provided examples of divine ingenuity for the natural theologies of John Wesley and, later, William Paley in England.[33] Paley's *Natural Theology* (1802), with its celebration of organic adaptation, was one source of inspiration to the young Darwin, focusing his mind on a mystery he would eventually solve.[34] Even after articulating his theory of natural selection, Darwin found it difficult to discard an assumption that had been integral to physico-theology—that every part in animals has a use to which it is fitted.[35]

If physico-theology could capture the imagination of those drawn to an intensive study of nature, it could also be a vehicle for the popularization of scientific knowledge. Inferences drawn from nature had a clarity sometimes lacking in biblical exegesis, giving them a potentially wide appeal. As John

Ray expressed it in the preface to his *Wisdom of God*, "These proofs, taken from effects, and operations, exposed to every man's view, are most effectual to convince all that deny or doubt of it. Neither are they only convictive of the greatest and subtlest adversaries, but intelligible also to the meanest capacities." In their introduction, Ann Blair and Kaspar von Greyerz have referred to the Frenchman François de Fénelon, whose physico-theology, which appeared in Hamburg in German translation in 1714, was deliberately targeted at an audience of lower social standing as a theology fit for everyone.[36]

The invocation of the sciences for a theological purpose meant that physico-theology could easily graduate into an apologia for science as well as for theology. Crucially, the fact that a theological justification could be given for scientific inquiry was arguably of critical importance in ensuring that an enduring culture of science became possible in Western Europe. Stephen Gaukroger has made that case, contrasting the durability of the scientific movement in Europe with the boom-and-bust patterns in other cultures.[37] There is important work to be done on the place and character of natural theology in Eastern Orthodox Christianity, where its low profile appears to coincide with recurrent indifference to the interrogation of nature.[38] By contrast, for much of the eighteenth century physico-theology in Britain and Western Europe helped to make the study of nature itself relevant to a wide audience. In Germany, Lorenz Heister, who was professor of anatomy, surgery, and medicine at Helmstedt, found in physico-theology a religious justification for practicing dissection.[39] It was dissection that revealed the "amazing fabric" of the human frame, testifying to the glory of the Creator—a testament that Heister declared to be the great and primary end of anatomy. The science was "highly useful to everyone who studied true wisdom and theology."[40] Friedrich Christian Lesser in Germany and the great taxonomist Carl Linnaeus in Sweden even constructed a deductive argument for science as a religious duty. This is Linnaeus: "If the Maker has furnished this globe with the most admirable proofs of his wisdom and power; if this splendid theatre would be adorned in vain without a spectator; and if man the most perfect of all his works is alone capable of considering the wonderful economy of the whole; it follows that man is made for the purpose of studying the Creator's works that he may observe in them the evident marks of divine wisdom."[41] Looking ahead to the nineteenth century, the natural theology of William Paley became a target for radical medical reformers in London; but even as it became increasingly outdated, it provided a platform from which the case could be made for introducing more science into university curricula.

As Jonathan Topham has argued, the natural theology of the eight *Bridgewater Treatises*, published in the 1830s, was more successful as a means of promoting politically safe science than it was as a rational defense of Christianity.[42]

Was physico-theology bad science? Certainly it could put constraints on what was conceivable to its exponents, its influence, as Darwin conceded, difficult to annul.[43] In anatomy, where structureless functions and functionless structures existed, it celebrated too restrictive a view of the interdependence of structure and function.[44] But it provided an apologia *for* science that was in many respects advantageous. It is worth remembering that despite his rejection of what Kant called the physico-theological proof, Kant acknowledged in his *Critique of Pure Reason* that it never fails to commend itself to the popular mind and imparts life to the study of nature, receiving new vigor from it. Within limits, it could even help to clarify what it was for God to be God.[45]

Was Physico-theology Bad Theology?

Whether physico-theology was bad theology is a trickier question to answer than whether it was bad science. One reason for this is that many of the criticisms directed against it have reflected intra-Christian controversies. Another derives from the complex social and political parameters that have interpenetrated those controversies. In her discussion of physico-theology as a form of mass literature in eighteenth-century Germany, Trepp has noted how it coincided with a stabilization of social conditions. She observes that authors repeatedly emphasized the regularity and functionality with which God had guaranteed the stability of the creaturely world for the use of humankind.[46] When physico-theology had first materialized in Britain, it arguably reflected a quest for social and political stability that had been jeopardized by religious disputes and the strife of civil war. A theology that set its face against religious enthusiasm at one extreme, and skeptics and scoffers at the other, could be attractive to figures like Boyle who found rational support for a moderate Christian theism in the architecture of nature. Boyle had been dispirited by the proliferation of Puritan sects, each claiming its own hotline to God. Let a man come to London, Boyle had written, and he will come near to losing his faith, so fragmented had Christianity become by competing claims for divine illumination.[47] A common jibe of the scoffers was to ask what was in store for the countless numbers who had never heard of Jesus Christ. Were they destined for eternal damnation; how could they be fairly judged? Physico-theology sometimes had a role to play in responding.

One reply was an exegesis of Paul's Letter to the Romans 1:20: "Ever since the creation of the world [God's] invisible nature, namely his eternal power and deity, has been clearly perceived in the things that have been made. So they are without excuse." To that extent physico-theology was biblical, even as it focused attention on the things God had made. It gained a higher profile, too, as it addressed an intellectual challenge from atomists and materialists inspired by Lucretius, that "witty villain," as he was denounced by the physician Walter Charleton.

Because of his combative style in the defense of providence, Charleton is a particularly engaging figure. Lucretius had to be refuted because, by the 1650s, he had been widely read and because he had dispensed with a creator altogether; he asserted that worlds simply come into, and pass out of, existence by a random aggregation and eventual disaggregation of atoms. As with his French precursor Pierre Gassendi, Charleton insisted that the order, architecture, and beauty of nature could not have arisen spontaneously in this way.[48] It was the Deity who had first created the atoms, arranged them in a preordained pattern, and set them in motion. Although opinion is divided over whether Charleton's *Darknes of Atheism* should be included in the canon of physico-theology, it was a book in which evidence for design was certainly adduced.[49] There was more industry in the proboscis of a flea, "in its delicate and sinuous perforation, than all the costly aqueducts of Nero's Rome."[50] Charleton aimed to show that a theistic interpretation of even a potentially dangerous science was perfectly possible. It would be harsh to label as bad theology such an aim. One of the grounds on which Lucretius had dismissed the idea of a deity having special regard for humankind was that humans were too fragile, too prone to destruction from storm and disease. The absence of protection implied the absence of providence. To Charleton this was a mere quibble for anyone who understood that there was both a general and a special providence active in the world with purposes that transcended the particularities of human discomfort.[51] Contrary to conventional critiques of physico-theology, this was not a manifestation devoid of references to special providence or to scripture. Charleton insisted that the true Christian was blessed with the gift of revelation and a spiritual state beyond that of the "natural man."[52] Like Ray, Charleton should not be reproached for failing to *prove* the existence of God, which both presupposed. The exposure of defects in the reasoning of one's opponents was valuable in itself and reassuring for those who already believed. As Katherine Calloway maintains, it is hard to accept the charge leveled against physico-theology that advocates such as

Charleton and Ray unwittingly brought about the decline of religion through an overambitious rational defense of Christianity.[53]

A large question remains, however, concerning the limitations of natural theology and physico-theology in particular. Mandelbrote has observed that at the end of the seventeenth century, natural theology could not provide for salvation, it did not represent a sufficient basis for overcoming doctrinal division either among Christians or between Christians and professors of other faiths, and it could not be used to propagate doctrine.[54] One could add that among its proponents, there was often a lack of clarity concerning the precise relationship between truths derived from revelation and those from natural reason. An enticing research project would be to explore whether there was a significant difference between those physico-theology writers who brought the Bible into their investigations of nature and those who, irrespective of their own views of revelation, left it out. In this context, a systematic examination of the last chapters of the classic natural theology texts becomes a desideratum. That was where references to revelation often appeared, together with remarks that could not always be predicted from the preceding text. In the last chapter of John Wilkins's *Principles and Duties of Natural Religion* there was reference to the necessity of revelation, given man's degenerate state; but, as Calloway wryly observes, readers, and perhaps even Wilkins himself, "are surprised to find that he has not jettisoned the more reformed doctrine of his youth in favor of the latitudinarian ideals he has come to cherish."[55] Many years later, in the last chapter of his *Natural Theology*, Paley declared revelation indispensable for particularities concerning God's designs as a moral governor.[56] This was where he also said that his project was designed to facilitate belief in the "fundamental articles of Revelation." It was a way of establishing the character of a deity who could reasonably be expected to deliver a revelation.

Marginalized or not, there can be no simple test to decide whether this was good or bad theology. Connections were made by physico-theologians with the leading of a Christian life. Praise and gratitude, fear and obedience were seen by William Derham as appropriate responses to the divine power displayed in nature. There were also strategic reasons why Christian apologists might choose to downplay revelation in rhetoric designed to engage atheists of various kinds. One was stated by Henry More in his *Antidote against Atheism*, which has been described as the first text of physico-theology: "I did not insist upon any sacred History mainly because I know the Atheist will boggle more at whatever is fetch'd from established Religion, and fly away from it, like a wild Colt in a Pasture at the sight of a bridle snuffing the Aire and smelling a Plot afarre off, as he foolishly fancies."[57] When physico-theologians glossed over the necessity of revelation, this often reflected a strategic belief that a rational defense of the faith, largely independent of revelation, would be the most effective weapon against unbelief. It is because of the unintended consequences of this strategy that natural theology, and physico-theology in particular, have enjoyed a poor theological reputation. A rational theology, independent of revelation, could so easily be detached and co-opted by deists as a contrary, even a contradictory, form of religion. It did not require a great mental shift to bring about this change. John Tillotson, the latitudinarian archbishop of Canterbury who had overseen the posthumous publication of John Wilkins's treatise on natural religion, could say that there was nothing more incredible than that divine revelation should contradict the clear and unquestionable dictates of natural light.[58] By "natural light," Tillotson probably meant something akin to the divine inspiration that could enable the attainment of a rational understanding. But it was not far from his position to the dangerous principle of the deists that no proposition in scripture should be accepted unless it had independent rational support. It is not difficult to see why physico-theology, especially when transposed into a lay theology, should be regarded as heterodox or subversive. It would prove ambiguous in its religious significance, coexisting with dangerous Unitarian heterodoxy and with more orthodox pieties. It was an ambiguity that could even have advantages for those wishing to advance the study of nature without having to bare their theological souls.[59]

For Unitarians like Newton and Joseph Priestley, a *Trinitarian* creed was bad theology, grounded in the corruption of biblical texts. For Christian orthodoxies, Unitarian dissenters were politically dangerous perpetrators of bad theology. For the secular historian there is no privileged standpoint from which to make absolute judgments. But while the labels "good" or "bad" can be distractions, there certainly was a deficiency in physico-theology that generated its own set of problems. This relates to different understandings of the doctrine of the Fall. As Peter Harrison has shown, this was a doctrine that featured prominently in the scientific movement of seventeenth-century England. Experimental inquiry, according to Francis Bacon, could help restore some at least of the knowledge that God had intended for Adam but which had been lost through his disobedience. The practical application of this knowledge, in fields such as agriculture and medicine, promised improvement, a kind of redemption of nature, for the glory of God and the relief of man's estate.[60] Nature bore the marks of the Fall. It was currently far from perfect. But here was the problem. In reading the texts of physico-theology, one could be forgiven for thinking that the natural world is more or less perfect—the kind of world that revealed the perfect wisdom of God. Not all advocates of physico-theology stated that this is the best of all possible worlds, but some came very close. By contrast, in more conservative Protestant theologies, nature was not even supposed to be perfect.

How, then, did physico-theologians deal with pain and suffering, earthquake and flood, noxious creatures and toxic plants? Unsurprisingly, they often floundered. There were responses, of course. It was surely inappropriate to question divine wisdom. But that move sat uncomfortably with claims to be demonstrating the wisdom of God. Confronted with venomous rattlesnakes, Derham reassured his readers that they do at least issue a warning: they rattle before they strike.[61] Toxic plants like hemlock were poisonous to humans but made excellent food for goats. Creatures that could injure humans were admissible because, as Derham put it, "in greater variety, the greater art is seen."[62] On the question of pain and disease, references to divine punishment were still an option. For what might be seen as defects in creation, there was another resource for those, like Ray, whose theology of nature was infused with elements of Platonism. Ray drew on the philosophy of the Cambridge Platonist Ralph Cudworth (1617–88), whose God delegated activity in nature to a nonmaterial agency, a "plastic principle." Because this agent was not omnipotent, in its interaction with recalcitrant matter, there would be occasions when nature was frustrated.[63] For Ray it was the outright mechanists such as Descartes whose theism was most unrealistic and vulnerable to objections from nature's flaws.

Whatever judgment we make about the status of physico-theology, there must, of course, be questions about its effectiveness as an apologetic resource. I suspect that these questions are extremely difficult to answer. I do, however, concur with Mandelbrote's judgment that "wonder at the bounty and the regularity of nature drew people powerfully to faith more often than materialism or mechanism encouraged them to doubt."[64]

Conclusion

When physico-theologies were constructed by scientific thinkers, one motive was to educate churchmen who were at best indifferent, and at worst hostile, to scientific innovation. Commenting on this in 1960, Dillenberger observed that "it is ironic that their defence against the attack of churchmen should have been the very thing which transformed theology, namely, the elaboration of an independent natural theology."[65] There is irony, too, in the fact that this independent natural theology managed, eventually, to dig its own grave.[66] If theological doctrine and a particular scientific understanding of nature are fused together—if, for example, God's providence is illustrated from Newton's analysis of the solar system—then the doctrine becomes a more sharply defined target for attack.[67] It was in France in the late eighteenth century that the God required by Newton periodically to reform the solar system was shown by Laplace to be redundant.[68] It is in France, too, where we can perhaps see most vividly another irony: how the crucial concept of laws of nature, originally embedded in a theistic framework, could become a weapon for deists wanting an autonomous natural world. For Newton, Clarke, and other theological voluntarists, the laws of nature expressed the will of God in producing certain effects in a regular, constant, and uniform manner. But it was not difficult to embrace the laws as legislation over which the Deity, in effect, no longer presided. Voltaire had no difficulty in enlisting Newton's natural philosophy for his anti-Catholic purposes.

In Britain, where physico-theology proved resilient until the mid-nineteenth century, another irony was played out. Advocates of design usually claimed that their argument was cumulative. Every instance of design added to its weight. For Derham and Paley, the eye was proof of design without the ear, the ear without the eye. But in multiplying example after example, the defense of theism began to sink under a weight of tedium and triviality. Writing in a post-Darwinian world, Alfred Russel Wallace mocked a claim he had just read that the soft scar on the coconut was a display of divine wisdom because, if it were not there, a young shoot would be unable to emerge. Wallace was scathing. This was inane and degrading. It imputed to the Supreme Being "a degree of intelligence only equal to that of the stupidest human beings." It was like praising an architect for remembering to put a door in his house.[69]

Perhaps the greatest irony in Britain was that so many advocates of design had put on a pedestal the very features of living things that would prove most susceptible to an alternative explanation once it was found. We find Darwin writing in the late 1850s that by "nature" he means "the laws ordained by God to govern the universe."[70] Nature was still a creation. There were still theological issues to be seen in, and extracted from, the natural world. Inferences to a divine legislator were still possible. But in these echoes of physico-theology, Darwin was no longer writing as a Christian theist. Nor was nature to be analyzed as a catalog of largely disconnected, separate instances of design. For Darwin, as for that other great explorer Alexander von Humboldt (1769–1859), it was the connections between living things, seeing nature as a web of interrelated species and their ecological niches, that offered the greatest insights.[71]

This chapter has been designed to show why a simple question (Was physico-theology bad theology and bad science?) cannot be given a simple answer. But was it perhaps bad, or at least doomed, philosophy? It was, after all, the philosophers Kant and Hume who subjected the argument for design to the most telling critiques. There is one feature of physico-theology in particular that would help in making that case. This was the prevalence of circular arguments in much of the rhetoric. As Hume insisted, it was impossible to infer God's infinite power and wisdom from features of a finite world. The conclusion of the design argument was more like a premise, the analogies on which it rested fragile. The circularity is decisive for philosophers who like to treat the classic proofs as formal, but failed, demonstrations of God's existence. But this can easily misrepresent how advocates themselves conceived their purposes. Design arguments had many functions, including strengthening the faith of religious believers as well as reassuring waverers, challenging skeptics, and justifying scientific enquiry.[72] Charleton and Ray, for example, knew they were taking for granted a preexisting Christian commitment. But this did not prevent them from exposing what they saw as implausible and defective in the views of their opponents. This is not entirely to exonerate them. When physico-theologians turned to the latest science to strengthen their case, they *were* sometimes blind to the circularities involved. One example will have to suffice. It concerns the physico-theology of eighteenth-century popularizers of Newton, in particular the universality of the inverse square law of gravitation. How might its universality be established before there was empirical proof from the behavior of stars? Newton's justification reflected his theological commitment to the unity and omnipresence of God: "If there be an universal life and all space be the sensorium of a thinking being who by immediate presence perceives all things in it the laws of motion arising from [that] life or will may be of universal extent."[73] There was, however, a catch. In post-Newtonian physico-theology, the argument was inverted. The unity of God was deduced from the universality of God's laws. This inversion can be seen in William Whiston's *Astronomical Principles of Religion* (1717). Whiston was Newton's heretical successor in the Lucasian Chair of Mathematics in Cambridge. The universe, he wrote, appears to be "evidently One Universe; govern'd by One Law of Gravity through the whole; and observing the same Laws of Motion everywhere So that this Unity of God, is now for ever established by that more certain Knowledge we have of the Universe."[74] It was not to be; but that is another story.

NOTES

1. Derham, *Physico-Theology*.
2. Charleton, *Darknes of Atheism*.
3. Gould, *Rocks of Ages*, 47–95.
4. Gillespie, *Charles Darwin*, 82–123; Richards, "Darwin's Theory"; Topham, "Biology in Service."
5. Dillenberger, *Protestant Thought* (1961), 133–62.
6. Buckley, *Modern Atheism*, 194–250.
7. Brooke, "Scientific Thought."
8. Jorink, *Book of Nature*.
9. Boyle, *Disquisition about Final Causes*; Shanahan, "Teleological Reasoning."
10. Newton, "Letter to Bentley," 48–49.
11. Trepp, "Natural Order and Divine Salvation," 137–42.
12. Ray, *Wisdom of God*, 7th ed. (1717), 30, 57–59; Hooke, *Micrographia*, 1–2; Wilkins, *Principles and Duties*, 80–81.
13. Harrison, *The Bible*, 174.
14. Harrison, *The Bible*, 176.
15. Mandelbrote, "Uses of Natural Theology."
16. Principe, *Aspiring Adept*, 212.
17. Boyle, *Disquisition about Final Causes*; Shanahan, "Teleological Reasoning."
18. Ray, *Wisdom of God*, 7th ed., 42.
19. Boyle, *Excellency and Grounds*.
20. Ray, *Wisdom of God*, 7th ed., 63–64.
21. Clarke, *Discourse*, 108–9.
22. Leibniz and Clarke, *Correspondence*.
23. Bentley, *Confutation of Atheism*, sermon 3, 11–13.
24. Hoskin, "Universe of Stars."
25. Iliffe, *Priest of Nature*, 310–13.
26. Derham, *Astro-Theology* (1715), lviii.
27. Derham, *Astro-Theology* (1715), 39.
28. Brooke, "'Wise Men Nowadays Think Otherwise.'"
29. Brooke, "'Wise Men Nowadays Think Otherwise,'" 206–8.
30. Trepp, "'Nature' as Religious Practice," 97.
31. Trepp, "Natural Order and Divine Salvation," 133.
32. Blair, "Jansenist Natural Theologian," 92.
33. Eddy, introduction to Paley, *Natural Theology*, xxiv.
34. Fyfe, "Paley's Natural Theology."
35. Darwin, *Descent of Man*, 42.
36. Blair and von Greyerz, introduction to this volume, note 30.
37. Gaukroger, *Scientific Culture*, 3, 4, 7, 23–24.
38. Nicolaidis, *Eastern Orthodoxy*, 170–75.
39. Vidal, "Extraordinary Bodies," 68.

40. Brooke, *Science and Religion* (1991), 216–17, or (2014), 294–95.

41. Linnaeus, *Reflections*, 18.

42. Topham, "Biology in Service."

43. Darwin, *Descent of Man*, 42.

44. Bynum, "Anatomical Method."

45. Kant, *Critique of Pure Reason*, 578–83.

46. Trepp, "Natural Order and Divine Salvation," 138.

47. Rattansi, "Social Interpretation," 21.

48. Osler, *Divine Will*, 160–62.

49. Charleton, *Darknes of Atheism*, 60–67.

50. Charleton, *Darknes of Atheism*, 66–67.

51. Charleton, *Darknes of Atheism*, 90–95.

52. Charleton, *Darknes of Atheism*, 47.

53. Calloway, *Natural Theology*, 113.

54. Mandelbrote, "Early Modern Natural Theologies," 89.

55. Calloway, *Natural Theology*, 92.

56. Paley, *Natural Theology*, 280.

57. More, *Antidote*, pref.; Calloway, *Natural Theology*, 37.

58. Calloway, *Natural Theology*, 92.

59. Brooke, "Natural Theology of the Geologists."

60. Harrison, *Fall of Man*, 186–244.

61. Derham, *Physico-Theology* (1713), 57.

62. Derham, *Physico-Theology* (1713), 55.

63. Ray, *Wisdom of God*, 7th ed., 51.

64. Mandelbrote, "Early Modern Natural Theologies," 91.

65. Dillenberger, *Protestant Thought* (1961), 186.

66. Brooke, "Fortunes of Natural Theology."

67. Funkenstein, *Theology and the Scientific Imagination*, 89–97.

68. Hahn, "Laplace."

69. Brooke, *Science and Religion* (1991), 220, or (2014), 298.

70. Richards, "Darwin's Theory," 61.

71. Wulf, *Invention of Nature*, 5.

72. Brooke, *Science and Religion* (1991), 192–225, or (2014), 261–306; Brooke and Cantor, *Reconstructing Nature*, 143–45; Mandelbrote, "Uses of Natural Theology."

73. Westfall, *Newton's Physics*, 397.

74. Whiston, *Astronomical Principles*, 130–31.

CHAPTER TWO

What's in a Name?

"Physico-theology" in Seventeenth-Century England

PETER HARRISON

The word "physico-theology" does not appear in any English book in the seventeenth century. It is unlikely, then, that there will be a distinct genre in the seventeenth century that historical actors understand as "physico-theology." That said, we do encounter several instances of the adjective "physico-theological" in this earlier period. But unhelpfully this term seems to be used in different ways, suggesting that the handful of authors who describe their approach as "physico-theological" may not share a single agenda. This leaves us with the question what, if anything, the term "physico-theological" is picking out during the seventeenth century.

One general trend of this period that might shed light on the meanings of "physico-theological" is the use of other compound terms with the prefix "physico-." The most common of these terms, ranging across European sources, is "physico-mathematical," followed by "physico-mechanical." Examination of the general logic of these hyphenated forms might indirectly illuminate uses of the term "physico-theological." In this chapter I examine this "hyphenation phenomenon," then turn to the specific uses of "physico-theological" in the seventeenth century, and finally draw some conclusions about whether there is an identifiable physico-theological genre in seventeenth-century England.

The Hybridization of Disciplines in Early Modern Europe

During the seventeenth and eighteenth centuries we encounter more than twenty different combinations of two, and occasionally three, disciplinary approaches, with the prefix "physico." The long list includes:

physico-astronomical	physico-logical
physico-astrological	physico-mathematical
physico-logic	physico-medical

physicomedicus	physico-chemical
physico-mechanical	physico-anatomical
physico-theological	physico-ethicum.
physico-theosophical	

There are, in addition, three-word combinations such as "physico-magico-medica" and "physico-medico-chymica." In the eighteenth century we find further combinations such as "physico-geographical," "physico-intellectual," "physico-mental," "physico-miraculous," "physico-philosophical/physico-philosophy," "physico-psychical," and "physico-psychological." Virtually all of these expressions are adjectival.

We can compare these hyphenated expressions with a second and subsequent set of compound terms that have a wide range of prefixes but with the same second term—"theology." Each of these appears in a book title (listed with a year of publication):

Astro-theology (1715)	Testaceo-theology (1744)
Pyro-theology (1732)	Bronto-theology (1745)
Hydro-theology (1734)	Akrido-theology (1748)
Litho-theology (1735)	Melitto-theology (1767)
Insecto-theology (1738)	Phyto-theology (1851)
Petino-theology (1742)	

These latter combinations, unlike the terms with "physico-" prefixes, are almost invariably in the noun form and, it is fairly safe to say, represent subdivisions of what now appears like the identifiable activity—physico-theology. In fact, looking backward from this later grouping of disciplines, it seems as if one of a range of physico-theological approaches in the seventeenth century begins to cohere into a single distinctive field. William Derham's *Physico-Theology* (1713)—the first title in which the noun form occurs—thus appears to set the pattern for the genre. Derham's *Astro-Theology* (1715), following in short order, opens up the way for a proliferation of similar titles. The equivalence is abundantly clear from the subtitles, both of which reproduce the formula "A Demonstration of the Being and Attributes of God from. . . ." This, then, is one of the dominant forms of physico-theology in the eighteenth century.

But what of the earlier physico-theological works? Here it is useful to consider what the range of hyphenated descriptors might have in common and see if this can shed light on the specific case of the "physico-theological." My suggestion is that the hyphenated forms signal an explicit desire to break down traditional disciplinary boundaries, along with attempts to raise the status of hybrid approaches along with that of their practitioners.

The case of one of the most common earlier combinations, "physico-mathematical," is instructive. Dutch philosopher Isaac Beeckman seems to have been the first to have used the term in a journal entry in 1618, following a meeting with the young Descartes.[1] Thereafter, the expression "physico-mathematical" begins to appear with some regularity. Instances include the Jesuit mathematician Paul Guldin's *Dissertatio-phisico-mathematica de motu terrae* (1622); Athanasius Kircher's *Ars Magnesia, hoc est, disquisitio bipartite-emperica seu experimentalis, physico-mathematica de natura* (1631); and Marin Mersenne's *Cogitata Physico-Mathematica* (1644).

This new term might be regarded simply as a synonym for a traditional category—that of mixed mathematics. The "admixture" here is an artifact of long-standing distinction, inherited from Aristotle, between the two theoretical sciences of natural philosophy and mathematics (the third theoretical science was metaphysics or theology).[2] Promiscuous crossing of boundaries— *metabasis*—was theoretically prohibited, although provision was made for "subordinate" or "mixed" sciences, such as optics or mechanics, that combined mathematics with natural philosophy.[3]

What, then, if anything, is the difference between "physico-mathematics" and mixed mathematical sciences? My proposal is that it has to do partly with the status of this "mixed" approach and partly with the status of those who employ it. The rise of physico-mathematics can be seen as an attempt by practitioners of the mathematical sciences to colonize territory traditionally occupied by natural philosophy—"a kind of muscle flexing on the part of mathematicians," as Peter Dear has described it.[4] Nicolò Fontana Tartaglia (ca. 1500–1557), Bernardino Baldi (1553–1617), G. B. Bennedetti (1530–90), and Galileo (1564–1642), for example, had offered mathematical demonstrations of flaws in Aristotle's physics of falling bodies, suggesting that mathematics could play a corrective role in the sphere of natural philosophy.[5] In the sphere of astronomy, Johannes Kepler (1571–1630) had brought together "geometrical and physical arguments" to provide a comprehensive account of the motions of the universe, underpinned by Lutheran theological convictions.[6] "In these chapters I will have the physicists against me," he conceded, "because I have deduced the natural properties of the planets from immaterial things and mathematical figures."[7] Kepler thus self-consciously moved from mathematical astronomy into the territory of "physicists" or natural philosophers (and theologians). These novel approaches were accompanied by an insistence that those who deployed them be granted an accordingly higher status than that accorded to mere mathematicians.[8] The culmination of this trend in the work of Kepler and Galileo famously came with the publication of Isaac Newton's *Philosophiae Naturalis Principia Mathematica* (1687), the title of which—"Mathematical principles of natural philosophy"—nicely embodies the principles of the physico-mathematical approach.[9] Newton's work can been seen as the final stage of the triumph of mathematics over scholastic natural philosophy.[10]

Generalizing from these examples, the proliferation of hyphenated categories can be regarded as symptomatic of a breakdown of traditional scholastic disciplinary boundaries and a renegotiation of the Aristotelian prohibition of *metabasis*. Similar categories such as physico-mechanical, physico-anatomical, and physico-astronomical could thus represent not merely the transgression of standard disciplinary divisions but also the aspirations and ambitions of those who promoted them.

On this principle, "physico-theological" approaches, rather than being simply a synonym for what we would now call "natural theology," would involve the kind of boundary crossing that we witness in the case of physico-mathematics. And just as mathematicians were to claim authority in the sphere of natural philosophy, some proponents of new mathematical natural philosophy or experimental natural philosophy came to see themselves as equipped to make claims in the sphere of theology.

Indirect evidence for this contention comes from the fact that in some cases it was the same individuals involved in both kinds of boundary transgression: physico-mathematical approaches could be combined with physico-theological. Kepler is a case in point, with his mathematical realism directly linked to his theological views about divine creation. Mathematical ideas, he insisted, "are the cause of natural things (a teaching that Aristotle rejected in so many places) [because] God the Creator had mathematicals with him as archetypes from eternity in their simplest divine state of abstraction, even from quantities themselves, considered in their material aspect." This option was unavailable to Aristotle, who "denied the existence of a Creator, and decided that the universe was eternal."[11] Kepler thus thought that expertise in mathematical astronomy was a sufficient qualification for making pronouncements in the sphere of theology, and he said so explicitly.[12] The offices of astronomer and theologian could thus overlap (reflecting also, to some degree, the Protestant collapse of the boundaries between the estates of clergy and laity).

This kind of boundary crossing was not solely the province of Protestants, however. Galileo was an inveterate boundary crosser, insisting, for example, that the book of nature was written in the language of mathematics and that mathematicians (rather than natural philosophers) were uniquely placed to interpret it.[13] Notoriously, in the *Letter to the Grand Duchess Christina* he set about correcting traditional interpretations of various passages of scripture, seeking to align them with Copernicanism. The implication was that mathematicians were qualified to read not only the book of nature but the book of scripture as well. This latter move contributed to his collision with the Holy Office.

These physico-theological trespasses represent what Amos Funkenstein has referred to as "a unique approach to matters divine, a secular theology of sorts" that appears for the first time in the sixteenth and seventeenth centuries.[14] Its practitioners were Galileo, Descartes, Leibniz, and Newton.[15] Physico-theological approaches are thus exemplifications of this "secular theology."[16] On analogy with the other hybrid methods, what we are looking for is less a coherent genre than a conscious recognition of the merits of flouting traditional disciplinary boundaries accompanied by attempts to redefine vocational roles around remade disciplinary boundaries. In the case of physico-theological approaches, then, we would expect to see new connections forged between theology and natural philosophy or natural history, along with the demand that natural philosophers or natural historians be permitted to contribute to theological discourse and debate.

Physico-theological Writings in Seventeenth-Century England

The label "physico-theological" also appears in the work of four writers in seventeenth-century England. Walter Charleton and Samuel Parker, though, do not conform to the pattern sketched out here. However, Robert Boyle and John Ray easily fit the pattern. In addition, a few other writers do not specifically employ the term but nevertheless exemplify some common elements. Henry More is important here, as too are Thomas Burnet and William Whiston.

The works of the first two authors just mentioned are not an obvious fit. Walter Charleton's *The Darknes of Atheism Dispelled by the Light of Nature. A Physico-Theologicall-Treatise* (1652) is, to the best of my knowledge, the first book to use the term "physico-theological." What Charleton means by this approach maps more or less on what we would now understand as natural theology, and indeed he later refers to this book as "a natural theology."[17] His aim is "the *Demonstration of the Existence of God*, by beams universally de-radiated from that Catholick *Criterion*, the *Light of Nature*."[18]

The occasion for Charleton's work is significant. If Kepler and Galileo were interested in the fit between new natural philosophies and theology, Charleton's self-declared concern is the belief that both "religion" and "the sacred authority of the Church" had been shattered by "our fatal Civill Warre." This resulted in heresies, enthusiasms, and even atheism. To counter these worrying developments, Charleton sets out what we would call the three classical arguments for God's existence. There is little contemporary natural philosophy here. Much of the argument rests upon a humanist style of argumentation that draws upon a variety of authors from antiquity to the present.

We do get a few hints of novel disciplinary transgression. For Charleton, atheism is to be subverted by "a Countermine of arguments purely Physicall."[19] These are to be provided not by theologians but by philosophers. The truths of natural theology, he says, "are to be demonstrated by *Philosophers*, rather then [*sic*] *Divines*."[20] Charleton elsewhere deploys the hyphenated "physico-anatomicae" to construct new medical knowledge based on mechanical explanations.[21] This latter category gives a better insight into what is going on in Charleton's *Darknes of Atheism*, and that is, as Holger Steinmann has suggested, more medico-theological than a strictly physico-theological argument. So overall, while Charleton is a boundary crosser of sorts, I am inclined to agree with Steinmann and Katherine Calloway, both of whom, for different reasons, decline to identify Charleton as progenitor of physico-theology.[22]

Like Charleton, Samuel Parker, in his *Tentamina physico-theologica de Deo* (1665), defends Epicurus and, in particular, Epicurean moral theory, although he is less eager to endorse atomism.[23] He also treats versions of design—the motions of the heavenly bodies and the uses of various parts of living bodies.[24] For the latter he draws on Galen's *De usu partium* and the more recent anatomical discoveries of Thomas Willis (1621–75).[25] Generally, Parker argues for evidence of providential design in animal and human bodies, although the form of the argument goes back to Galen, and the modern data are not different in kind. He attacks Descartes's version of the ontological argument, consistent with a general pattern of English seventeenth-century natural theology that looks to empirical, rather than metaphysical, argumentation.[26] To that extent, he is part of a trend to deploy a posteriori arguments that will draw upon natural philosophy. Parker also insists that we have no innate ideas of God and appeals to historical and cross-cultural data to support this claim. On the issue of boundary transgression, Parker elsewhere comes down on the side of observing the traditional divisions, critiquing not only other kinds of physico-theological combination but also the physico-mathematical approach.[27]

Parker's *Tentamina* might seem to undermine the idea that there is single coherent enterprise in seventeenth-century England called physico-theology.[28] But we could say alternatively that Parker's work simply exemplifies the possibility for the promiscuous deployment of the expression "physico-theological," and that his work points to contemporary disagreement about the proper way to deploy philosophy or natural philosophy in defense of religion. In fact, Parker's book helpfully identifies a competing and discrete form of physico-theological practice that was already well developed in Henry More's work and that would be taken up by other members of the so-called Cambridge Platonists, by Thomas Burnet and William Whiston, and by Continental Pietists such as Johann Jacob Zimmermann. This genre of physico-theological writing brought together natural philosophy and aspects of scripture and revealed religion in precisely the way that Parker seemed to oppose, and it often relied heavily upon the principles of Platonism and Cartesianism, which Parker also found uncongenial. At the same time, *Tentamina* also exhibits what Dmitri Levitin calls the "humanist mindset" that sought to prioritize scholarly, textual endeavors over more empirically oriented natural philosophical approaches that rely on "scientific" evidence (although these two strategies are by no means incompatible).[29]

Henry More, Joseph Glanvill, Robert Boyle, Thomas Burnet, William Whiston, and John Ray all sought to bring together natural philosophy and revealed or biblical truths—precisely the things that for Parker could not be known "but by revelation." Moreover, these thinkers would argue that the tasks of theology and natural philosophy significantly overlap. In short, they held that inquiry into the truths of revelation was not off-limits to natural philosophers.

Before looking at the content of these later physico-theological approaches, I briefly consider the general "vocational" or disciplinary aspects of this broad tendency—the philosophy-theology combination. Henry More, for example, proposes that every priest should strive to be "a philosopher."[30] He also maintains that scripture itself contains "a various Intertexture of Theosophical and Philosophical Truths, many Physical and Metaphysical Theorems."[31] The method for interpreting scripture must include an approach that he calls "physico-theosophical."[32] Biblical exegesis and reflection upon revealed truths thus require physico-theological work in a much stronger sense than what Parker would allow. In the same vein, Thomas Sprat maintains that the praises of the informed modern natural philosopher will prove more acceptable to God than "the blind applauses of the ignorant."[33] Joseph Glanvill similarly declares that "there was never more need [in our day] that the Priests should be Philosophers."[34] Finally, Robert Boyle comes at this from the other side, maintaining that philosophers need to act as priests.[35]

This advocacy of a theological role for natural philosophers was matched in the works of these writers by their attempts to apply physical or natural philosophical approaches to specific theological themes or sacred history. Again, the case of Henry More, one of Parker's chief targets, is instructive; More uses the analogous term "physico-theosophy," and his *Antidote against Atheism* (1653) was arguably the first systematic English natural theological work to draw directly upon new natural philosophy for its arguments. He was a significant influence on a remarkable physico-theological work, Johann Jacob Zimmermann's *Exercitatio Theoreticorum Copernico-Coelestium Mathematico-Physico-Theologica* (1689).[36]

I will not dwell on the familiar "design arguments" in More's *Antidote,* but the general idea is that the "exquisite contrivance" of the parts of animals is "an undeniable Demonstration that they are the effects of Wisdome."[37] This relatively uncontroversial theme was augmented with two other related strands: first, the specific deployment of contemporary natural philosophy (in this case that of Descartes, and note that this is different from contemporary *natural history*); and, second, the treatment of biblical themes and topics that had traditionally been part of revealed theology. These two emphases are most evident in *Conjectura Cabbalistica* (1653) and subsequently in *An Explanation of the Grand Mystery of Godliness* (1660). In the former, in which he deals with the creation of the world, More notes that the divine revelations of scripture convey many natural philosophical truths.[38] Revelation and natural philosophy are entwined from the beginning. These natural philosophical doctrines turn out to be, as More puts it, "strangely agreeing with the most notorious conclusions of the Cartesian philosophy."[39] These are arguments not about the conformity of Cartesian natural philosophy with the general idea of creation but about its conformity with the specific details of the Creation as set out in the Hebrew Bible—in other words, revealed truths. In the *Grand Mystery of Godliness*, More turns his attention to biblical prophecies concerning the end of the world. Again, the physical mechanisms that lead to the world's end are said to be explicable in terms of Cartesian natural philosophy.[40] As noted earlier, all of this is entirely compatible with "humanist endeavors," and thus More also wishes to argue that, while pagan writers may have not known the precise physical details of the earth's demise, nonetheless we encounter some version of a final conflagration in the Sybills, Pythagoras, Heraclitus, Ovid, and the Stoics.[41]

One reason that Henry More is a key figure in the English natural theology tradition is that he combines all of these elements—the familiar arguments from contrivance or design, "humanistic appeal" to common theistic notions, novel combinations of natural philosophy and truths of revelation, and, related to this, the admixture of natural philosophy and biblical interpretation. It is significant, then, that John Ray in his *Wisdom of God* places More at the head of a list of his predecessors: "Dr. More, Dr. Cudworth, Dr. Stillingfleet now Bishop of Worcester, Dr. Parker, late of Oxon, and to name no more, the Honourable Robert Boyl, Esquire."[42] Ray's own construction of an English tradition of natural theology thus recognizes the seminal role played by More.

This particular strand of physico-theological writing was further developed by Thomas Burnet (1681, 1689) and William Whiston (1696), who, respectively, apply Cartesian and Newtonian cosmology to the biblical narrative of the Flood and to biblical prophecies of the future destruction and renovation of the earth.[43] John Ray takes up similar themes in both the classic *Wisdom of God Manifested in the Works of Creation* (1691) and his *Physico-theological Discourses* (1692, 1693). In the latter he treats the origin of the earth, the Flood, the destruction of the earth by water or fire, and the postmortem fate of human souls. Ray certainly canvasses the opinions of the ancients, but he also considers the view of contemporary philosophers, including Descartes, Pierre Gassendi, and More.

Robert Boyle, in his *Some Physico-theological Considerations about the Possibility of the Resurrection* (1675), moves from the cosmological to the personal dimension of eschatology. Boyle proposes that while the resurrection will be brought about by God's power—that is, supernaturally—he nonetheless thinks that there may be "a Plastick Power in some part of the matter of a diseased Body; whereby, being divinely excited, it may be enabled to take to its self fresh matter, and so subdue and fashion it, as thence sufficiently to repair or augment itself."[44] So at the very least, contemporary natural philosophy throws up no objections to this revealed truth and, at best, provides some hints about how it might be accomplished.

Boyle reflects more generally on the relationship between theological and natural philosophical reasoning in his later *Disquisition about the Final*

Causes of Natural Things (1688). Here he broaches the issue of the legitimacy of mixed sciences, coming out clearly in favor: "And to me 'tis not very material, whether or no, in Physics or any other Discipline, a thing be prov'd by the peculiar Principles of that Science or Discipline; provided it be firmly proved by the common grounds of Reason."[45] He goes on to distinguish "physical arguments" from "physico-theological arguments."[46] "But if the revelations contain'd in the *Holy Scriptures*, be admitted, we may rationally believe More, and speak less Haesitantly, of the Ends of God, than bare Philosophy will warrant us to do."[47] A complete natural philosophical explanation of phenomena, he had already argued, might well require recourse to theology.[48]

The more general question of the relations between contemporary natural philosophy and theology had already been broached by Boyle in *The Excellency of Theology, compar'd with Natural Philosophy* (1674):

> The Gospel comprises indeed, and unfolds the whole Mystery of Man's Redemption, as far forth as 'tis necessary to be known for our Salvation: And the *Corpuscularian* or Mechanical Philosophy, strives to deduce all the *Phoenomena* of Nature from Adiaphorous Matter, and Local Motion. But neither . . . seems to be more than an Epicycle (if I may so call it) of the Great and Universal System of God's Contrivances, and makes but a part of the more general Theory of things, knowable by the Light of Nature, improv'd by the Information of the Scriptures.[49]

In respect of the subordinate parts of theology and philosophy, both the gospel and mechanical philosophy are constituents of a general hypothesis the objects of which are "the *Nature, Counsels, and Works of God*."[50] But the partiality of our knowledge—and its distribution into truths of reason or truths of revelation—is to be attributed both to the limitations of our present earthly condition and to the progressive nature of natural philosophical enquiry, which has the capacity to change the boundary between truths of reason and of revelation. These liminal areas of knowledge include such topics as the beginning of the world, the approximate age of the earth, that the earth will come to an end, and other "discoveries" about angels, the universe, and our souls.

Conclusion

A consideration of the uses of the expression "physico-theological" in seventeenth-century England may seem an unpromising way of picking out a distinctive genre of natural theology: the term seems to be used in too great a variety of ways. My suggestion has been that if we look instead to other common disciplinary hybridizations such as physico-mathematics we get some important clues about at least some physico-theological approaches of the period. What the latter share with other hyphenated approaches are an explicit rejection of inherited disciplinary divisions, along with an attempt to rethink the identity of the practitioners of the various disciplines. In their more radical moments, these trangressionary attempts bring together natural philosophy and truths of revealed theology in quite novel ways. The most frequent topics of these speculations are creation, generation, and eschatology, and we see instances in More, Burnet, Whiston, Boyle, and Ray—although, unhelpfully, they mostly avoid using the term "physico-theological." Were we to extend the range beyond England, and beyond the seventeenth century, we would get further confirmation of the currency of this kind of physico-theological approach. Examples would include various Continental physico-theologies such as that of Zimmermann, *Exercitatio theoricorum Copernico-coelestium* (1689), and other works including Zimmermann's translation of Burnet, Jean Pierquin's *Dissertations Physico-théologiques Touchant la Conception de Jésus-Christ dans le Sein de la Vierge Marie sa Mere* (Amsterdam [Paris], 1742), and Charles Bonnet's remarkable *Palingénésie philosophique* (Geneva, 1769). This view of things is consistent with Amos Funkenstein's notion of "secular theology," with a consensus among historians of science about the distinctive discipline-busting approaches of the period, and to some degree with Fernando Vidal's suggestions of there being two kinds of physico-theology during this period.[51]

With the introduction of the noun form in Derham's *Physico-Theology* (1713), a more stable pattern of usage began to emerge along with the development of a recognizable genre of natural theology that was focused primarily on the principle of contrivance or design. Necessarily, this form was also dependent upon contemporary natural history and, to a lesser extent, natural philosophy. It was in this form that natural theology was critiqued by Hume and Kant and was ultimately thought to have fallen victim to Darwin's theory of evolution by natural selection. This idea of natural theology as an *argument* and the new terminology of "physico-theology" and "*the* physico-theological argument" were thus destined to eclipse the variety of physico-theological approaches described in this chapter, and this overshadowing of the earlier traditions has complicated our attempts to understand seventeenth-century meanings of the term "physico-theological."

NOTES

1. Dear, *Discipline and Experience*, chap. 6; Schuster, *Descartes-Agonistes*, 31–58, 108–66.

2. Aristotle, *Metaphysics* 1025b–26a. Cf. Aristotle, *Posterior Analytics* 75a–b; Aristotle, *Metaphysics* 989b–90a; Aristotle, *On the Heavens* 299a–b.

3. See Funkenstein, *Theology and the Scientific Imagination*, 35–37, 303–7; Livesey, "Divine Omnipotence and First Principles," 14–15.

4. Dear, "'Mixed Mathematics,'" 152.

5. See discussion in Dear, *Discipline and Experience*, 174–79.

6. Jardine, *Birth of History and Philosophy of Science*, 250; Westman, "Astronomer's Role in the Sixteenth Century."

7. Kepler, *Mysterium Cosmographicum*, 37.

8. Bagola, *Galileo Courtier*, esp. 3. Cf. Feldhay, "The Simulation of Nature."

9. Cunningham, "How the Principia Got Its Name."

10. Dear, *Discipline and Experience*, 247.

11. Kepler, *Mysterium Cosmographicum*, 38.

12. Kepler, *Gesammelte Werke*, 13:40.

13. Galilei, *The Assayer*, 237–38.

14. Funkenstein, *Theology and the Scientific Imagination*, 3. Also see Grant, *History of Natural Philosophy*, 239–73; Shank, "Natural Knowledge in the Latin Middle Ages," 103.

15. Funkenstein, *Theology and the Scientific Imagination*, 3–5.

16. Harrison, "Physico-theology and the Mixed Sciences."

17. Charleton, *Three Anatomic Lectures*.

18. Charleton, *Darknes of Atheism*, Advertisement to the Reader.

19. Charleton, *Darknes of Atheism*, Advertisement to the Reader.

20. Charleton, *Darknes of Atheism*, Advertisement to the Reader.

21. Charleton, *Exercitationes physico-anatomicae*. Cf. Crooke, *Mikrokosmographia*, 57.

22. Steinmann, *Absehen—Wissen—Glauben*, 19–20; Calloway, *Natural Theology*, 34.

23. Parker, *Tentamina*, 41.

24. Parker, *Tentamina*, 45–64.

25. Parker, *Tentamina*, chap. 3. For praise of Galen, p. 77.

26. Parker, *Tentamina*, 164; Charleton, *Darknes of Atheism*, sig. br, B3r–v; More, *Antidote*, 6–8.

27. Parker, *Censure of the Platonick Philosophie*, 79, 82; Parker, *Disputationes de Deo*, 281.

28. See Levitin, "Rethinking English Physico-theology."

29. See Levitin, "Rethinking English Physico-theology."

30. More, *Collection of Several Philosophical Writings*, v.

31. More, *Conjectura Cabbalistic a*, 104–5.

32. More, *Divine Dialogues*, 565.

33. Sprat, *History of the Royal Society*, 349–45. Cf. Stubby, *A reply unto the letter written to Mr. Henry stubble*, 25.

34. Glanvill, *Essays*, essay 4, p. 42.

35. Boyle, *The Christian Virtuoso*, 538 (cf. 531); Boyle, *Some Considerations touching the Usefulnesse of Experimental Naturall Philosophy*, 18–20. Cf. Ray, *Three physico-theological Discourses*, pref.

36. For Zimmerman's physico-theology see Zuber, "Copernican Cosmotheism."

37. More, *Antidote*, 5. Also see Harrison, *The Bible*, 171–72.

38. More, *Conjectura Cabbalistica*, pref.

39. More, *Conjectura Cabbalistica*, 135, 150, 161–62. See also Harrison, "Cartesian Cosmology in England."

40. More, *Grand Mystery of Godliness*, 240.

41. More, *Grand Mystery of Godliness*, x.

42. Ray, *Wisdom of God*, pref.

43. Burnet, *Telluris theoria sacra*; Whiston, *New Theory of the Earth*.

44. Boyle, *Some Physico-theological Considerations*, 10–11.

45. Boyle, *Disquisition about Final Causes*, 23–24.

46. Boyle, *Disquisition about Final Causes*, 105.

47. Boyle, *Excellency of Theology*, 80.

48. "[W]e shall perhaps finde the more Catholick and Primary causes of Things . . . of all which it will be difficult to give a satisfactory Account, without acknowledging an intelligent Author or Disposer of Things." Boyle, *Some Considerations touching the Usefulnesse of experimental naturall philosophy*, 67.

49. Boyle, *Excellency of Theology*, 51.

50. Boyle, *Excellency of Theology*, 52.

51. Vidal, "Extraordinary Bodies"; Funkenstein, *Theology and the Scientific Imagination*.

CHAPTER THREE

The Form of a Flower

JONATHAN SHEEHAN

> The heart is ever gladdened by the beauty, the exquisite spontaneity, with which life seeks and takes on its forms in an accord perfectly responsive to its needs. It seems ever as though the life and the form were absolutely one and inseparable, so adequate is the sense of fulfillment.
>
> Louis Sullivan, "The Tall Office Building Artistically Considered" (1896)

Form is a curious and powerful idea. On the one hand, it names something simple, the shape of a thing, how it appears to our eyes. On the other hand, it conveys something much deeper about the world. What is the form of a perfect skyscraper? asked the modernist American architect Louis Sullivan in 1896. We might list some common features—taller than they are wide, for example, windows on the sides, and so on. How it looks does not, however, exhaust the form. The form is more, "an outward semblance that tells us what they are"; it perfects a thing, gives it "life" for everyone to see.[1] This was an old thought. Aristotle too thought that form exceeds appearance, that form determines what the thing is, a statue determined as a statue by its form. For ancient and modern alike, form tells us about the purpose of things, the ends toward which they are oriented. A statue might look like a statue but *be* a door knocker, or a hat pin, or an improbably shaped sport of nature. What makes a skyscraper a skyscraper—or an apple blossom an apple blossom—is not merely how it looks but the purpose that it serves and the end that it perfects.

That a version of Aristotle's purposes might persist even in the high-modernist Chicago skyline testifies to the durable creativity of the teleological imagination. This was no less true in the first century of modern science, when physico-theology emerged as a partner to the scientific revolution and invested nature with purposes visible in its parts and coordinated into systems that witnessed the divine authorship of things. In the 1691 formulation of English botanist and Royal Society member John Ray, there is an "admirable art and wisdom that discovers itself in the make and constitution, the order and disposition, the ends and uses of all the parts and members of this stately fabrick of Heaven and Earth."[2]

From the perspective of a crude scientism, this seems a mistake. The destruction of the Aristotelian physical framework from Descartes onward was supposed to eliminate formal and final causes from scientific explanation. Natural laws are the province of efficient causes; they describe observed regularities without feigned hypotheses about *why* nature is ordered in this particular way and not some other. In this view, the unflagging attention to purposeful nature that still attended the scientific revolution, whether in the form of physico-theology or otherwise, might be deemed either a mistake or a kind of atavism that will be overcome when the natural sciences finally attain independence from the Christian-Aristotelian worldview.

From another perspective, however, physico-theology was interesting less for its piety than for the consequential problems that it attempted to resolve. These problems were deeper than that of reconciling God and science or of proving God's existence in the face of new empiricism. At the root of physico-theology lay questions fundamental to the philosophical (and theological) enterprise itself: Why is there something rather than nothing? Why is this something ordered and not simply a chaotic aggregate? What is the relationship between individual things and the larger order to which they belong? These were *the* questions motivating physico-theology and indeed much of Enlightenment natural science, as well as its philosophy, psychology, economics, and politics, for that matter.[3] We see these problems at play already in John Ray, with his interest in the origin of the various orders that we observe and inhabit, the nature of these orders, the relationship between individual parts and organizing wholes, and so on.

Among them was the question of purposes, the "ends" that Ray mentioned. Already in 1688, in his *Disquisition about Final Causes*, Robert Boyle developed a measured defense of final causes, at least insofar as they cannot be *ruled out* in some domains of natural explanation. As he put it, "There are some effects that are so easy and so ready to be produced that they do not infer any knowledge or intention in their causes, but there are others that require such a number and concourse of conspiring causes, and such a series of motions and operations, that 'tis utterly improbable they should be produced without the superintendency of a rational agent, wise and powerful."[4]

At issue was whether chance was enough to explain the origins of complex and organized bodies. The conviction that it was not—that as complexity grew in scope from inanimate to animate nature it would demand a "concourse of conspiring causes"—drove the mind, in Boyle's view, to the conviction that the world was filled with the purposes of the divine author. Physico-theology generalized these convictions across many intellectual terrains, as the chapters in this volume testify. But the issue of purposes assumed real conceptual clarity only in 1728 with the German metaphysician Christian Wolff, who gave the science of purposes the name that it still retains today: "teleology."[5]

Even before he named it, Wolff showed robust interest in the science of divine intentions. His 1725 work on the "parts of man, animals, and plants," for example, sought to demonstrate how the parts of a plant (roots, bark, wood, and so on) function in service of God's ultimate purpose—namely, to nourish and sustain vegetable life. Flowers "exist for the sake of the seed," fertile seeds serve to preserve plant species "as long as the world continues," excess seeds are made to feed humans and animals, everything tied (*verknüpfft*) together as God intended.[6] *Consona qui diversa sonant* (the harmony of diversity), as the frontispiece proclaimed, gives us a world knit together sustaining God's highest creation, human beings.

Flowers, their forms, and the play of purposes: in them, this chapter discovers some of the imaginative power of physico-theology, the questions it found urgent, and the legacies it left for eighteenth-century thought. In a lovely essay on flowers and the imagination, the literary critic Elaine Scarry remarks how "a blossom lends itself to be imagined, to being mentally captured in nearly the same degree of extraordinary vivacity it has in the perceptual world." Flowers "lift us above the material world," she continues, disencumbering us of the givenness of things, freeing us in a way from the chains that nature lays upon us.[7] On the face of it, Wolff could not have disagreed more, seeing in flowers convincing witnesses to divine necessity and order. If flowers are reduced to the instrumental services they provide plants and those nourished by them, however, what are we to make of the perceptually evident feature of flowers, their beauty, their "exquisite spontaneity," as Louis Sullivan put it? The eighteenth century would take up questions just like this. In the form of the flower, natural scientists, poets, and philosophers discovered clues to the greatest mysteries of all: the purposes of God, the organization of the world, and the nature of human freedom.

In the wake of Wolff's essay, fellow German physico-theologists grew curious about flowers too. The Hamburg-based literary society, the Deutsche

Gesellschaft, for example, offered a prize in 1737 for essays addressing the question that Wolff ignored: What purposes of the Creator can be found in the figure (*Darstellung*) of a flower? "Shouldn't their beauty be bound to usefulness?" asked the young mathematician Georg Friedrich Bärmann in his winning entry. Praising Wolff, William Derham, and Ray, Bärmann wrote about the importance of the petals in the protection of the seeds, the importance of the stamen in the process of plant reproduction, and so on.[8] Unlike these other pillars of the physico-theological tradition, however, he took the question of usefulness in a new direction, asking in effect why, among all the possible means God might have used to sustain the vegetable economy, he chose such a beautiful one. What, in other words, is the use of beauty?

Flowers pose this question most sharply, because, as Bärmann wrote, there is "no fashion among them."[9] They are uniquely beautiful, appealing to all people at all times. We discover in the flower, this universality tells us, *another* universality—namely, the universality of God's beautiful creation, "the most beautiful order, the most beautiful harmony [*Zusammenstimmung*]" condensed in the form of a flower.[10] "Are you not astonished, O mankind, [to see] in us, as small as we are, the greatness of his power?" the flowers ask us.[11]

The flower isolated, in other words, the beautiful from the general economy of nature. Perhaps for this reason, authors like Johann Benemann, in his *Gedancken über das Reich derer Blumen* (1740), paid special attention to flowers that seemed useless to the conservation of vegetable life. Flowers that "hold no fruit, nor even seed in themselves"—these must have "an entirely different purpose" than the other parts of plants.[12] In his immensely popular *Spectacle de la nature* (1732), Noël-Antoine Pluche had thought along similar lines. Perhaps the "the flowers' ultimate end was to generate the seed of the plant," but we must recognize that "they are at the same time intended to adorn our travels with . . . the radiance of their colors." There are many flowers—the double hyacinth beloved in the eighteenth century, for example—which have "no other merit than their finery."[13]

Aesthetics was in the air in the 1730s and 1740s, as Simon Grote has shown, emerging at the intersection of moral philosophy, theology, and natural philosophy.[14] Already from the early eighteenth century onward, in fact, rumination on the order of things often turned to beauty, understood to name the coherence of individual things into perfected and united wholes. Beauty was a powerful antidote, in a sense, to chance, to the problem of the chaotic aggregate, as I put it earlier. What a difference there is, the Earl of Shaftesbury exclaimed in 1711, "between an organized body and a mist or cloud driven by the wind!" Order, proportion, unity: a "system of parts" coheres in a beautiful whole testifying to the grand design of things.[15]

If beauty depends on a certain relation of part to whole, then the beauty of a flower depends on its subordination to, and thus revelation of, some larger design. Different designs might be revealed. For Shaftesbury and Bärmann, it was the order of the universe itself that was ultimately mirrored in all its parts. For Benemann and Pluche, it had more to do with God's kindliness and love toward his special creation, mankind, whose pleasure in floral beauty mirrors the pleasure that God takes in the things that he loves. "Divine wisdom resembles a tender mother to whom all of the needs of her children are dear, [and] who, without demeaning herself, deigns to play with them, and interest herself in their pleasures," Pluche wrote.[16] In either case, though, the purpose of a beautiful flower is ultimately the purpose of another, usually God, whose ultimate purposes are revealed in the pleasures afforded a grateful humanity.

Could a flower, however, have its *own* purposes? Put another way, can we imagine a flower not just as a "part," or in some relationship of subordination, but as an autonomous thing, with its own purposes? For flowers, the stakes might seem small, but if we substitute humans as the subject, they grow quickly. Physico-theology imagined a world of interlocking parts, all operating smoothly to further God's wise and grand design. However, in the context of what were, from the late seventeenth century onward, urgent discussions about freedom, necessity, theodicy, divine justice, and human autonomy, such causal tightness could threaten as much as defend the moral economy of the world.

Few people were as invested in this question of autonomy as the enthusiastic amateur botanist, and floral enthusiast, Jean-Jacques Rousseau. Flowers and plants might not be self-legislating, but they were interesting to Rousseau to the extent that they *resisted* an instrumental calculus, and especially the calculus of human benefit that Wolff, Bärmann, Benemann, Pluche, and many physico-theologians emphasized. Let doctors confine themselves to the benefits of the plant kingdom, but for *botanists*, he wrote, "as soon as [the plant's] form is destroyed and ground in a mortar it is no longer anything." A true botanist has eyes not for instrumental use, but delights in the "most elegant forms, the most lively colors, the charming flowers, the delicious perfumes."[17]

Flowers were especially interesting to him. "The fields covered with flowers are the sole laboratory of the botanist," Rousseau wrote in a fragment, the place where nature showcases the elegance of its "forms" and "distributions."[18] The "flower . . . comes first. It is in this part that nature has enclosed the summary of her work," he wrote in a 1771 letter to his botanical pupil Madeleine-Catherine Delessert.[19] Far more than just an instrument of reproduction, the flower is notable for its complexity, for its beauty, for its fragrance, all things of principal interest to any botanist worth the name. This complexity is the proper subject for the science of botany, which should interest itself above all in the "mutual correspondence" among the parts of the plant that make it a unity. The botanist must not subject these parts to any instrumental idea of purposiveness, as "ingredients for enemas," as he put it.[20] "These forms [*figures*], these colors, this symmetry," Rousseau wrote about flowers, "were not put here for nothing"; their purposes are just that—not *our* purposes, but rather their own.[21]

Put in Aristotelian terms, the purpose of a plant is internal, what Aristotle would have called its *entelechy* or *energeia*, the realization of the *telos* to which a plant is oriented. *Entelechy* and *energeia* were, in turn, intimately connected to *form*: the "actuality [*energeia*] is the end, and it is for the sake of this that the potentiality is acquired . . . matter exists in a potential state, just because it may attain to its form; and when it exists actually, then it is in its form."[22] It is unsurprising then that the language of form recurs in Rousseau as well, naming as it does both the external *shape* of the flower and the actuality that a living plant embodies: "The botanist studies in plants their tissue, their shape, their organization, their generation, their birth, their growth, their life, and their death. He can also consider them by their color, their taste, their order, their flavor . . . but this is only an analogical and secondary study for clarifying and confirming that of forms."[23]

Form was not a well-specified concept for Rousseau. Perhaps it would have become so if he had written a complete treatise on botany, but even in these vague statements, we can see form working as an alternative to "part," a way of giving "an identity to natural beings," in Jean Starobinski's apt phrasing.[24]

As it turned out, form would become a concept of considerable use in the mid-eighteenth century for natural historians trying to address the *same* questions so urgent to physico-theology but casting around for *different* kinds of answers. Rousseau botanized with Linnaeus tucked under his arm, and it is in Linnaeus that we find some of the most sustained commitment to a neo-Aristotelian language of form. The "continuity of form" was visible, he thought, in the development of plants, whose different parts emerge out from the interaction of a form-generating medulla and a form-constraining cortex.[25] In the flower, "the inner structure of a plant, its *real* structure, becomes evident to the human eye," because there we see most clearly the "form-generating" operation of the medulla.[26] The French naturalist Georges Buffon objected to the entire taxonomic project of Linnaeus, but he too found form useful to think with. In explaining how living beings become the beings that they become, and not something else, he developed the idea of the "interior mold," an "inner form" (as a contemporary German critic translated it) to which particular organic development strives.[27] This was an awkward claim for Buffon, who insisted that in nature there is nothing but "individuals and successions of individuals" and thus would seem to have had little patience for an empirically undetectable "mold" that determines which individual each thing will become.[28] But the determination of an individual *as* an individual—and not some other kind of thing, or just an example of a generic species—seemed to demand the language of form.

The most influential biological "form" of the later eighteenth century, though, came from Johann Friedrich Blumenbach, professor of comparative anatomy at Göttingen and one of the most influential natural scientists of that era. In his work on the forces of nature, Blumenbach discovered what he called the *nisus formativus*, the *Bildungstrieb*, a "particular drive" found in every organized body "to assume at the beginning its determined form [*Gestalt*], and then to preserve it throughout its life."[29] For Blumenbach, "form" (both *Bildung* and *Gestalt*) responded to what we can call the physico-theology of the seed.

Already we have observed how important the seed was to the discussion of flowers—the seed gave the flowers a purpose in the larger economy of nature. But seeds were everywhere in physico-theological speculation from the late seventeenth century onward. Macro-systematics, the planet-wide coordination of natural systems, both inorganic and organic, paralleled a micro-systematics that coordinated the living organisms with the logic of seeds. "An entire tulip is seen in the seed of a tulip bulb," the Oratorian Nicolas Malebranche wrote in 1674, collapsing together the problems of individuation (why one thing becomes *that* thing) and speciation (how individuals are gathered into collectives).[30] For Gottfried Wilhelm Leibniz, Jan Swammerdam, and the generations of natural historians that followed, seeds offered both a metaphor for, and an embodiment of, God's direction of natural purposes. God makes the seeds from which every organism comes, every seed contains within it every future seed, and so every individual is both expression and bearer of the species. The "purpose" of a plant is simply to become the thing that it was already designed to become. Put another way, the seed collapses the distinction between potentiality and actuality, between matter and form, so important to Aristotle.

Blumenbach found seeds inadequate to explain, however, the hybridization experiments of Joseph Kölreuter (1733–1806), who managed a "completely perfected transformation of one natural plant species into another" by artificial pollination of tobacco flowers.[31] The "entire form of the maternal seed was destroyed and transformed into another," form succeeding form in plastic yet constrained manners.[32] In abandoning seeds, Blumenbach by no means abandoned purposes. Seed theory outsourced purposes to a divine beyond; the *nisus formativus* made purposes immanent in the organisms. Skeptics, Blumenbach later wrote, mock purposes, but anyone who has studied nature cannot help noticing the "preestablished harmony . . . between the purposive form of the creature and its way of life."[33] Yet form and purpose should not be exported to God's secret domain. Rather, they are visible here and now, visible in the process of generation and growth, in the anatomical structures of natural organisms, and in the codetermination of these structures and lived behaviors.

Form had quite a resurgence in the later eighteenth century—in Germany, we can find the concept structuring inquiry into aesthetics, logic, music theory, grammar, and probably many other disciplines, and there are some analogues in France as well, at least in the biological sciences. Goethe's later science of morphology, as well as his interest in the *Urformen* of plants, is only the most well-known example of this wider phenomenon. Constraints here forbid a fuller exploration of how a very old idea got such vigorous new life in the late Enlightenment. Although reports of the death of Aristotle have long been exaggerated in my view, what we are seeing here is more than just a return to the Greeks. Rather, we see instead a return to the *questions* that Aristotle tried so systematically to answer: Why is there something rather than nothing? Why is this something ordered and not simply random? And so on.

Physico-theology was a remarkably coherent effort to answer these. It tightly coordinated mechanist efficient causation with divinely ordered formal and final causation. As a result, it had to answer neither macroscopic questions about the origins of natural systems nor microscopic questions about the origins of natural organisms. The doctrine of seeds made the distinction between matter and form analytically uninteresting—hence, the well-known language of "preformationism"—and the purposes that form used to compass were given over to divine wisdom. But for those who, like

Rousseau, Buffon, and Blumenbach, were interested in *immanent* teleology, in seeing purposiveness within rather than beyond the natural world, form assumed conceptual importance so that they could articulate both the difference between simple matter and complex organisms and the coordination of organisms with the natural systems they inhabit.

As a final example of form's new power and range—and a final apostrophe to the flower—we will end with the man who piqued Goethe's morphological curiosity.[34] From at least 1763 onward, Immanuel Kant was interested in the physico-theological project. His essay "The Only Possible Argument in Support of the Demonstration of the Existence of God" saw the physico-theological proof of divine existence as essentially *formal*—just as Aristotle "derived not the matter or stuff of nature, but only its form, from God," so too can only the formal unity of natural systems (the ways, say, that the parts of a plant formally cohere) testify to God's existence.[35] By 1781, however, Kant declared the physico-theological proof of God impossible. Although "the present world discloses to us such an immeasurable showplace of manifoldness, order, purposiveness, and beauty . . . that our judgement upon the whole must resolve itself into . . . astonishment," to proceed from this astonishment to God's existence would be to "bend" nature to our own ends.[36] The problem is that we can make inferences about this showplace with regard only to its *form*, not to its underlying substance. As Kant put it:

> The purposiveness and well-adaptedness of so many natural arrangements would have to prove merely the contingency of the form, but not of the matter, i.e., of substance, in the world; for the latter would further require that it be able to be proved that the things of the world would in themselves be unsuited for such an order and harmony according to universal laws if they were not in their substance the product of a highest wisdom; but entirely different grounds of proof from those provided by the analogy with human art would be required for this. Thus the proof could at most establish a highest *architect* of the world, who would always be limited by the suitability of the material on which he works, but not a *creator* of the world, to whose idea everything is subject.[37]

In other words, the evident purposes that we see in natural systems—the way that, for example, teeth, saliva, and stomach coordinate into a digestive system—can show us only that these systems might have been ordered in a different way. Their "forms" are, in Kant's terms, contingent. But from this contingency of form, we can infer nothing about the substantial *creation* of nature. As proof of the existence of God, then, physico-theology must necessarily fail.

Kant's critique of physico-theology laid an important foundation for his 1790 *Critique of Judgment*, which one might justly understand as one long and complex investigation of form (the aesthetic judgment) and purpose (the teleological judgment)—that is, the two halves of the critique. Space is brief, so let us take just one example from the first half. What Kant called the "pure" judgment of taste arises, he argued, when a viewer judges something beautiful without any sense of what the object is supposed to be, what use it might serve, what end it might accomplish.

We find "free beauty," as he called it, above all in flowers: "Flowers are free natural beauties. Hardly anyone other than the botanist knows what sort of thing a flower is supposed to be; and even the botanist, who recognizes in it the reproductive organ of the plant, pays no attention to this natural end if he judges the flower by means of taste. Thus this judgement is not grounded on any kind of perfection, any internal purposiveness to which the composition of the manifold is related."[38] What Kant means is reasonably clear. Insofar as we are concerned to distinguish the aesthetic judgment from the activity of reason, we cannot subordinate the former to the latter. Once concepts are in play—for example, the flower is for making seeds—the aesthetic judgment has been usurped by judgments of reason. Understood as "free," then, flowers invite Kant to define beauty this way: "the form of purposiveness [*Form der Zweckmäßigkeit*] of an object, insofar as it is perceived in it without representation of an end." The tulip, he writes in a note, is held to be beautiful because of a "certain purposiveness" unrelated to a distinct end, or *telos*.[39]

Although form is everywhere in Kant's thought, here it interests especially in light of our organizing theme.[40] The experience of flowers, for Kant, models a way of seeing the world *freely*, free especially from the subordination of purpose, divine or otherwise. The flower offers to our view the "*form of purposiveness*," that is, the autonomy of an object understood as having its *own* purpose free of any end to which it might be put. As in Rousseau earlier, form here designates this autonomy. The form of a flower does not mean merely the flower's shape; rather, it is what determines the flower as *its own* thing, having ends of its own. And it is precisely in this formal freedom that, Kant says, we take pleasure in beautiful flowers, because they show us the "free play" of our *own* cognitive capacities. The purposiveness without subordination that we experience in the flower gives us an experience of our own purposiveness without subordination. The law of form, Sullivan much later wrote, opens up "the airy sunshine of green fields and gives to us . . . freedom" and clears room for our "own characteristic individuality."[41] Kant and others in the later eighteenth century would have agreed. We are set free of physico-theology, set free from the logic of divine determination, by the form of the flower.

NOTES

1. Sullivan, "Tall Office Building," 408.
2. Ray, *Wisdom of God* (1691), 11–12.
3. For a broader exploration of these questions, see Sheehan and Wahrman, *Invisible Hands*.
4. Boyle, *Disquisition about Final Causes*, 45.
5. Wolff, *Philosophia rationalis sive Logica*, 38. For Wolff on teleology, see van den Berg, "Wolffian Roots of Kant's Teleology"; and Buchenau, "Die Teleologie zwischen Physik und Theologie."
6. Wolff, *Vernünfftige Gedancken von dem Gebrauche der Theile in Menschen, Thieren, und Pflanzen*, 727, 734, 735.
7. Scarry, "Imagining Flowers," 91, 92.
8. Bärmann, "Abhandlung von den Absichten des Schöpfers, bey Darstellung der Blumen," 285.
9. Bärmann, "Abhandlung von den Absichten des Schöpfers, bey Darstellung der Blumen," 298.
10. Bärmann, "Abhandlung von den Absichten des Schöpfers, bey Darstellung der Blumen," 306.
11. Bärmann, "Abhandlung von den Absichten des Schöpfers, bey Darstellung der Blumen," 304.
12. Benemann, *Gedancken über das Reich derer Blumen*, 30.
13. Pluche, *Le Spectacle*, 3:494. On the contemporary love of the hyacinth, see Hyde, *Cultivated Power*, 69–71.
14. Grote, *Emergence of Modern Aesthetic Theory*.
15. Shaftesbury, *Characteristics*, 274.
16. Pluche, *Le Spectacle*, 3:494.
17. Rousseau, "Fragments on Botany," in *Collected Writings*, 8:251. For Rousseau on flowers, see Scott, "Rousseau and Flowers."
18. Rousseau, "Fragments on Botany," 251.
19. Rousseau, "Letters of Mme Madeleine-Catherine Delessert," in *Collected Writings*, 8:131.
20. Rousseau, "Fragments on Botany," 252.
21. Rousseau, "Fragments on Botany," 252. On Rousseau and botanical form, see also Kuhn, "'A Chain of Marvels,'" 5.
22. Aristotle, *Metaphysics*1050a9.15, in *Complete Works*, 2:1658.
23. Rousseau, "Fragments on Botany," 250.
24. Starobinski, "Rousseau's Happy Days," 150.
25. Here I am following the remarkable article of Stevens and Cullen, "Linnaeus."
26. Stevens and Cullen, "Linnaeus," 209.
27. Reill, *Vitalizing Nature in the Enlightenment*, 47.
28. Sloan, "The Buffon-Linnaeus Controversy," 374.
29. Blumenbach, *Bildungstrieb*, 24.
30. Malebranche, *The Search after Truth*, 27.

31. Blumenbach, *Bildungstrieb*, 68.

32. Blumenbach, *Bildungstrieb*, 69.

33. Blumenbach, *Beyträge zur Naturgeschichte*, 1:128.

34. On Goethe's relation to Kant here, see Lenoir, "Eternal Laws of Form." More generally, see Richards, *Romantic Conception of Life*.

35. Kant, "The Only Possible Argument, 165–66.

36. Kant, *Critique of Pure Reason*, 579, 581.

37. Kant, *Critique of Pure Reason*, 581.

38. Kant, *Critique of the Power of Judgment*, 114.

39. Kant, *Critique of the Power of Judgment*, 120.

40. More generally, see Pippin, *Kant's Theory of Form*.

41. Sullivan, "Tall Office Building," 409.

PART TWO

NATIONAL TRADITIONS

CHAPTER FOUR

What Was Physico-theology For?

SCOTT MANDELBROTE

The first use of the adjective "physico-theological" in print in England was by the physician and Christian synthesizer of Epicurean ideas Walter Charleton (1619–1707) in the title of his book *The Darknes of Atheism Dispelled by the Light of Nature: A Physico-Theologicall Treatise* (London, 1652).[1] Similar adjectives are to be found in the expanding vocabulary employed at the time to describe the scope of natural philosophy. Many of them were deployed by Robert Boyle (1627–91), in particular "physico-mechanical."[2] In Latin, French, and later English, terms of this kind were widely used by correspondents of the intelligencer Samuel Hartlib (d. 1662), who was based in London during the 1640s and 1650s ("physico-technico-mysticus"; "physico-mechanicus"; "physico-magicus"; "physico-astrologicus," later "physico-astrological"; "theologo-physico-mathematique").[3] Interestingly, in one translation that Hartlib published, the term "physico-technico-mysticus" was rendered as part of the phrase "In which harmonicall One-triple, *viz.* Naturall, Artificiall, Mysticall Systeem, or Body, all Arts and Sciences . . . are . . . implicitly imbosomed."[4] Boyle also used the adjective "physico-theological," again in the title of a book: his 1675 discussion of the chemical and other natural philosophical effects that might make plausible in natural philosophy the literal interpretation of the revealed truth that there would be a resurrection of the same body.[5]

Boyle later defined what might be meant by the term "physico-theological" in his 1688 *Disquisition about the Final Causes of Natural Things.* He began by arguing:

> And, to make way for what I am to offer by a Distinction, the want of which seems to have contributed to the Obscurity of my Subject; I shall observe to you, that there are two ways of Reasoning from the Final Causes of Natural Things, that ought not to be Confounded.

He then established that:

> Sometimes from the Uses of things Men draw Arguments that relate to the
> Author of Nature, and the General Ends he is suppos'd to have intended in
> things Corporeal: As, when from the manifest Usefulness of the Eyes, and all
> its parts, to the Function of Seeing, Men infer, that at the Beginning of Things
> the Eye was fram'd by a very Intelligent Being, that had a particular care, that
> Animals, especially Men, should be furnish'd with the fittest Organ of so nec-
> essary a Sense as that of Sight.

From this, he went on to distinguish his second form of reasoning:

> And Sometimes also, upon the supposed Ends of things Men Ground Argu-
> ments, both Affirmative and Negative, about the peculiar Nature of the Things
> themselves; and Conclude, that This Affection of a Natural Body or Part ought
> to be granted, or That to be denied, because by This, and not by That, or by
> This more than by That, the End design'd by Nature may be best and most
> conveniently attain'd.

Finally, Boyle set out a clear distinction between physico-theological and
physical arguments, based on the two forms of reasoning he had outlined:

> This latter sort of Arguments I am wont to call purely or simply, *Physical* Ones;
> and those of the former sorts may, for distinctions sake, be styl'd *Physico-
> Theological* Ones; or (if we will with *Verulamius* refer Final Causes to the
> Metaphysicks,) by a somewhat shorter name, *Metaphysical* Ones.[6]

Both Charleton's usage and that of Boyle allow one to place the term
"physico-theological" within a much older history of natural theology. Charleton was explicit about the debt that he owed to older authors, in particular
to the natural theologies of Philippe Duplessis-Mornay (*De Veritate*, 1583 and
many reprints), Juan Luis Vives (*De Veritate fidei Christianae*, 1543), and Raymond Sebond (*Liber naturae sive creaturarum*, written 1434–36, printed
from 1487 and published in French by Montaigne in 1569). Here and elsewhere
in his diverse publications, Charleton shared with such writers a key component of natural theology that was more broadly understood: apologetics.

Natural theology was initially successful as a response to classical and Renaissance ideas of skepticism. A dialogue with classical interpretations of
nature and divinity remained critical in the formation of early modern interpretations of natural theology. While the importance for early modern writers of countering ancient forms of atheism has often been recognized, the sur-

vival of classical writers as genuine interlocutors in this period does not always receive the attention that it deserves.[7] Moreover, religious choices were responses to a divine call, rather than being simple acts of human rationality. Remnants of common belief thus also justified attempts to convert pagans and made it possible to propose accommodation between Christian worship and the practices of other faiths. In debate with the competing religious traditions of Islam, such as that engaged in by Catholic missionaries in Persia and the Middle East, moreover, natural theology offered points of easy agreement. Its claim that knowledge of the natural world gave authority to speak about God provided the basis from which to advance to more contested points of positive truth.[8] Access to traditions of natural religion explained the glimmers of truth found even in materialistic classical philosophy, whereas direct knowledge of the Judaeo-Christian tradition made plausible the deeper, spiritual insights that some detected in ancient Platonism.[9]

These were themes that Charleton picked up in his later consideration of the harmony of natural and positive divine laws, which explicitly treated the relationship between Christianity and Judaism.[10] But Charleton's initial apologetic intention had immediate contexts too: the upheavals of the English Civil Wars, which "at this unhappy day foster more swarms of *Atheisticall monsters.*" It aimed to demonstrate the existence of God, the truth of his two cardinal attributes as creator and moderator of the universe, and the existence of the human soul as *"a substance perfectly distinct from [the] body, and endowed with Immortality."*[11] Katherine Calloway has argued that Charleton at this point was not writing physico-theology, yet it would seem strange not to take his claim to be a physico-theologian seriously.[12] It is true that his method was to concentrate on the philosophical arguments of ancient writers, rather than on contemporary scientific observation. But the topics that he treated included those that became characteristic of physico-theology, as a distinctive branch of natural theology: above all, the mechanisms of Creation, the universal Deluge, and what Charleton called "physiology," according to which "the speculation of Natural Causes hath a power to raise the mind of man to a generous height" from which it might dispel superstition.[13] Moreover, what Boyle claimed marked out "physico-theological" argument was drawing attention to "the General Ends [God] is suppos'd to have intended in things Corporeal." In a succession of works, Charleton repeatedly did precisely this, considering in particular the providential ("physico-anatomical") function of parts of the human body or the moral lessons to be taught by animal behavior.[14]

As Boyle remarked, the consideration of such ends was "metaphysical." "Metaphysics" was the usual term in early modern England for what we would now call "natural theology" (i.e., "theology based upon reasoning from observable facts rather than from revelation" [*Oxford English Dictionary*]). Francis Bacon (1561–1626) had tried to separate out "Divine Philosophie or Natvrall Theologie" from metaphysics, "which heretofore hath been handled confusedly." The contrast that he drew dealt with "that knowledge or Rudiment of knowledge concerning GOD, which may be obtained by the contemplation of his Creatures. . . . The boundes of this knowledge are, that it sufficeth to conuince Atheisme; but not to informe Religion" and "the inquirie of FORMALL and FINALL CAVSES," which makes "the wisdom of God more admirable."[15] This distinction embodied an emerging gap between what others would refer to as "natural religion" and the form of natural theology that Boyle later called "physico-theology."

For many contemporaries, however, Bacon's division of terms within natural theology was premature. When Thomas Hobbes attacked natural theology in his *Leviathan* (1651), he called it "metaphysics," and his target was the Aristotelian philosophy of the universities. As the Independent minister and tutor Theophilus Gale (1628–79) acknowledged, "*Aristotle's Metaphysicks* passe in the *Scholes* under the splendid title of *Natural Theologie*."[16] The place of the *Metaphysics* in the curriculum had certainly changed, however, in the context of the long-running debate over the teaching of Aristotle, which began in the late thirteenth century. Doubts about the success of the synthesis of Aristotelian philosophy and Christian theology achieved by Thomas Aquinas (1225–74) focused much debate on metaphysical concerns (such as the limits of the body or location of God with regard to the blood shed by Christ on the cross). At this stage, metaphysics served, as Aristotle himself had intended, as a theoretical science, which dealt with unchanging things that existed separately from matter. As such it was akin to theology and different from both mathematics and physics (the science of changeable things). Its impact on the teaching of natural philosophy was therefore strictly circumscribed.[17] Despite the reservations expressed about metaphysics by Petrus Ramus (1515–72), both Lutheran and Reformed writers of the late sixteenth and early seventeenth centuries, such as Bartholomaeus Keckermann (c. 1572–1609), Johann Heinrich Alsted (1588–1638), or Clemens Timpler (1563–1624), came to accept the value of combining theoretical with practical knowledge of God, thus adapting natural philosophy to theological ends in part through awareness of metaphysical concerns regarding divine will, freedom, and consistency. What might be conceivable in terms of metaphysics helped to shape what might be knowable about both God and nature through natural theology.[18]

Topics that might once have been confined to metaphysical theology and that related particularly to the form and intentions of God and human beings were now taught as part of the propaedeutic to the study of physics (itself a part of natural philosophy) within the arts curriculum.[19] At the same time, the philosophy and method of post-Tridentine reformulations of scholastic metaphysics continued to be relevant to Protestant arguments about the distinctions between final and efficient causes and the necessity of cognition and rationality (in God, angels, and men) in order to comprehend what might be observed in nature. This background helped to make problematic many of the assumptions of Cartesian natural philosophy with which both Charleton and Boyle were in critical dialogue.

On the one hand, "physico-theological" simply implied a form of apologetic, historical, and metaphysical argument that was not very different in either content or purpose from a much longer history of natural theology.[20] On the other, Boyle wished to make an important distinction, which would be clarified by the later development of the term "physico-theological" and by the definition of a genre of "physico-theology" by John Ray (1627–1705) and his disciple and editor, William Derham (1657–1735). According to that distinction, "meer Reason" was insufficient properly to comprehend the purposes of God simply through observation and reflection, irrespective of the wonder that nature declared and for which "humble Thanks" were required. This was why Bacon had argued that natural religion was not the same as natural theology. For Boyle too, the speculative "physical" arguments of much traditional natural theology had to be distinguished from more conclusive and properly "metaphysical" ones. In addition, Boyle suggested that arguments about the lawfulness of the universe might "reach to Prove anything about the determinate Nature of particular Bodys" and would be able to extend the conclusions of metaphysics.[21]

According to the *Oxford English Dictionary*, "physico-theology" is simply a synonym for "natural theology," but what has been said so far should give us pause before endorsing this view. In particular, one should be careful before associating the growth of "physico-theology" with the development of natural philosophical arguments about the structure of the universe and about the lawfulness of God's general providence framed in an explicitly Cartesian or Newtonian manner. Instead, "physico-theology" is best understood as a new term developed within a growing lexicon of practices in natural philosophy that specified a particular kind of natural theological argument drawn from an assumed knowledge of final causes. In the preface to his *Physico-Theology: Or, A Demonstration of the Being and Attributes of God, from His Works of Creation* (initially delivered as the Boyle Lectures in 1711–12), William Derham claimed that the approach that he was taking was both "Mr. *Boyle's* own . . . Way" and one that had not been followed by any of the earlier Boyle Lecturers, several of whom had presented metaphysical natural theological arguments against atheism.[22] In practice the sort of argument most closely associated with "physico-theology" as envisaged by Boyle and adapted by Ray and Derham is an appeal to the special providence of God: showing how God has taken care of incidental aspects of the creation of the natural environment and of the animals and people who inhabit it, in such a way as to work toward their particular benefit and to ensure that a rational observer informed by the Christian religion might therefore interpret those incidentals as being demonstrations of divine omnipotence and goodness.

This distinction can be stated too firmly: arguments cross over between categories, and the formulation of a division between general and special providence is not always present in the original texts. It is, however, explicit in Derham's *Physico-Theology*, which draws repeated attention to God's special providence.[23] A feature of the development of "physico-theology" that supports this claim may be explained with reference to the list of authors whom Derham cited as having done something similar to him (but whom he also claimed not to have consulted before writing). This list included the French Minim (and one of the correspondents of Hartlib mentioned in a note at the start of this chapter) Marin Mersenne (1588–1648), in particular his *Quæstiones celeberrimæ in Genesim* (Paris, 1623); the Scottish Episcopalian John Cockburn (1652–1729), in particular *An Enquiry into the Nature, Necessity, and Evidence of Christian Faith* (London, 1696); Ray, whose reworked Cambridge exercises, *The Wisdom of God* (London, 1691), were greatly expanded by their author until a fourth edition appeared in 1704; the Cambridge divine and classicist Richard Bentley (1662–1742), whose *Folly and Unreasonableness of Atheism* (London, 1692) represented the first Boyle Lectures; and the French archbishop and sympathizer with Quietism François Fénelon (1651–1715), whose *Demonstration of the Existence of . . . God* had been translated from French in the year of its original publication.

With the partial exception of Bentley, all of these authors were concerned with specific moral as well as physical conclusions. Their concern was in that sense with human salvation and an awareness of divine goodness, as much as it was with the proof of divine omnipotence and governance of the frame of nature. Fénelon was clear here, in a manner that also indicated a broader change of emphasis in natural theological literature. He made a distinction between "metaphysical proofs of God's existence," which he said were not within the reach of every observer, and moral proofs, which were more accessible: "But there is a less perfect Way, Level to the meanest Capacity. Men the least exercised in Reasoning, and the most tenacious of the Prejudices of the Senses, may yet with one Look discover Him who has drawn Himself in all his Works. The Wisdom and Power he has stampt upon every Thing he has made, are seen, as it were in a Glass, by those that cannot contemplate Him in his own Idea. This is a Sensible and Popular Philosophy, of which any Man, free from Passion and Prejudice, is capable."[24]

The background to the change of emphasis that Derham's book embodied can be found in the thinking of John Ray and his contemporaries in the early 1690s. In 1692 Ray expanded the *Wisdom of God* and hurriedly reworked another surviving sermon from his days in Cambridge to meet the demands of his publisher, Samuel Smith, and in fear that he was dying.[25] Smith then wanted a new edition of the second book even before the first had sold out. Ray's *Three physico-theological Discourses* (London, 1693) was, its author acknowledged, an imperfect work, rushed to completion in May 1692 and then revised further before appearing near the end of the year. Initial translations into Dutch (*De werelt van haar begin tot haar einde*, Rotterdam, 1694) and German (*Sonderbahres Klee-Blätlein der Welt Anfang, Veränderung und Untergang*, Hamburg, 1698) obscured the change of emphasis in the new title while testifying to its immediate influence.

Part of the occasion for interest in Ray's work was controversy over the writings of Thomas Burnet (ca.1635–1715) and the resulting debate about philosophical interpretations of Genesis, Cartesian readings of Mosaic physics, and their compatibility with syncretic interpretations of pagan and Christian philosophy, in which truth might be hidden from the comprehension of the vulgar by the use of a language of accommodation.[26] For this reason, Ray was urged to prepare a Latin translation of his *Discourses*, which he resisted not least because he did not himself see what he was doing as answering Burnet.[27] The way in which Ray eventually developed his text took it away both from the historical form of his original sermon and from Burnet's work. It built on the activities in which he and his friends were engaged in the early 1690s in collecting fossils, plants, and insects toward other publishing aims that

Ray had, in particular his *Synopsis methodica animalium* (London, 1693), his *Synopsis stirpium* (London, 1696), and his planned synopsis of insects (published posthumously in 1710). Ray's interests at this time were increasingly in objects and specimens as testimonies to divine wisdom, a theme that he had developed in his earlier *Wisdom of God*.[28] By contrast. the idea of special providence, and of punishment for sin in the ordering of the world, received a particular boost from the revisions that the physician Hans Sloane (1660–1753) brought to the text after its original submission to Smith, especially through the account given by Ray of the earthquake that utterly destroyed Port Royal, Jamaica, on June 7, 1692.[29] Ray appears to have settled on the new title for his work only at the last minute in December 1692. A year later, he was considering a similar providence in the plague of locusts that his friend Edward Lhwyd had identified in Pembrokeshire.[30] For Ray, "physico-theology" represented a choice that allowed him to reposition work that he had originally drafted in the late 1650s and early 1660s more thoroughly in a contemporary intellectual climate. Its use enabled him to differentiate his work from styles of natural theology based primarily on the wondrousness of the natural and spiritual worlds that had characterized the approach of mid-seventeenth-century Cambridge writers like Henry More (1614–87).[31] At the same time, an adventitious concentration on the special providence of God allowed Ray to indicate how different his view of the history of the earth was from that of Burnet, who, according to his critics, was excessively wedded to processes of "ordinary" or general providence.[32]

Derham was closely involved with the posthumous development of Ray's reputation and publications: he saw the third edition of *Three physico-theological Discourses* through the press in 1713 and edited Ray's letters.[33] In addition, his work in *Physico-Theology* clearly revealed a debt not only to Boyle but also to his former Oxford tutor, Thomas Willis (1621–75), whose accounts of human anatomy formed the basis for much of what Derham argued.[34] *Physico-Theology* was later supplemented by a new work, *Astro-Theology* (London, 1715). This provided a third section of argument about the heavens (examined through the telescope) to go with Derham's existing claims about the visible creation (particularly anatomy and the habits of men and animals), which had made up the majority of the first two sections, delivered in *Physico-Theology*. *Physico-Theology* itself was remarkably successful: two large impressions sold out immediately, and a third edition appeared within a year. Plans were quickly afoot for translations into two or three other languages.[35] Like Ray, Derham was concerned to draw implications for the worship of God and the moral behavior of human beings from "physico-theology," not simply to concentrate on an account of divine attributes. This was because consideration of final causes brought out the moral meaning of the efficient working of nature and the suitability of God's creation and his creatures for the world in which they found themselves. Like Ray again, Derham was critical of Burnet, who was faulted for suggesting that the current state of the world might be imperfect.[36] Nevertheless, he was reluctant to convict Descartes of atheism, even if he remained convinced that both Epicureanism and Cartesianism might lead to impiety.[37]

Derham's work was very quickly taken up by journals such as the *Spectator*.[38] The moral universe of plants and animals suited to a divine purpose and fit for their habitats and for the use of man contrasted starkly with a mechanical view of animals as little better than automata. Unaware though animals might be of the true ends that governed them, their behavior in physico-theology might nevertheless betray positive moral characteristics. In this way, Derham's ideas reached a wide audience, perhaps less interested in the detail of natural philosophical observation and debate that appeared in his footnotes. That audience was broader than the "Young Gentlemen at the Universities" for whom he had originally thought his work most suitable.[39] The evangelical hymnographer William Cowper (1731–80) suggested that Derham's work was "very intelligible even to a child," although the artist and naturalist Mary Delany (1700–1788) felt that "many things are too abstruse for me in [Derham's lectures]."[40]

In his *General Magazine*, the Newtonian lexicographer and instrument maker Benjamin Martin (1705–82) characterized physico-theology as a science arising from "the Consideration of *meer Matter*." He enumerated fifteen parts to the practice of physico-theology: Helio-theology, Planeto-theology, Seleno-theology, Cometo-theology, Astro-theology, Aero-theology, Hydro-theology, Geo-theology, Phyto-theology, Zoo-theology, Ornitho-theology, Ichthyo-theology, Insecto-theology, Herpeto-theology, and Anthropo-theology.[41] Physico-theology had come to mean any particular study of nature that shed light on divine design.

Yet the fundamental point of physico-theology for Derham had been to show that nothing in the world was accidental. As Immanuel Kant later recognized, that conclusion depended on assumptions both about divine action and about human understanding: "Physico-teleological theology, is the cognition of the Deity, as being the author of that order and perfection in the natural world of sense, which is every where discoverable."[42] Physico-theology might be said to have been for the moral improvement of mankind by making rational beings think about the moral consequences and physical mechanisms, whereby final causes could be considered to act through efficient means in created nature. As such, it modified both the apologetic and the metaphysical nature of earlier forms of natural theology.

NOTES

1. See *OED* under the relevant terms: "physico-"; "physico-mechanical"; "physico-theological."

2. Boyle, *New Experiments Physico-Mechanicall*; see also Boyle, *Correspondence*, 1:437–38 (Robert Sharrock to Boyle, Nov. [24], 1660).

3. Variously: Sheffield University Library, Hartlib Papers: 1/33/16A (Cyprian Kinner to Hartlib, Oct. 9, 1647) and 1/34/4B (from Kinner, *Cogitationum didacticarum diatyposis summaria*, sig. A4v); 39/3/20A (Henry Oldbenburg to Hartlib, Oct. 28, 1658); 56/3/14B (from Abraham von Frankenberg's "Cum Deo & die plus ultro"); 8/60/2B; 42/1/9A (Benjamin Worsley to Hartlib, Oct. 14 & 20, 1657); 51/42B (John Beale to Hartlib, Dec. 14, 1658); 18/2/30B (Marin Mersenne to Hartlib, Nov. 23, 1640); see also Boyle, *Correspondence*, 1:243–44 (Hartlib to Boyle, Dec. 8, 1657).

4. Kinner, *Continuation of M. John-Amos-Comenivs School-Endeavours*, 8; cf. Hartlib Papers, 14/1/7A–B. Kinner's threefold distinction relates to his parallel enumeration of "Nature, Art and God himself" (3) and his notion of the relationship between the human being as microcosm and God as macrocosm (4).

5. Boyle, *Some Physico-theological Considerations*.

6. Boyle, *Disquisition about Final Causes*, 104–5. Boyle refers ultimately to Bacon, *Aduancement of Learning*, bk. 2, fols. 27–29.

7. Cf. Charleton, *Darknes of Atheism*, e.g., 39–49.

8. Heyberger, *Les Chrétiens du Proche-Orient*, 319–26; Hillgarth, *Ramon Lull*, 5–27.

9. Charleton, *Darknes of Atheism*, 103–6; cf. Creech, *Lucretius, De Natura Rerum, Done into English Verse*, sig. c1r–v.

10. Charleton, *Harmony of Natural and Positive Divine Laws*.

11. Charleton, *Darknes of Atheism*, sig. a1r, a4v.

12. See her contribution to this volume and Calloway, *Natural Theology*, 34.

13. Charleton, *Darknes of Atheism*, 94–198, esp. 153.

14. Charleton, *Oeconomia animalis*; Charleton, *Onomasticon zoicon*; Charleton, *Enquiries into Human Nature*.

15. Bacon, *Aduancement of Learning*, bk. 2, fols. 22r–30v.

16. Gale, *Court of the Gentiles. Part II: Of Philosophie*, 415.

17. Grant, *Foundations of Modern Science*, 135; Grant, *Nature of Natural Philosophy*, 91–118.

18. Daston and Stolleis, *Natural Law and Laws of Nature*; Freedman, *Philosophy and the Arts*; Friedrich, *Die Grenzen der Vernunft*; Hotson, *Commonplace Learning*; Lohr, "Metaphysics and Natural Philosophy."

19. Brockliss, *French Higher Education*, 205–16.

20. Cf. Roling, *Physica sacra*.

21. Boyle, *Disquisition about Final Causes*, sig. I8v–K1r.

22. Derham, *Physico-Theology*, sig. A4r–v.

23. Derham, *Physico-Theology*, 69, 78.

24. Fénelon, *Demonstration of the Existence . . . of God*, 2–3.

25. Keynes, *John Ray*, 91–99, 107–10; cf. Ray, *Miscellaneous Discourses*.

26. See Poole, *World Makers*; Levitin, *Ancient Wisdom*, 183–89.

27. Ray, *Further Correspondence*, 236–37 (Ray to Edward Lhwyd, Mar. 22, 1692/3).

28. Cf. Ray, *Wisdom of God* (1691), 74–86.

29. See Ray, *Correspondence*, 249–51 (Ray to Sloane, May 25, 1692); Ray, *Three physico-theological Discourses* (1693), 186–94.

30. Ray, *Further Correspondence*, 234 (Ray to Lhwyd, Dec. 28, 1692), 240–42 (Ray to Lhwyd, Dec. 26, 1693).

31. Cf. the attack on More mounted by the scholastic Oxonian reader of Charleton Samuel Parker, *Tentamina*; see also Levitin, "Rethinking English Physico-theology."

32. See the defensiveness on this issue in Burnet, *Answer to the Exceptions made by Mr. Erasmus Warren*, 2–4.

33. Ray, *Three physico-theological Discourses*, 3rd ed. (1713), xvii–xix; Derham, *Letters between Mr. Ray and Correspondents*.

34. Derham, *Physico-Theology*, 147.

35. Steinmann, *Absehen—Wissen—Glauben*, 110–26.

36. Derham, *Physico-Theology*, 47; cf. Ray, *Three physico-theological Discourses*, 35–42.

37. Derham, *Physico-Theology*, 271.

38. Steinmann, *Absehen—Wissen—Glauben*, 127–49.

39. Derham, *Physico-Theology*, sig. A8r.

40. Cowper, *Letters*, 1:388–90 (Cowper to William Unwin, Sept. 7, 1780); Swift, *Correspondence*, 4:414–16 (Delany to Swift, Nov. 19, 1735).

41. Martin, *General Magazine*, 5:10–59.

42. Willich, *Elements of the Critical Philosophy*, 137.

CHAPTER FIVE

Physico-theology in the Seventeenth-Century Dutch Republic

The Case of Willem Goeree (1635–1711)

ERIC JORINK

As is well known, the Dutch Republic played an important role in the physico-theological offensive that swept across Europe around 1700. In 1715 Bernard Nieuwentijt published his highly influential *Het regt gebruik der werelt beschouwingen*, and this book, as Rienk Vermij describes in the present volume, was only one example of the seemingly acute interest in employing the latest scientific discoveries and the "argument from design" to combat the danger of atheism. Nieuwentijt's book, explicitly addressed against Spinozists, coincided with similar publications. It also went hand in hand with the oration "On Certainty in Physics" that Herman Boerhaave delivered at Leiden University in 1715, the subsequent appointment of the Newtonian Willem Jacob 's Gravesande as the professor of physics at the same academy, and the publication of the pirated Amsterdam second edition of Isaac Newton's *Principia*. In the eighteenth-century Dutch Republic, physico-theology, Newtonianism, and the empirical-experimental approach to nature were more or less synonymous. Building on the long tradition of natural theology, Dutch scholars challenged the dangerous rationalistic and materialistic ideas of Descartes and Spinoza, whom they viewed as their common enemy.[1] The physico-theological offensive of 1715 was a well-orchestrated action, involving many scholars, publishing houses, and institutions.[2]

In earlier literature, the physico-theological offensive in the Netherlands is most often seen as the adaption of an originally British movement to local circumstances.[3] Indeed, looking at the editions and translations published in the Netherlands by authors such as William Derham, John Ray, Robert Boyle, George Cheyne, and Newton gives the impression that this movement was just the implementation of ideas and practices invented in the British Isles. However, underlying, longer local traditions in the Netherlands suggest that the 1715 physico-theological offensive had a firm root in Dutch intellectual culture. This interpretation, of course, depends upon our conception of physico-theology. One of the characteristics of Nieuwentijt's influential work was its basic claim that recent scientific research not only proved the existence of the divine Architect but also confirmed a literal reading of the Bible. This approach was explicitly addressed against Spinoza. However, with this two-fold approach of trying to reunite scripture and the latest scientific discoveries and trying to show that the order of nature proved the existence of the almighty Creator, Nieuwentijt was elaborating on some earlier, partly overlapping practices in the Netherlands.

In this essay I first consider the 150 years preceding Nieuwentijt's book and exploring the various strategies in accommodating natural philosophy with the word of God. Then my focus turns to the now largely forgotten figure of Willem Goeree (1635–1711), a prolific writer who in lavishly illustrated and highly successful books tried to synchronize the most recent scientific discoveries and ideas with a literal reading of the Bible—once again with the explicit aim of combating Spinoza.

Some Backgrounds

In the wake of the starting of the Dutch revolt (1568) some developments relevant to our theme took place. First, the advance of Calvinist thought proved to be of great importance for the perception and study of God's creation.[4] As Calvin had stressed, God made himself known in his great theater or book of nature, and man was created to find him and honor him here. Second, many Protestant artisans, artists, scholars, and publishers fled from Antwerp—then an international hub of trade, art, and science—to the northern provinces of the Low Countries. A new status quo soon emerged, resulting in the establishment of permanent Protestant institutions, such as Leiden University in 1575. Although the Calvinist Church never was the official state religion, public incumbents all had to subscribe to the Reformed articles of faith. The Dutch Republic became a haven for Protestants from all over Europe; for a still-large minority of Catholics; and, around 1600, for Jews. Members of the Reformed Church had to subscribe to the articles of faith of the Belgic Confession (1569). Article 2 is worth quoting in full:

> We know him [God] by two means. First, by the creation, preservation, and government of the universe, since that universe is before our eyes like a beautiful book in which all creatures, great and small, are as letters *to make us ponder the invisible things of God: his eternal power and his divinity,* as the apostle Paul says in Romans 1:20. All these things are enough to convict men and

to leave them without excuse. Second, he makes himself known to us more openly by his holy and divine word, as much as we need in this life, for his glory and for the salvation of his own.[5]

The idea that God revealed himself to mankind through both the Bible and the book of nature had been a strong undercurrent in Christian thought, going back to Augustine.[6] The Reformation laid new stress on God's revelation in the order of nature. In sixteenth- and seventeenth-century Protestant countries, we find many references to the notion of the *liber naturae*. However, seen from European perspective, the Dutch Reformed Church was the *only* denomination that explicitly included this dogma in its confession.

However, the concept of the "two books" was not without its internal discussions or contradictions. Of relevance are the following three points. First, the conviction that God revealed himself in nature had been firmly rooted in Dutch culture since the late sixteenth century. Second, the Belgic Confession stimulated a very lively debate on exegetical principles in relation to natural phenomena. Many Dutch theologians, especially of the orthodox wing of the Reformed Church, subscribed to a strict form of biblical literalism, what aptly has been called Mosaic physics. Ministers, scholars, natural philosophers, and laymen alike praised God's glory in his creation (see Ps. 104); tried to identify Behemoth, Leviathan, and Jonah's "wonder-tree"; made efforts to describe and reconstruct Solomon's Temple; and opposed the Copernican hypothesis. Third, the relatively open intellectual culture in the Netherlands was fruitful for debates on the relationship between the Bible and the book of nature, and that between revealed theology and natural theology. The most notable cases are, of course, René Descartes and his radical offspring, Baruch Spinoza. This open intellectual culture was noted in the rest of Europe, and many scholars traveled to the republic—including, for example, John Ray. Leiden University especially attracted many students from Protestant countries.

Descartes lived most of his life in the Dutch Republic. In the wake of the publication of his *Discours de la méthode* (Leiden, 1637), a furious debate took place, lasting for decades. The orthodox wing of the Reformed Church was outraged by Descartes's materialist approach toward nature and his implicit rejection of the value of the Bible in *rebus naturalibus*. For example, the debate on Copernicanism after 1640 was essentially not about astronomy but about exegetical principles and about the relation between theology and philosophy.[7] The orthodox wing, led by the Utrecht professor Gisbertus Voetius

(1589–1676), feared that philosophy, now the handmaiden of theology, soon would become her master. Indeed, Descartes conspicuously had remained silent on the relation between the word of God and the book of nature.

However, moderate Reformed scholars and natural philosophers influenced by Descartes's mechanical conception of nature ignored his epistemology and metaphysics and increasingly left the Bible out of their study of nature, instead putting full emphasis on the structure, order, and beauty of God's creation. In their view, the traditional "argument from design," already expressed by ancients such as Cicero, Seneca, Pliny, and Galen, now gained a powerful impulse from Cartesian physics. One example is Johannes Swammerdam (1637–80), who brought the ideas of Descartes to their logical conclusion: if nature is stable and uniform, the "lowest" creatures, such as insects, are worth attention. Setting aside the emblematic tradition and biblical framing of insects, Swammerdam now placed full emphasis on the creatures' incredible inner anatomy. His book on insects was published in Dutch in 1669. The purpose of studying insects, Swammerdam repeated on page after page, was to provoke piety. All creatures, great and small, have one single cause: "Namely the inscrutable God and the unfathomable Creator, wondrous and matchless in his works; in which all are based on few rules and unsearchably coherent with one another; he is Good, Wondrous and Worthy of Adoration."[8] This sounds, of course, familiar to any reader of the works of Derham and Ray, published some twenty years later. God can be seen in the louse or the bee as well as in the elephant. According to Swammerdam, God was an artist rather than a legislator, an architect rather than a writer. Swammerdam died in 1680, but his most important work was published as the *Biblia naturae* only in 1737–38.

The Emergence of Radical Biblical Criticism

Whereas pious Cartesians such as Swammerdam no longer considered the Bible an unproblematic key to the book of nature and now, instead, studied God's creation in its own right, others started to attack overtly the Bible's status as God's revelation to mankind. The Dutch Republic not only provided fertile ground for natural philosophy but was highly stimulating for biblical hermeneutics. Joseph Scaliger (1540–1609), Hugo Grotius (1583–1645), Isaac Vossius (1618–89), and others made contributions to a historical-critical approach to the Holy Scripture.[9] Decades before the publication of Spinoza's *Tractatus Theologico-politicus* in 1670, the conviction was held in certain circles that the Bible was not a book full of timeless truths but the product of an earlier culture, with all the problems of interpretation that that entailed. The publication of Isaac la Peyrère's *Preadamitae* in Amsterdam in 1655 was a bombshell.[10] La Peyrère rejected the notion that the Bible contained the history of all humanity, advanced rational explanations for biblical miracles, and doubted the universality of the Flood. In 1668, influenced by Descartes and the young Spinoza, a fellow student of Swammerdam at Leiden, Adriaan Koerbagh, published a book in Dutch, *Een Bloemhof.* In this work he ridiculed not only Reformed orthodoxy but in fact all established Christian institutions. "The word Bible," he wrote, "is a bastard Greek word which means a book in general, no matter what kind of book it may be, whether Renard the Fox or Till Eugenspiegel."[11] Getting his teeth into the story of the Flood, of which a literal account was becoming increasingly problematic, Koerbagh commented on the dimensions of Noah's Ark. "*Ark*," Koerbagh wrote: "a box to store things. The Bible mentions such a box, in which all kinds of animals, male and female, were stored as the water was covering the earth."[12] Moses, Koerbagh wrote, obviously had thought the earth was flat, as he wrote that a few days of rain were enough to cover the earth with water: "Moses was not present when this happened, and he was not an able mathematician, because he failed to reconstruct the right dimensions of the Ark: How could so many animals fit in this box measuring 300 cubits by 50 cubits by 30 cubits? How could food for more than a year be stored here?"[13] For this and the many other heretical ideas in *Een Bloemhof* (and in its unpublished sequel, *Een light*), Koerbagh was arrested and sentenced to a heavy penalty and forced labor—to which he succumbed within a year.[14] Only a few months after Koerbagh's death, in April 1670, Spinoza published his infamous *Tractatus Theologico-politicus*, in which he, among other things, maintained that the Old Testament was just a history of a certain tribe in the Middle East. Spinoza denied the historical reality of the biblical miracles—an idea that caused general outrage. In a less-discussed but equally devastating remark, Spinoza ridiculed those who believed that the Bible provided the sole key for understanding nature and science. Commenting on the description of the "molten sea" outside Solomon's Temple ("He made the Sea of cast metal, circular in shape, measuring ten cubits from rim to rim and five cubits high. It took a line of thirty cubits to measure around it," 1 Kings 7:23),[15] Spinoza remarked: "Because we are not bound to believe that Solomon was a Mathematician, we are allowed to affirm that he did not know the ratio between the circumference of a circle and its diameter, and that he thought, like ordinary workmen, that it is 3 to 1. But if it's permitted to say that we do not understand that text, then I certainly don't know what we can understand from Scripture." Spinoza's work—both the anonymous *Tractatus* and the *Opera posthuma* (1677)—had a tremendous impact, both in the Dutch Republic and beyond. Spinoza's monism, his biblical criticism, and his claim to absolute mathematical certainty were felt as immediate threats to the whole of Christendom. They forced theologians and scholars to reconsider the relation between theology and philosophy, between revealed and natural theology, and between the Bible and the book of nature. How was the Bible to be interpreted? How could new conceptions of nature be set within a religious framework? In the light of philological discoveries and new natural philosophies, how could a debatable biblicism be avoided without lapsing into materialism or deism?

The Case of Willem Goeree

As noted, the 1715 offensive against Spinoza is often seen as the starting point of physico-theology in the Netherlands. Here I contend that if we take a slightly broader view, many themes and approaches anticipating Nieuwentijt's work can already be detected in the long seventeenth century. Physico-theology was rather more a *basso continuo* than a wake-up call. A case in point is the work of the architect, draftsman, collector, theoretician of art, and publisher Willem Goeree.[16] In the decades before Nieuwentijt, he published many lavishly illustrated books—all in Dutch—which enjoyed great popularity and went through reprint after reprint. Now largely forgotten, his work gives an interesting view of the ways in which an educated layman tried to square a literal reading of the Bible with the latest developments in the natural sciences.

Goeree was born in Middelburg, in the province of Zeeland, in 1635. He was raised in a highly intellectual milieu in the Zeeland capital. His father, Hugo, was trained as a physician as well as a Hebraist; he was a friend of Adam Boreel (ca. 1602–65), known for his translation of the Mishna as well as for his alchemical and optical experiments. The two men enabled the Middelburg rabbi Jacob Jehuda Leon (ca. 1602–75) to construct a model of Solomon's Temple. This was by nature a project that required as much knowledge of the Old Testament as of mathematical principles. Rabbi Leon "Templo," as he was nicknamed, became famous after he moved to Amsterdam around 1650, where he received many visitors. In 1675 he made a trip to London and there demonstrated his model to members of the Royal Society.

Young Willem Goeree was a product of this learned environment. He too moved to Amsterdam, where he published many books on the theory of drawing, painting, and architecture. Although having no academic background, he had a critical mind and was well aware of the latest developments both antiquarianism and natural philosophy. Of great significance is the fact that the learned Goeree stressed the importance of history and historical knowledge. He argued, for example, that in painting historical and biblical scenes—the highest form of art—the artist should be extremely careful in checking the historical facts. What robes and togas were worn? Which animals and plants were indigenous in the Holy Land? Goeree advised artists who wanted to represent the Jewish captivity in Egypt to adorn their paintings with depictions of the recently found shabti and other ritual objects; he scorned those who painted Romans soldiers with scimitars.

In 1683 he published a book by the late Leiden professor Petrus Cunaeus (1586–1638), *De republyk der Hebreen, of het gemeenebest de Joden* (The republic of the Hebrews, of the commonwealth of the Jews), a historical description of the ancient Jews based on both the Bible and other sources, including visual and material objects. The small booklet included many illustrations and was an immediate success. On the basis of unpublished work by his own learned father, Goeree compiled three sequels, which were equally successful. Somewhat anachronistically, these books could be qualified as attempts to write a cultural history of the biblical Jews. Elaborating on this concept, Goeree embarked on two related projects, *Voor-bereidselen tot de bybelsche wysheid, en gebruik der heilige en kerkelijke historien* and *Mosaize historie der Hebreeuwse kerke,* published in 1690 and 1700, respectively. Especially the latter work was impressive. It was published in four massive volumes and lavishly illustrated with splendid engravings, partly designed by Goeree himself, partly by Jan Luiken (1649–1712), by far the best Dutch engraver at that time. The work is a massive display of erudition. Goeree discusses every episode, every chapter, and every verse of the Pentateuch in astonishing detail in an effort to explain them to the reader. Using many extra-biblical sources—not only ancient writings as well as contemporary works on exegesis, church history, travel accounts, and natural history, but also visual and material objects including coins, inscriptions, and archaeological findings—Goeree's aim is to clarify the historical world of the Bible while also proving that the Bible is true and must be understood literally. The use of illustrations underlines this point, as it allows the reader to become a witness.

Relevant to our theme is the fact that the erudite Goeree also used the latest scientific and scholarly works to illuminate the Bible. In doing so, he was working in the overlapping traditions of Mosaic physics and physico-theology.

His account of Creation, for example, heavily borrows—both conceptually and visually—from Thomas Burnet's *Sacred Theory of the Earth*. The explanation of Genesis 1:6–8, concerning "the lights in the firmament of heaven," consists of elaborate discussions of the most recent astronomical discoveries, made possible by that great Middelburg invention, the telescope. In the discussion of the third plague described in Exodus 8:16, "Then the Lord said to Moses, 'Tell Aaron, "Stretch out your hand with your staff over the streams and canals and ponds, and make frogs come up on the land of Egypt."'" Goeree refers to the world recently revealed by the microscope and gives detailed descriptions and images of all kinds of lice, heavily borrowing from Francesco Redi's *Esperienze intorne del' insetti* (1668; Swammerdam's work on the louse was already written but not published until 1737–38). The astonishing anatomy of these creatures clearly proved that these were the works of the Almighty Architect.

Throughout the work, Goeree repeatedly stresses that one of his main motives is to demolish the ideas of Spinoza. Convinced as he is that the Bible should be interpreted literally, one of the bottom lines is that extra-biblical sources (be they pagan works, archaeological findings, or contemporary scientific discoveries) all prove that Moses was divinely inspired and that everything in the Bible is true. In this sense, Goeree operates in the same spirit as Joseph Scaliger, who heavily relied on extra-biblical sources in order to reconstruct the biblical chronology. Scaliger struggled with the *Tomoi* by Manetho, with its table of Egyptian dynasties that seemingly went back further in time than the accepted date of Creation. Goeree would stumble over similar problems.

One of the topics that fascinated Goeree was Noah's Ark. In the *Voorbereidselen tot de bybelsche wysheid* of 1690 as well as later in the *Mozaische historie der Hebreeuwse kerke*, Goeree devoted many pages and many illustrations to the description and reconstruction of the Ark. The section was explicitly addressed against Adriaan Koerbagh, "this impudent and awkward liar, who in his filthy book has dared to throw up the idea that Moses has forgotten the dimensions of the Ark."[17] In one of the illustrations, Luiken depicted the Ark according "to the fancy of the Artists" and to the "True shape according to Scripture." Subsequently, Goeree firmly tried to settle this issue, which already had been the subject of a long scholarly debate involving Origen, Hugh of Saint-Victor, the French Jesuit mathematician Johannes Buteo, and the omnipresent Kircher.[18] Goeree had to tackle a range of problems. The first was the most obvious: given the exact dimensions of the Ark provided in Genesis, how long was a cubit? Goeree, as an architect well aware of the problem of measurement as well as of many relevant sources from antiquity, knew that there had been, among others, Lebanese, Syrian, Babylonian, and Egyptian cubits. He was very cautious on this issue, which—as we will see— was of great importance to other biblical reconstructions as well.

The second problem was that Noah had no examples to work from. Goeree's imaginative approach is illustrated by the following: "Facing the prospect of the Deluge, and the task assigned to him, Noah must have panicked and have prayed to God. He must have asked himself: 'How will I be able to construct, out of these huge pieces of wood, a ship? How can I make it stable enough to withstand the forces of the ocean? How will I be able to bring all the animals in?'"[19] But Goeree, the down-to-earth artisan, imagined that God had implanted a blueprint in Noah's mind and that Noah subsequently worked this out in a drawing, format specifications, and a scale model. Building a scale model himself, Goeree found that experiments confirmed what he already feared: without a keel the ark would be highly unstable. Further practical issues arose as well. Others had already addressed the questions which animals had been on the Ark and how all these creatures had fitted in a floating object of 30 by 50 by 300 cubits—be it Babylonian or Egyptian ones. The matter had gained new urgency in the light of the steadily increasing number of previously unknown animals brought in from all parts of the world. Goeree addressed the issue and came, as Kircher did, to the comforting conclusion that these animals fitted in nicely. Displaying an impressive knowledge of the more recent literature on natural history, Goeree also drew up a scheme on where which pairs of animals must have been housed. Working his way through the format specifications, Goeree, the experienced *bouwmeester*, calculated that a year was sufficient to build the vessel—which contradicted the traditional reading of the relevant passage in Genesis, from which it was concluded that Noah had spent 120 years in constructing the Ark. More traditional was the question how, once the Ark was ready, all animals had boarded (Goeree imagined that Noah had also built some wooden cranes).

The practical architect raised other questions no scholar had posed before. Given the fact that each elephant eats about a hundred kilos of hay every day and produces eighty kilos of excrement, Goeree made extensive calculations of how much excrement all the animals together must have produced on a daily basis and designed an elaborate sewer system. This was a grandiose effort to reconstruct the Ark—it had been done before but never on a scale

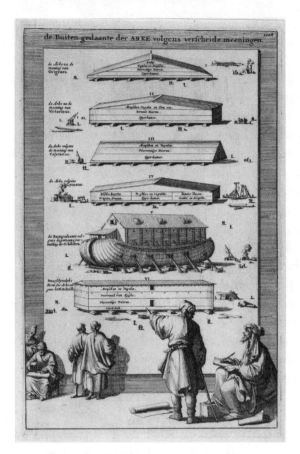

Jan Luiken, *Various opinions on the outward shape of the Ark*. This etching, included in Willem Gocree, *Voor-bereidselen tot de bybelsche wysheid* (Amsterdam, 1690), presented reconstructions that according to "the imagination of the painters" were the "most likely according to Scripture." Courtesy of Rijksmuseum, Amsterdam. Object number RP-P-1896-A-19368-833.

and with an eye for detail like this. But all Goeree's efforts to prove that the Bible's description of Noah and his Ark were right, and to emphatically confront his readers with the problems Noah must have faced, had some uncomforting outcomes: that according to experience and calculation the Ark must have had a keel, and that Moses obviously got it wrong when he suggested that Noah spent more than a year in building the vessel.

Even more ambitious was Goeree's reconstruction of Solomon's Temple, a subject that equally had occupied the finest spirits of the age, including Goeree's own father and Leon Templo. Goeree's work on the Temple has been recently studied in some detail.[20] Of relevance here is that in this case, Goeree pushed his historicizing approach to its limits. Studying all the sources available, he could elaborate as well on the kinds of wood and the origin of the copper as on the grander structure of the Temple. Goeree once again addressed the thorny question of the length of a cubit. Whereas in the case of the Ark he was more or less able to ignore this problem, in the case of the Temple it was at the heart of the matter. Like many before him, Goeree was puzzled by the phrase in Ezekiel 40:5 "in the man's hand a measuring reed of six cubits long, each of which was a cubit and a handbreadth." He tried to resolve the question by postulating that this should be understood as a *ratio* between the cubit and the hand breath and not an absolute length. And the cubit certainly was the Egyptian cubit, as Solomon's knowledge came from the Egyptians. Goeree was, of course, not the first to praise the wisdom of the Egyptians, but the implicit message of his work was that he seemed to deny divine inspiration, opting for a historical transmission of knowledge from Egypt to Jerusalem. More generally, the excessive calculations he made (in the case of the Ark, he drew up extremely detailed format specifications, including the costs of the materials, the wages the artisans must have earned, and the total contract price) seemed to turn the biblical Temple into an all-too-human real estate project. The astonishing foldout engraving in the *Mozaische historie*, intended to convince the reader of the historical truth of Solomon's Temple, basically conveyed the same message. Whereas visitors to Jehuda Leon must have gazed with their mouths open at the evocation of the Temple, readers of Goeree witnessed the rather messy work in progress.

Ironically, Goeree's pious enterprise to destroy Spinoza's rejection of the biblical truth turned against him. It should be noted, for example, that Goeree ignored Spinoza's devastating argument on the circumference of the "molten sea." Although explicitly addressing that most horrible (to him) of all philosophers, Goeree seemed not to be aware of the fact that his historical and computational approach bore at least some resemblance to the target of his ire. Orthodox theologians noted that Goeree too seemed to deny divine inspiration, relied heavily on mathematics, and used a questionable number of extra-biblical sources to clarify the word of God. Accusations of Spinozism were the result, as Jetze Touber recently has shown.[21] Goeree's use of ancient languages, natural history, and mathematics demonstrates that physico-theology could sometimes lead to unorthodox ideas.

More generally, his work demonstrates that already decades before the 1715 offensive in the spirit of Nieuwentijt, Dutch scholars sought different strategies to reconcile the word of God with contemporary scholarship and science. They were as much rooted in local traditions as inspired from abroad. And vice versa. Given the fact that the republic attracted many students and scholars from other Protestant countries—especially from England, Germany and Scandinavia—it would be worth taking a comparative approach in studying the rise of physico-theology.

NOTES

1. Bots, *Tussen Darwin en Descartes*; Wall, "Newtonianism and Religion."
2. Jorink and Maas, *Newton and the Netherlands*; Jorink and Zuidervaart, "Newton's Reception in the Low Countries."
3. Notably Vermij, *Secularisering*.
4. Jorink, *Book of Nature*.
5. Bakhuizen van den Brink, *De Nederlandse belijdenisgeschriften*, 73.
6. Van Berkel and Vanderjagt, *Book of Nature in Antiquity and the Middle Ages*; van Berkel and Vanderjagt, *Book of Nature in Early Modern and Modern History*; Methuen, "Interpreting the Books of Nature and Scripture"; Augustine, *Enarratio in Psalmum XLV*, 6–7.
7. Vermij, *The Calvinist Copernicans*.
8. Swammerdam, *Historia insectorum generalis*, 28.
9. Grafton, *Joseph Scaliger*; Grafton, "Isaac Vossius, Chronologer."
10. Popkin, *La Peyrère*; Grafton, *Defenders of the Text*, 204–13; Jorink, "'Horrible and Blasphemous'"; Jorink, "Noah's Ark Restored (and Wrecked)."
11. Koerbagh, *Een bloemhof*, 95.
12. Koerbagh, *Een bloemhof*, 95.
13. Koerbagh, *Een bloemhof*, 95.
14. On Koerbagh: Wielema, *Adriaan Koerbagh*, and the introduction in that volume by van Bunge.
15. Quotations from the Bible are from the New International Version.
16. Roemer, "Regulating the Arts"; Jorink, *De Ark*; Touber, "Right Measure."
17. Goeree, *Voor-bereidselen tot de bybelsche wysheid*, 1026.
18. Cf. Allen, *The Legend of Noah*.
19. Goeree, *Mosaize historie*, 986.
20. Touber, "Right Measure"; Jorink, *De Ark*.
21. Touber, "Right Measure."

CHAPTER SIX

Back to the Roots?

J. A. Fabricius's "Register of Ancient and Modern Writers" of 1728

KASPAR VON GREYERZ

Older German research has regularly identified Johann Albert Fabricius (1668–1736), the philologist, theologian, and professor at the Akademisches Gymnasium in Hamburg, as the *spiritus rector* of German physico-theology. In their influential works, Otto Zöckler (1879) and Wolfgang Philipp (1957) unanimously attributed to him this crucial role.[1] Manfred Büttner and Anne-Charlott Trepp, on the other hand, have dated the inception of German physico-theology to the late sixteenth and early seventeenth centuries.[2] In what follows I will diverge from these interpretations. The close examination of Fabricius's "Register" of 1728 reveals that we must distinguish two phases in the development of German physico-theology. The first, beginning in the 1670s, consisted almost exclusively of academic disputations concerned with the argument from design, particularly with different parts of the human body—hands, heart, ears, and eyes. Only in the second phase, beginning in the 1710s, did Fabricius seize the role of an indefatigable promoter of vernacular physico-theological work, particularly as a translator of (among other works) William Derham's seminal physico-theological treatises. The shift from Latinate to vernacular physico-theology, which generally distinguishes the two phases of German physico-theology, can be explained to a large extent by the increasing affinity of the movement with the early Enlightenment and its didactic aims.

Fabricius's "Register," that is, his "Verzeichnis derer alten und neuen Scribenten," is a foreword to his German translation of William Derham's *Astro-Theology*, first published in 1728.[3] The *Astrotheologie* was republished four times in 1732, 1739, 1745, and 1765, including Fabricius's "Verzeichnis" or "Register." All editions appeared in Hamburg. The "Verzeichnis" in the 1728 edition consisted of fifty-one pages in quarto. In the 1765 edition it ran from page xv to xc; that is, it had grown to a length of seventy-five pages in the same format. New references to recent titles were added to the editions of 1732 and

1739, in the latter case posthumously by the unknown editor. There are no further addenda in the 1745 and 1765 editions.

Johann Albert Fabricius was born and raised in Leipzig.[4] There he was trained above all as a literary scholar. After heading westward as a student in 1693, he was stranded in Hamburg when he ran out of money. However, his luck changed when he found a job as an amanuensis of the controversial Hamburg pastor and professor at the University of Kiel, Johann Friedrich Mayer (1650–1712). He was mainly in charge of Mayer's huge library, work that seems to have been the first stepping-stone toward a career that eventually made him one of the eighteenth century's most remarkable bibliographers. His first major work (of many to follow) was the *Bibliotheca Latina*, comprising, as he noted in the subtitle, "notes about Latin authors of antiquity whose writings have been preserved to this day." It was published in 1697. Two years later Fabricius received a doctorate in theology at Kiel. Only a few months after gaining his doctorate, he was elected to a professorship at the Akademisches Gymnasium in Hamburg, a position he kept until he died.

After his appointment in 1699, Fabricius did not take long to get thoroughly integrated into Hamburg's social and cultural elite. From 1715 he was a member of two consecutive local Enlightenment *sodalitates*. The second one, founded in 1723, was the Patriotic Society, a small, elitist club that, in spite of its exclusive character, was committed to educational and didactic aims. Among the members were the poet Barthold Hinrich (or Heinrich) Brockes and the polyhistor Michael Richey. During Fabricius's later years, this circle provided the encouragement and the sociocultural background for his engagement in the promotion of physico-theology, be it as a correspondent with like-minded virtuosi in other parts of Germany, as a translator of English works, or as an author of works on hydro-theology and pyro-theology.[5]

The whole title of Fabricius's text that I examine here reads (in English translation): "Register of those ancient and modern writers, who have made it their concern to lead men to God by an examination [*Betrachtung*] of nature and [its] creatures." As noted, it is a lengthy preface to Derham's German *Astrotheologie* and essentially constitutes a bibliography with commentary. The bibliography runs from page xii to lxiv of the foreword and includes an impressive number of authors and titles. Of course, this number looks minuscule compared to the growing size of Fabricius's private library. When it was auctioned off after his death in 1736, it contained 33,275 volumes.[6] This was not exceptional for Hamburg in the first half of the eighteenth century, where Martin Mulsow has counted six private libraries that each held between 20,000 and 30,000 books.[7]

In 2003 Fernando Vidal drew attention to the "Register of those ancient and modern writers" and declared it to be "of considerable historical interest."[8] Six years later, Anne-Charlott Trepp pointed more explicitly than Vidal to the lines of continuity suggested by Fabricius's "Verzeichnis" and demanded that these should be taken seriously.[9] However, in what follows, I will not so much focus on this continuity, which I consider to be more rhetorical than factual in the case of Fabricius's "Register," but rather attempt to categorize Fabricius's entries in his bibliography in terms of their relationship with the most characteristic aspects of physio-theological writing during its heyday, between about 1650 and 1760. For brevity's sake, I refer back to the list of these characteristics in the introduction to this volume.

Fabricius's "Register" is divided into seven sections of very unequal length. The shortest is section 5, concerned with authors who have made their treatises on "the essence, qualities, and strengths" of the human soul a ground for admiring and praising the Creator. Its length is one page only. Here, our author refers only to the Dutchman Caspar Barlaeus's *Oratio de animae admirandis* of 1635 and to the Marburg professor Franz Ulrich Ries's treatise *De existentia Dei ex stupenda mentis cum corpore unione demonstrata* of 1726, and otherwise directs his readers to his own *Delectus Argumentorum et Syllabus Scriptorum qui veritatem religionis Christianae adversus Atheos, Epicureos, Deistas, Seu Naturalistas, Idololatras, Judeos et Muhammedanos ... asseruerunt* of 1725.[10] This is a massive tome in both its length and the author's stupendous learnedness, to which is added a long, unpaginated *Index Scriptorum* (index of authors). Its contents span authors from antiquity to the early eighteenth century much like the "Register" discussed in this chapter.

The longest of the seven sections, section 2, runs from page xxv to lx. It begins with a mention of Athanasius's first book against the gentiles and ends with a general reference to dissertations defended at the University of Altdorf under the presidency of Johann Christoph Sturm that point to the moral significance of the study of nature.[11] The whole section is dedicated to authors "who do not touch upon the six days of Creation [this is the theme of section 1] but have otherwise used the examination [*Betrachtung*] of nature for proving God's existence and for a recognition of his superb qualities and of the duties we owe him."[12] This section is less extensive on ancient authors than the first one and advances to writers of the Middle Ages fairly quickly. Here Isidore of Seville and the Venerable Bede are mentioned, and so is the Dominican Johannes Nider, otherwise mainly known as an early demonologist. Jean Gerson and Marsilio Ficino also get their due, before early modern authors are addressed. Among these Isaac Barrow and Samuel Parker get special attention. In particular, Fabricius focuses on Barrow's interpretation of specific biblical passages, such as Jeremiah 10:12 and 51:15, as well as Genesis 1:27, and on Barrow's critique of Descartes's physics, which is criticized as deficient in explaining the most important things we see in nature.[13] Fabricius also offers a short overview on what I assume must be Parker's *Disputationes de Deo et providentia divina* of 1678, although he calls them *Cogitationes* in error.[14] To this, he adds a reference to François Fénelon's *Démonstration de l'existence de Dieu*, including the English, Dutch, and German translations of this treatise, followed by a roll call of (among other references) the authors well known today for their physico-theological works, such as Bernard Nieuwentijt, Robert Boyle, Barthold Hinrich Brockes, the Scottish poet James Thomson, John Ray, and John Woodward. Among these luminaries he also lists many other authors, such as Daniel Straehler, the philosopher who became one of Christian Wolff's earliest opponents at Halle,[15] and Samuel Fabricius, whose *Cosmotheoria Sacra, oder heilige Betrachtung über den CIV Psalm* appeared in 1626. Neither this particular Fabricius, who was a Lutheran pastor at Zerbst, nor Straehler can be counted among the physico-theologians, nor can several other authors in this section. We should note, however, that Fabricius also praises the fourth volume of the Jansenist Charles Rollin's *De la manière d'enseigner et d'étudier les Belles-Lettres par rapport à l'esprit et au Coeur . . .* , first published in 1726, for its appreciation of the proper observation of nature and its beneficial effects in instilling respect for and veneration of the Creator.

Anne-Charlott Trepp has justly highlighted Fabricius's "undogmatic and supraconfessional attitude" with which he surprises the reader right at the beginning in invoking the Jesuit Athanasius Kircher's reference to 6,561 ways of demonstrating that there is a God.[16] Perhaps even more surprising, considering his academic training under the supervision of an orthodox theological polemicist such as Johann Friedrich Mayer, is Fabricius's thoroughly positive appreciation of Roberto Bellarmine's *De ascensione mentis in Deum per scalas creaturarum*, first published in 1618. It would be difficult to assume that he was uninformed about the cardinal's involvement in the procedures taken by the Roman Inquisition against Giordano Bruno and, later, against Galileo Galilei. Fabricius devotes a full two pages to Bellarmine's treatise, first with a short synopsis of its contents, then with an overview of all the translations it underwent, including a German Protestant translation published at Görlitz in 1705.[17]

Along with the amazing number of treatises he had obviously scrutinized firsthand, Fabricius also quotes works referred to in learned journals. It is clear from his "Register" that he regularly consulted the *Acta Eruditorum*, the *Journal des Savans*, the *Journal littéraire*, the *Journal de Trévoux*, the *Unschuldige Nachrichten*, *L'Europe Savante*, and others.[18] He almost overdoes it at the end of his foreword where he lists a real plethora of reviews of Derham's *Astro-Theology*.

I take comfort from Fabricius's occasional reliance on reviews rather than the original because it would be utterly impossible to consult all of the titles he lists within a reasonable period of time. This applies in particular to his fourth section, concerned with demonstrations of God's existence based on the examination of human eyes, ears, the heart, stomach, brain, the hands and legs, the tongue and speech,[19] emotional state (*aus den Affekten*), and consumption of food, as well as to section 6, concerned with the *Confessio Naturae contra Atheos* (nature's evidence opposing atheists). Here his categories are rain, snow, thunder, wind, monsters, the devil, minuscule worms, spiders, ants, the voices of animals, the diversity of forms of animals and their instincts, the magnet, movement and gravitation, and natural bodies and the proportion of size and number among them. As I have mentioned, section 5 in between is only one page long. In the fourth and sixth sections, Fabricius again refers to some Catholic authors, such as a work on monsters published by the Jesuit Georg Stengel at Ingolstadt in 1647, or another by Fabricius's contemporary Nani Falaguasta on demonstrations of the existence of God from the smallest, as well as largest, creatures. He also lists a treatise on ants by M. Publius Fontana published in Bergamo in 1594. He highlights, once again, Johann Christoph Sturm's pioneering role in dissertations written under his supervision, such as that of 1678 on the quality of the human eye in recognizing God. Sturm's later work (before his death in 1703) was clearly linked to the first stirrings of the early Enlightenment in Germany, but in chronological terms his commitment to basic tenets of physico-theology came first.

Fabricius's "Register" records practically no works written by well-known representatives of German Lutheran orthodoxy. His reference to a treatise published by Valentin Ernst Löscher in 1724 is an exception.[20] This general omission throws doubt on the assumption that there was some kind of affinity between Lutheran orthodoxy and physico-theology. Such an affinity was suggested, albeit in fairly oblique terms, by Wolfgang Philipp in 1957, in a work that was astonishingly influential in the germanophone world considering its shortcomings.[21] Philipp based his claim that there was a proximity between German physico-theology and Lutheran orthodoxy chiefly on the fact that Fabricius worked for a while (before 1699) as an assistant and librarian for the orthodox polemicist Johann Friedrich Mayer. However, Philipp was mistaken for the simple reason that, as far as I can see, hardly anybody among Mayer's former students, except for Fabricius, came to adhere to the physico-theological movement. More generally speaking, it is practically impossible to recognize any affinity between contemporary physico-theology and the writings of exponents of Lutheran orthodoxy. A different story can be told about the former pupils of Johann Christoph Sturm. Bona fide physico-theologians like Jakob Wilhelm Feuerlein or Johann Jakob Scheuchzer were among them, and so was the Lutheran pastor Christian Philipp Leutwein, who published *Theologia nivis physico-mystica . . . oder Geistliche Lehrschule vom Schnee* in 1693 (in short, a "snow theology"), one of the first genuinely physico-theological treatises written in the German vernacular.

I will end my account by looking briefly at the first and last sections of Fabricius's text. The first section concentrates on interpretations of the *hexaemeron*, that is, the six-days of Creation as described in Genesis 1:1–2 and 4. This section begins with references to Theophilus of Antioch and Basil of Caesarea, mentions Pico della Mirandola's *Heptaplus de opere sex dierum Geneseos . . .* of 1496, strongly recommends the fourth book of Johann Arndt's *Four Books on True Christianity*, lists Andreas Libavius's and Marin Mersenne's contributions to the theme, and finally refers the reader to reviews contained in the *Unschuldige Nachrichten* of 1705 and 1706 of hexameral works by Johann Heinrich von Schönau and Johann Sophron Kozak published in 1688 and 1686 respectively.

The seventh and final section is devoted to astronomy.[22] In chronological terms it runs from Marsilio Ficino's *De Sole et Lumine* (1493) via Athanasius Kircher and Caspar Barlaeus to William Whiston's *Astronomical Principles of Religion, natural and reveal'd* of 1717 while also covering lesser-known treatises and dissertations. It ends with a two-page celebration of the work of William Derham, quite appropriately because Derham's *Astro-Theology* in German translation is the text that follows.

What kind of insights can we gain from scrutinizing Fabricius's "Register"? First of all, we should recall that his text is an extended foreword to a German translation of William Derham's *Astro-Theology*. This foreword obviously

intends to legitimate physico-theological writing by providing it with historical depth. However, the historical panorama unfolded by Fabricius is largely
that of natural theology and its long history beginning with Plato, Cicero, and
the church fathers. He does not focus exclusively on the physico-theology that
emerged in the course of the seventeenth century as a new and specific branch
of natural theology. Thus, Fabricius does not offer support for the notion
underlying the present volume that physico-theology was more than a "manner of thought," a timeless *Denkform* that remained basically unchanged
from the church fathers to William Paley and beyond. In other words, Fabricius's "Verzeichnis derer alten und neuen Scribenten" does not establish a
pedigree for the physico-theology specific to the late seventeenth and early
eighteenth centuries but lists instead a "fair number of writers . . . who by an
industrious observation of creatures have endeavored with much diligence to
encourage their readers to the recognition of, as well as love and respect for,
the creator." These are Fabricius's own words in his letter of dedication addressed to his associate Barthold Hinrich Brockes. In part, the "Register" may
even reach beyond the confines of natural theology, for it is a matter of definition whether one wants to subsume homilies and commentaries on the *hexaemeron* under the label of natural theology.[23] Fabricius is happy to embrace all
these different ways of reaching God through nature under one very big tent.

Second, if we are willing to see in physico-theological writing after the
1650s the expression of an intellectual movement that begins to acquire European proportions from about the 1690s onward, then Fabricius's "Register"
provides a valuable confirmation of this view in highlighting what we would
today call the book-historical aspects of this development. Here, I can mention only a few examples: Fabricius points, for instance, to the several English
editions of Richard Bentley's *The Folly and Unreasonableness of Atheism* of
1692 but also to its Latin and German translations, by Ernst Jablonski (Berlin: Johann Michael Rüdiger,1696) and Mathaeus Seidel (Hamburg: Samuel
Heyl, 1715). He also touches upon the English and French translations of Bernard Nieuwentjit's classic of 1715. He notes that the "most diverse writings"
(*unterschiedlichste Schriften*) of Robert Boyle were translated into English
(when the original was in Latin) or into Latin and German, and he goes on
at some length about the different editions and translations of the works of
John Ray, noting that the first German translation of Ray's *Three physico-
theological Discourses* was unsatisfactory and much improved upon by a
new translation that appeared in 1713. The bibliographic focus of the German

practice of *historia litteraria* led Fabricius to pay more attention to the diffusion of physico-theology than many a modern historian.

The third insight to be gained from Fabricius's foreword has to do with possible stages in the appearance of physico-theology in Germany. To my mind, there can be no doubt that the beginning of an actual *movement* of physico-theology in Germany had to do with the inspiration provided by British examples, first perhaps by treatises written by Samuel Parker and Robert Boyle, but almost certainly by the works of John Ray. They became available in German translation from 1698. However, it is of some interest that Fabricius points to earlier texts, published in Germany during the last decades of the seventeenth century, that argued from design. We must note, though, that these are invariably dissertations, some of them lengthy, like a demonstration of God's existence by an examination of the human hand presided over by Christian Donat (or Donati), published at Wittenberg in 1686,[24] some much shorter, such as an interpretation of Acts 14:17 (God "gave us rain from heaven, and fruitful seasons . . .") under the direction of Gerhard Meyer (who was, like Fabricius, a professor at the Akademisches Gymnasium in Hamburg), published in 1697. Unlike these two texts, another dissertation supervised by Gerhard Meyer on spiders as witnesses of God's existence, also published in 1697, explicitly uses physico-theological language when it highlights that the spider's composition and the accurate symmetry of its parts reveal the superb wisdom of its creator far better than human wisdom could reach on its own ("quo aptius per sapientiam humanam nequeat excogitari").[25] And, last but not least, there were the dissertations supervised by Johann Christoph Sturm at Altdorf, mentioned by Fabricius on three different occasions, beginning with a thesis on the human eye as an organ capable of recognizing God, published as early as 1678.[26] Considering their manifest interest in promoting the argument from design, it is safe to claim that Sturm and Meyer were forerunners of the physico-theological movement in Germany. However, we should bear in mind that we are dealing with dissertations written in Latin that were never reedited and probably appeared in few copies. It is difficult to say whether their physico-theological orientation was self-inspired or the result of extraneous influences. Fabricius also lists older treatises with similar subjects, such as one by Caelius Secundus Curione on the ant, published in Basel in 1544, and that by M. Publius Fontana on the spider (mentioned previously), which appeared at Bergamo in 1594. But to my mind they are part of a different story and historical context.

In my fourth and final observation, I want to address a significant lacuna in Fabricius's bibliography. In discussing physico-theological texts, he largely concentrates on the argument from design and almost completely ignores treatises dealing with the connection between the biblical Flood and fossils. He mentions only two treatises related to this particular debate: John Ray's *Three physico-theological Discourses Concerning I. The primitive Chaos . . . , II. the general deluge . . . , III. the dissolution of the world and Future Conflagration* of 1693, and the English version of John Woodward's *Natural History of the Earth, Illustrated, Inlarged, and Defended* of 1726. It is interesting to note that Woodward's partner in this particular interpretation of the origin of fossils, Johann Jakob Scheuchzer, is listed exclusively in the index, where Fabricius adds a handful of references to authors who are not mentioned in his text. These special references appear in the 1728 edition of Fabricius's text (of which the index seems to be an integral part and not added at a later date) and did not undergo any changes in the subsequent editions. We find here allusions not only to Scheuchzer's *Jobi Physica sacra* of 1721 but also to a "copper-Bible" ["Kupffer-Bibel zur Erläuterung der Physica sacrae"] allegedly published from 1727 onward.[27] This is rather mysterious at first sight because Scheuchzer's famous *Physica sacra* was published only in four folio volumes between 1731 and 1735. What likely happened is that Fabricius took at face value the call for subscriptions of 1727, where the first installment of the *Physica sacra* was overoptimistically announced for the same year.[28] However, the fact remains that, as noted, Fabricius's "Verzeichnis derer alten und neuen Scribenten" puts a premium on publications concerned with the argument from design while neglecting almost completely the debate about the role of the biblical Flood in the emergence of fossils that was strongly connected to the work of Woodward and Scheuchzer.[29] Intentionally or not, Fabricius's text suggests that this latter discussion took place on a different stage, even though we now see it as an integral part of contemporary physico-theology. In addition to attending to actors' categories, historians can also appreciate their limitations. Fabricius blended physico-theology within a broad and long-running tradition of natural theology, while omitting from it many contemporary treatises on fossils and the Flood.

NOTES

1. Zöckler, *Geschichte der Beziehungen*, 87–88; Philipp, *Das Werden der Aufklärung*, 33–35.

2. Büttner, "Zur neuen Epochen-Gliederung." None of the authors (Danaeus, Mercator, Lütkemann, and others) considered by Büttner to represent the initial German phase were physico-theologians. Trepp, *Von der Glückseligkeit*, 33–60, dates this inception to the publication of the fourth volume of Johann Arndt's *Four Books on True Christianity* in 1610. This takes much inspiration from Raymund of Sabunde's *Theologia naturalis seu liber creaturarum* (1434–36; Natural theology, or, the book of creatures). *Pace* Trepp, I share Peter Harrison's argument that "early modern talk of contrivance and design in nature was not simply the continuation of a medieval tradition of natural theology with a better and more objectively established data set." Harrison, *The Territories*, 113. See also Stebbins, *Maxima in minimis*, 63–64, and Krolzik, *Säkularisierung der Natur*, 18 and 151–53.

3. I would like to thank Ann Blair and Rienk Vermij for their helpful comments on an earlier draft of this article.

4 For the following, see Häfner, "Literaturgeschichte und Physikotheologie"; Raupp, "Fabricius, Johann Albert"; Petersen, "Brockes, Fabricius, Reimarus"; Krolzik, *Säkularisierung der Natur*, 133–82; Verner, "Johann Albert Fabricius." Raupp and Verner offer extensive bibliographies of Fabricius's works.

5. *Hydrotheologie*, 1734; *Pyrotheologiae sciagraphia*, 1732. In the latter case, only an extremely detailed table of contents has come down to us.

6. Krieger, *Patriotismus*, 79.

7. Mulsow, "Johann Christoph Wolf," 81.

8. Vidal, "Extraordinary Bodies," 62, n. 5.

9. Trepp, *Von der Glückseligkeit*, 328.

10. Fabricius, *Delectus Argumentorum et Syllabus Scriptorum*. A new edition of this text by Walter Sparn is forthcoming (Stuttgart: Fromann-Holzboog).

11. On Sturm, see Gaab, Leich, and Löfflandt, *Johann Christian Sturm*.

12. Fabricius, "Scribenten" (1728), xxiii.

13. The reference is to the *Works of the Learned Isaac Barrow*.

14. "Sam. Parkers . . . Gedanken von Gott und der Vorsehung Gottes, Cogitationes de Deo & Providentia, in sechs Discoursen. . . ." Fabricius, "Scribenten" (1728), xxxii–xxxiii. He must also be mistaken about the date of publication, which he notes as 1672.

15. On Straehler, see Zedler, *Universal-Lexicon*, 40: cols. 482–83. The larger context of this debate was the fundamental criticism voiced against Wolff by the Halle Pietists, which resulted in his banishment from the Prussian lands by King Frederic William I in 1723.

16. Trepp, *Von der Glückseligkeit*, 313.

17. Fabricius, "Scribenten" (1728), xxix–xxxi.

18. Fabricius, "Scribenten" (1728), sec. 2, references to Fénelon.

19. On this, see Maehle, "'Est Deus ossa probant.'"

20. Fabricius, "Scribenten" (1728), sec. 2.

21. Philipp, *Das Werden der Aufklärung*. In light of the work's problems, it suffices here to refer to the introduction to this volume, note 4, and to J. Steiger, *Bibel-Sprache*, 55–56.

22. Fabricius, "Scribenten" (1728), lx–lxiv.

23. See Sparn, "Natürliche Theologie"; Topham, "Natural Theology and the Sciences"; Mandelbrote, "Early Modern Natural Theologies." See also the comments on this question in the introduction to this volume.

24. Christian Donat was a professor of logic at the University of Wittenberg from 1672 to 1694. He obviously is the author of this treatise (rather than his student). This does not invoke the typical physico-theological vocabulary but confirms that not only is the human hand the work of God but that man is entirely God's creation (sec. 39). This is an unusual treatise because the University of Wittenberg was only just beginning to leave behind the period of Lutheran high orthodoxy—no friend of physico-theology.

25. Meyer and Mentzer, *Disputationum Hamburgensium Decima aranearum*, 104 (sec. 14).

26. Sturm and Volland, *Oculus theoscopos.* See also note 18 in this chapter.

27. "Biblical coppers as illustrations of the Physica sacra." In the eighteenth century the word *Kupfer* (copper) also meant engraving, as engravings were made on copper plates.

28. Scheuchzer's call for subscriptions is printed in Müsch, *Geheiligte Naturwissenschaft,* 185–87.

29. On this, see Greyerz, "Protestantism, Knowledge, and Science."

PART THREE

STYLES OF RELIGIOSITY

CHAPTER SEVEN

Miracles, Secrets, and Wonders

Jakob Horst and Christian Natural Philosophy in German Protestantism before 1650

KATHLEEN CROWTHER

In 1569 Jakob Horst (1537–1600), a German Lutheran physician, published a thick book enthusiastically titled the *On the wonderful secrets of nature, and on the fruitful contemplation of these, which are not only useful but also pleasant to read about*.[1] Horst offered his readers extensive descriptions of the natural world, including the human body, plants, animals, and the heavens. The information ranged from the philosophical to the practical, from the serious to the comical. A reader could find descriptions of the four elements and the structure of the cosmos, simple cures for headaches and techniques for getting rid of vermin, stories about the great physician Hippocrates, and warnings about magicians who made men impotent. Throughout the book he explicitly connected natural knowledge to both religious piety and practical utility. Though written in German and thus accessible to a nonscholarly audience, the text remained true to the author's university training, as it was peppered with Latin and bristled with references to classical and contemporary authorities. The book sold briskly, going through ten more editions during Horst's lifetime and five after his death.[2] The success of the book, as measured by its multiple editions, indicates that Horst had a keen sense of the contemporary book market and what would appeal to vernacular readers. Despite its contemporary popularity, the *Wonderful secrets of nature* has attracted little attention from historians of science, and Jakob Horst has long since faded into obscurity.

The *Wonderful secrets of nature* is not a text that reports new discoveries or ideas unique to the author; rather, it is an encyclopedic treatment of the natural world, gleaned largely from books and intended to entertain and edify a broad lay audience. For Horst, "contemplation" of the natural world was "fruitful," "useful," and "pleasant" because such contemplation could both lead a person closer to God and reveal practical "secrets" that could improve a person's life. Although a mixing of science and theology in which descriptions of the natural world are combined with moral exhortations is deeply strange to modern sensibilities, the *Wonderful secrets of nature* was a product of a period in which scientific and religious concerns were thoroughly intertwined. In his now-classic work *Theology and the Scientific Imagination*, Amos Funkenstein argues that science and religion were more closely connected in the sixteenth and seventeenth centuries than they were in the Middle Ages. "Never before or after," he writes, "were science, philosophy, and theology seen as almost one and the same occupation."[3] But if knowledge of nature could bring his readers closer to God, it could also (or so Horst claimed) bring them more immediate and tangible benefits, such as health and wealth. The *Wonderful secrets of nature* was one of hundreds of books in the sixteenth and seventeenth centuries to offer readers access to such "secrets." William Eamon's magisterial study, *Science and the Secrets of Nature*, makes clear that "books of secrets" were among the most popular genres of books across Europe in the early modern period and were an integral part of the development of modern science.[4] The conjunction of piety and practicality, of divine and mundane, that Horst offered in the *Wonderful secrets of nature* was enormously appealing to early modern readers.

In the second half of the sixteenth century, the number of scientific works in vernacular languages increased dramatically.[5] Some of these books were translated from Latin,[6] although "translation" in this period could involve substantial reworking of a text, and others were originally written in vernacular languages. These books were marketed to the growing numbers of readers literate in their native tongues but not necessarily in Latin.[7] I use the term "scientific" to refer to books on a range of subjects, including astronomy, astrology, alchemy, botany, zoology, geography, agriculture, and medicine. Horst covers all these topics, and although the encyclopedic breadth of his coverage is unusual, his choice of topics was not.

While Horst's *Wonderful secrets of nature* was shaped by the European-wide trends I have just sketched out, it was also shaped by the more specific context of German Lutheranism. The late sixteenth century was a period of fierce confessional disputes between Catholics and Protestants, between Lutherans and Calvinists, and between different factions of Lutherans. These conflicts led to the formation of distinct confessional identities, characterized not just by adherence to specific doctrines but also by different styles and genres of art, music, literature, and science.[8] The *Wonderful secrets of nature* was written by a Lutheran and intended first and foremost to appeal to a Lutheran readership. As such, it offers unique insight into the meanings and uses of natural knowledge in German Lutheran culture of the late sixteenth and seventeenth centuries.

In this essay, I highlight three distinct ideas about nature and about the usefulness of the study of nature in Horst's *Wonderful secrets of nature*. The first is that natural knowledge leads to awe and wonder at God's creation and inspires greater piety. Contemplation of nature is a means of access to the divine. The second is that an understanding of nature gives practical information on how to lead a "good" life—that is, a life that is both physically healthy and morally upright. The third is that natural knowledge is potentially dangerous. Although contemplation of the natural world *should* lead to greater piety and morality, natural knowledge can be used incorrectly and could lead people astray, both physically and spiritually. In isolation, all these themes can be found in the work of Horst's Catholic and Calvinist predecessors and contemporaries. The first theme, in particular the idea that the study of nature inculcated love and admiration for nature's creator, was ubiquitous in scientific texts of the Middle Ages and early modern period by writers of all three Abrahamic faiths. However, the combination of these themes, with the inherent tensions and ambiguities between them, was distinctive to Lutheran writing on the natural world in the late sixteenth and early seventeenth centuries.

My analysis of Horst's book builds on the work of historians who have demonstrated that natural knowledge played an important role in Lutheran intellectual culture.[9] Martin Luther himself extolled contemplation of the natural world as a means of access to the Divine. In Luther's view, contemplation of nature should inspire not only awe and wonder but also immense gratitude. The splendor of the physical world demonstrated God's omnipotence as well as his fatherly compassion and goodness. Like an earthly father, God supplied his children with a house and food and everything they needed. Reading about or directly observing the human body, the heavens, plants, animals, or any other aspect of the natural world could bring a devout Christian to a deeper knowledge and love of the Creator.[10] Many of Luther's colleagues and students shared his interest in and enthusiasm for natural knowledge. The most important and influential of these colleagues was Philip Melanchthon (1497–1560), a professor of Greek at the University of Wittenberg who redesigned the university curriculum to bring it in line with Lutheran theology. Although belief in divine providence was not new, nor was a sense that contemplation of the natural world was a pious activity, Sachiko Kusukawa has demonstrated that study of the natural world was emphasized at Lutheran universities in a way that it was not elsewhere.[11] And Anne-Charlott Trepp has argued that doctrines of providence and of salvation shaped Lutheran understandings of nature and the use of natural philosophy in texts intended for a broad lay audience, demonstrating that nature played an increasingly important role in Lutheran spirituality and devotional life beginning in the second half of the sixteenth century.[12]

Robin Barnes's study of astrology in Reformation Germany lends further support to the claim that Lutherans linked natural knowledge to religious piety. As Barnes points out, the cheapest, most plentiful texts on the natural world in the sixteenth century were calendars and almanacs that offered astrological predictions for the coming year. And nowhere were these texts more popular than in Lutheran areas of the Holy Roman Empire. Barnes argues that these texts fostered a view of the cosmos as a united and orderly whole, created and governed by an omnipotent and benevolent God. Popular astrology texts were "the most pervasive manifestation of the early modern mathematization and naturalization of time and space."[13] Far from promoting an irrational or superstitious world view, astrology trained people to see the cosmos as regular and governed by immutable laws. In sum, the notion that natural knowledge—of the stars, the earth, and the human body—lead to a deeper understanding and love of God was widely disseminated in Lutheran culture. Evidence for this view can be found in texts owned and read by men and women across the social spectrum, whether university professors, wealthy nobles and burghers, artisans, or peasants.

A Brief History of the *Wonderful secrets of nature*

Horst's book was a translation, albeit with significant additions and rearrangement, of the magnum opus of the Dutch Catholic physician Levinus Lemnius (1505–68), *De occultis naturae miraculis*, a book intended to display the wonders of God's creation in nature generally and especially in the human body. Lemnius was educated at the University of Leuven and spent most of his life practicing medicine in his native city of Zierikzee, in the province of Zeeland.[14] He published quite extensively, writing books on astrology, medicine, and botany, among other topics. Far and away his most successful book, just judging by the numbers of editions, was *De occultis naturae miraculis*. Lemnius published the work in Latin in Antwerp in 1559. He expanded the text in 1564, and this larger version went through numerous Latin editions, the last in 1666.[15] It was also translated into French, Italian, English, and German.

Lemnius does not attempt to give a comprehensive account of the natural world or the human body; rather, the book is about those facets of nature that are secret or occult or, as he puts it, "things which do not present a manifest demonstration to the sense and understanding, and therefore are called by physicians, hidden qualities." While he states that it is not always possible to provide certain explanations of such phenomena, he promises the reader "probable conjectures" about their causes.[16] Further, he asserts that "contemplation of nature," and particularly of occult phenomena, inspires "admiration and love of the Creator."[17] The work is divided into four books. The first book begins with a chapter on "Nature, God's Instrument," followed by a chapter on "Man's Worth and Excellency." Most of the rest of book 1 deals with human procreation, including such topics as how children are nourished in the womb and when the fetus is ensouled. It ends with a few chapters on odd topics such as why grapes will not ripen in moonbeams and how to kill weevils in corn. The second book deals with the causes and cures of various human diseases. The third and fourth books deal with a wide variety of natural phenomena, including the effects of the moon on the tides and the winds on human health, an explanation of why dogs go mad, and a new method for making salt. The work ends with an "Exhortation" on how to lead a good life, a section that gives advice on both bodily and spiritual health.

Jakob Horst was born in Torgau in 1537. He was educated, in both medicine and theology, at the Lutheran universities of Wittenberg and Frankfurt. After obtaining his degree, he worked as a physician in Schweidnitz, Sagan, Iglau, and Krems. He joined the medical faculty at another Lutheran university, Helmstedt, in 1584, and remained there until his death in 1600.[18] Horst made at least two different versions of his translation. In the editions he published between 1569 and about 1588, he followed the structure and content of the original more closely, although he added new material and rearranged the text somewhat. In 1588 he produced a substantially revised version that departed considerably from the original. This new and expanded version was divided into ten books, rather than four, and gave a comprehensive account of the entire natural world—heavens, earth, and human body. It was also significantly reorganized. Horst translated Lemnius's addendum on how to lead a good life and made it the first of the ten books. It was this second version that continued to be reprinted into the late seventeenth century.

In the process of translating, rearranging, and augmenting Lemnius's *De occultis naturae miraculis*, Horst transformed it from a Catholic to a Lutheran text. Although Horst generally steered clear of the specific doctrinal controversies that divided Lutherans, Catholics, and Calvinists, he made clear that he wrote as a devout Lutheran. The most striking evidence of this is his assertion that his translation of Lemnius's work was analogous to Martin Luther's translation of the Bible. Just as Luther made the word of God accessible to German readers, so Horst made nature, God's creation, accessible to the same audience.[19] Another example of the Lutheran character of the *Wonderful secrets of nature* is Horst's inclusion of a chapter titled "a little rosary of all kinds of hidden properties in natural things," which was not in Lemnius's original.[20] This was both a critique of and an alternative to the Catholic practice of praying the rosary. Horst offers ten pages of interesting nuggets of information about the hidden properties of natural objects, such as that beer made with well water tastes better and lasts longer than beer made with rainwater, egg whites mixed with caulk can be used to repair broken glass and pottery, and onions grow best when the moon is waning.[21] These little facts are strung together like beads on a rosary but suggest to the reader a different kind of devotional practice. Instead of prayers to the Virgin Mary, the reader should contemplate the myriad splendors of the natural world, not just in the grand structure of the cosmos, but in the common and mundane objects around them, like beer, eggs, and onions. Horst also singles out several Lutheran authors—Philip Melanchthon, David Chytraeus (1530–1600), and Conrad Bergius (1544–92)—for particular praise.[22]

Nature and God's Providence

Horst asserts that contemplation of nature will lead to deeper piety and that this was his main motivation for composing the *Wonderful secrets of nature*. "The entire world," Horst writes, "was created by God the Almighty out of His inexpressible wisdom, like an exquisite building, that reasonable people cannot admire enough. And [the world] is so wonderfully ruled and maintained by God that even the heathens, who knew nothing of God, by this omnipotent and omniscient ordering, firmly believed that there was a God and that he was an all-powerful, supremely wise and most just being."[23] Horst's description of the natural world as "an exquisite building" is highly reminiscent of Martin Luther's description of nature in his *Sermons on the First Book of Moses* as a "house" prepared by God for humankind.[24] Horst, like Luther, believed that contemplation of nature should inspire not only awe and wonder but also immense gratitude. The splendor of the physical world demonstrated God's omnipotence, but it also demonstrated his fatherly compassion and goodness.

These were not just pious sentiments in the prefatory material. Horst refers repeatedly to divine providence throughout his descriptions of the natural world. For example, in a section on the heavens in the 1588 edition, Horst not only explains the complex motions of heavenly bodies but also describes how they are perfectly arranged by divine providence to benefit human beings. He explains that the sun moves around the earth on an eccentric circle, that is, one not centered on the earth. This was arranged by divine providence "so that in the summer, when the rays of the sun stream directly down [on the earth], the sun on its eccentric is farther away from the earth, so that the great, unbearable heat is mitigated." And, conversely, in the winter, when the sun's rays hit the earth obliquely, God has arranged the universe so that the sun is closer to the earth in order that we do not become too cold.[25] Other planets also are closer to or farther away from the earth as they travel around it, and their effects on the earth—and on humans—are greater when they are closer. God has planned for this as well. Saturn, for example, is cold, so God has arranged the cosmos in such a way that Saturn is at its closest to the earth only during the summer.[26]

Horst acknowledges that nature can at times be harmful to human beings, but he ascribes the damage caused by the natural world to the providence of God as well. God, like an earthly father, punishes his children, not out of malice but out of love, so that they will learn and improve their ways. Most sixteenth- and seventeenth-century parents and educational theorists believed that strict discipline, including corporal punishment, was essential to raising a pious and obedient child.[27] Parents who did not discipline their children were seen as at best lax and at worst uncaring. Horst applies these ideas about fatherhood to God. When he describes various kinds of damaging weather, including snow and hail that harm crops and winds that bring disease in their wake, he writes that God sends these phenomena when people are "hardened in sin" so that they will learn to "obey his command."[28] Understood in this way, even destructive natural phenomena should be seen as signs of God's benevolence.

Practical Uses of Natural Knowledge

Horst asserts that knowledge of nature, particularly knowledge of human anatomy and physiology, helps people to lead better lives, both physically and spiritually. "Medicine," he writes, "prescribes a sober and moderate life."[29] The first book of his expanded ten-book version of the *Wonderful secrets of nature* was "an admonition on how every human being can achieve the very best life."[30] This book was based on the similarly titled addendum to Lemnius's book, but with substantial modifications. It contains advice on how to maintain the health of the body as well as the health of the soul. However, lessons about leading a sober, moderate, God-fearing life are scattered throughout all ten books of the *Wonderful secrets of nature*. The biggest lesson from human anatomy, physiology, and medical theory is that moderation and sobriety are the keys to health. For example, Horst asserts that "when we carefully observe the narrow *mesaraic veins*, through which food goes from the stomach to the liver, the smallness of the vessel of the stomach and the soft, tender members, we have to acknowledge that we were born to moderation, good order, and sobriety and not to gorging, swilling, and generally overindulging."[31] Horst's advice reflects the Galenic theory of digestion, in which food is processed in the stomach and then passes to the liver, where it is converted to blood and dispersed to the entire body. Horst insists that knowledge of anatomy makes clear that our bodies were designed by God to consume food and drink in fairly small quantities. Large quantities of food and drink overwhelm the small stomach and the narrow and delicate vessels that carry food from the stomach to the liver. When we gorge ourselves or drink to excess, we are damaging our bodies, and we are also violating God's law and abusing God's creation. Horst gives similar advice about moderation in sex. Both male and female bodies produce "seed" that joins together during sexual intercourse to form offspring. This seed is drawn from all parts of the body. "From healthy limbs healthy seed is made, and from unhealthy limbs unhealthy seed."[32] Hence, it is essential to refrain from sex while one is sick or drunk or otherwise indisposed. Further, because human "seed" is formed from blood drawn from all parts of the body and requires heat to produce, too much sex depletes the body.[33]

Secrets, Shame, and Danger

The third significant theme in Horst's book is concern about the misuses of natural knowledge when placed in the hands of the uneducated. Although the first two themes were certainly present in Lemnius's original text, this last one was not. Horst added a section at the beginning of the book on "the correct use of the hidden secrets of nature."[34] In this section he rails against the misuse of the material in his book. Substantial portions of the book deal with sex and reproduction. Horst insists that the book is only for "pious husbands, sensible wives, diligent students of nature, and lovers of health." It is acceptable for "pious, honorable unmarried women" to read the book (or for it to be read to them), but they need to skip over certain parts.[35] He denounces "rude people and coarse young men" who read the book in public and with young women. He promises that divine wrath will fall upon them.[36] Horst's attention to reproduction, as well as his anxiety about the "proper" use of this information, are both consistent with contemporary Lutheran writing on the subject. Several scholars have pointed out that Lutherans devoted particular attention to sex, sexual desires, and procreation.[37] Having denounced celibacy as an unnatural and immoral state, Lutherans were at pains to present sex and marriage as both natural and divinely ordained. But they were sensitive to accusations of licentiousness and to the ongoing ambivalence about clerical marriage.[38]

Horst is particularly caustic about the misuse of natural knowledge by women. In his section on female reproductive anatomy, he asserts that the reason God placed the uterus deep inside the female body instead of near the surface was to protect the uterus from harm, inflicted by women themselves "through carelessness, but especially through the malice of certain wicked women, who would then have been able to rid themselves of unwanted pregnancies."[39] In this extraordinary passage, Horst asserts that God himself chose to keep the uterus hidden from women because he did not trust them to use knowledge of their own bodies for good and not for evil: all the more reason that men, like Horst, should keep this knowledge secret from women.

As I noted earlier, Horst explicitly compares his translation of Lemnius's *Secrets of nature* to Martin Luther's translation of the Bible. The same objections that were made to Luther's translation could be made to Horst's translation of Lemnius—namely, that it was better that the sacred text remain in Latin so that the uneducated could not read it and be led into error.[40] Although Horst acknowledges this as a possibility, he asserts that the good that can come of his translation outweighs the possible harm. Nonetheless, he repeatedly warns his readers to use his book properly lest they fall into grave sin. Horst's comparison refers to the fact that at least parts of the book are translated from a Latin text by Lemnius and other material is culled from Latin writers not available in the vernacular. The allusion to Luther's translation of the Bible also evokes the well-known metaphor of the book of nature.[41] Horst claims that he is not just translating a particular human author but making the entire book of nature, authored by God, accessible to vernacular readers.

Horst does promise to leave certain passages, "which are entirely too shameful," in Latin.[42] And there are indeed several such passages in the book in Latin. As an example, let us look at the third chapter of the first book of the edition from 1580. This chapter is "On giving birth or fathering children, and how human beings should undertake this work with honor . . . ," and includes an extensive description of male and female reproductive organs.[43] There are scattered terms in Latin and a few in Greek throughout the anatomical section, and there are three substantial passages in Latin, the first on male ejaculation, the second on the different shapes and sizes of penises, and the third on the different textures and sizes of the cervix.[44] It is not clear why this particular material would have been considered unsuitable for some readers, especially female readers. These passages are no more graphic than other parts of the text, which contains explicit details of reproductive anatomy and function. Further, none of these passages was in Lemnius's original text. Horst, as I noted earlier, made significant additions to Lemnius. In the case of this chapter on procreation, about two pages out of fifty are an actual translation of Lemnius's writing. The rest of these passages, which Horst claims to have "left in Latin," were original material composed by him for this book. They are in a different typeface and accompanied by marginal notes (in German) that summarize the contents. The overall effect is to highlight these passages through a kind of textual striptease rather than to hide them. I would suggest that these Latin passages were added to create the impression of dangerous and forbidden knowledge, rather than because the material they contain is more problematic than other parts of the text. The *Wonderful secrets of nature* was almost certainly shaped not just by Horst's Lutheran faith and education but by what he and his publishers believed would sell books. And making natural knowledge seem forbidden and dangerous may have heightened the appeal of the book.

Clearly, Horst's *Wonderful secrets of nature* shares many of the characteristics of physico-theology as delineated by other authors in this volume: Horst conveyed to his readers a vision of God as rational and benevolent and of the study of nature as inspiring awe, wonder, and gratitude toward the Creator. He expressed considerable excitement about new scientific research, including empirical methods and direct observation of nature. His work had a strong moral dimension. He had, however, no concern to combat atheism or deism. His work encouraged readers to see and experience God in nature. He did not seek to prove the existence of God using the natural world. Interestingly, Horst's *Wonderful secrets of nature* stayed in print until the end of the seventeenth century and undoubtedly continued to be read and circulated long after that. The *Wonderful secrets of nature* shared space in bookstores and on bookshelves with works of physico-theology. In other words, although we might see Horst's book as a precursor to later physico-theological works, it was also a contemporary example of the genre.

NOTES

1. Horst, *Occulta Naturae Miracula: von den wunderbarlichen Geheimnissen der Natur* (1569).
2. A search of WorldCat and the Karlsruhe Virtual Catalog found editions from 1569, 1571, 1572, 1575, 1579, 1580, 1581, 1588, 1589, 1592, 1593, 1600, 1605, 1612, 1630, and 1672.
3. Funkenstein, *Theology and the Scientific Imagination*, 3.
4. Eamon, *Science and the Secrets of Nature*.
5. Giacomotto-Charra, "Entre traduction et vulgarisation"; Pantin, *La poésie du ciel*; Harkness, *The Jewel House*; Crowther-Heyck, "Wonderful Secrets of Nature"; and Johnson, *Astronomical Thought in Renaissance England*.
6. Van Hoof, *Histoire de la traduction en Occident*, 42.
7. Engelsing, *Analphabetentum und Lektüre*, and Chrisman, *Lay Culture, Learned Culture*.
8. On confessionalization, see Ehrenpreis and Lotz-Heumann, *Reformation und konfessionelles Zeitalter*; and Kaufmann, *Konfession und Kultur*.
9. Barker, "Role of Religion" and "Lutheran Contribution"; Bellucci, *Science de la nature et Réformation*; Crowther, *Adam and Eve* and "Lutheran Book of Nature"; Harrison, *Fall of Man*; Kusukawa, *Transformation of Natural Philosophy*; Methuen, *Kepler's Tübingen*; Trepp, "Natural Order and Divine Salvation."
10. Crowther, *Adam and Eve*, chap. 5.
11. Kusukawa, *Transformation of Natural Philosophy*.
12. Trepp, "Natural Order and Divine Salvation."
13. Barnes, *Astrology and Reformation*, 49.
14. Hoorn, *Levinus Lemnius*.
15. Margolin, "Vertus occultes et effets naturels"; Blair, "Mosaic Physics"; Crowther, "Sacred Philosophy, Secular Theology."
16. Lemnius, *De occultis naturae miraculis*, from pref. to reader.
17. Lemnius, *De occultis naturae miraculis*, from dedication to Eric of Sweden.
18. Biographical information from *Deutsche Biographie*, s.v. Horst.
19. Horst, *Occulta Naturae Miracula. Wunderbarliche Geheimnisse der Natur* (1580), 24.
20. Horst, *Occulta Naturae Miracula. Wunderbarliche Geheimnisse der Natur* (1605), bk. VI, 142–52.
21. Horst, *Occulta Naturae Miracula. Wunderbarliche Geheimnisse der Natur* (1605), 143, 145, 149.
22. Horst, *Occulta Naturae Miracula. Wunderbarliche Geheimnisse der Natur* (1588), sig.)(iiir.
23. Horst, *Occulta Naturae Miracula. Wunderbarliche Geheimnisse der Natur* (1580), 76–77.
24. Crowther, *Adam and Eve*, chap. 5; Crowther, "Lutheran Book of Nature."
25. Horst, *Occulta Naturae Miracula. Wunderbarliche Geheimnisse der Natur* (1588), 20.
26. Horst, *Occulta Naturae Miracula. Wunderbarliche Geheimnisse der Natur* (1588), 20.
27. Ozment, *When Fathers Ruled*.
28. Horst, *Occulta Naturae Miracula. Wunderbarliche Geheimnisse der Natur* (1605), bk. VI, 87.

29. Horst, *Occulta Naturae Miracula. Wunderbarliche Geheimnisse der Natur* (1588), 18.

30. Horst, *Occulta Naturae Miracula. Wunderbarliche Geheimnisse der Natur* (1588), 1.

31. Horst, *Occulta Naturae Miracula. Wunderbarliche Geheimnisse der Natur* (1580), 9; (1588), 120.

32. Horst, *Occulta Naturae Miracula. Wunderbarliche Geheimnisse der Natur* (1580), 165.

33. Horst, *Occulta Naturae Miracula. Wunderbarliche Geheimnisse der Natur* (1580), 165.

34. Horst, *Occulta Naturae Miracula. Wunderbarliche Geheimnisse der Natur* (1580), 1–26.

35. Horst, *Occulta Naturae Miracula. Wunderbarliche Geheimnisse der Natur* (1580), 5.

36. Horst, *Occulta Naturae Miracula. Wunderbarliche Geheimnisse der Natur* (1580), 3–4.

37. Crowther, *Adam and Eve*, chap. 4; Encarnación, "The Proper Uses of Desire"; and Stolberg, "A Woman Down to Her Bones."

38. Plummer, *From Priest's Whore to Pastor's Wife.*

39. Horst, *Occulta Naturae Miracula. Wunderbarliche Geheimnisse der Natur* (1580), 182.

40. Horst, *Occulta Naturae Miracula. Wunderbarliche Geheimnisse der Natur* (1580), 25.

41. Bono, *Word of God*; Van der Meer and Mandelbrote, *Nature and Scripture.*

42. Horst, *Occulta Naturae Miracula. Wunderbarliche Geheimnisse der Natur* (1580), 26.

43. Horst, *Occulta Naturae Miracula. Wunderbarliche Geheimnisse der Natur* (1580), 157.

44. Horst, *Occulta Naturae Miracula. Wunderbarliche Geheimnisse der Natur* (1580), 165, 173, 184.

"Rather Theological than Philosophical"

John Ray's Seminal *Wisdom of God Manifested in the Works of Creation*

KATHERINE CALLOWAY

I appeare now in the plaine shape of a mere Naturalist, that I might vanquish Atheisme. . . . For hee that will lend his hand to help another fallen into a ditch, must himself though not fall, yet stoop and incline his body. . . . So hee that would gaine upon the more weake and sunk minds of sensuall mortalls, is to accommodate himself to their capacity.

Henry More, *An Antidote against Atheism* (1653)

Thus far the Doctor [Cudworth], with whom for the main I do consent. I shall only add, that Natural Philosophers, when they endeavor to give an account of any of the Works of Nature by preconceived Principles of their own, are for the most part grosly mistaken and confuted by Experience.

John Ray, *Wisdom of God* (1691)

John Ray is justly celebrated—not least in this volume—for his seminal contributions to natural theology in an age of scientific change in England and on the European continent. Two works are normally singled out: his *The Wisdom of God Manifested in the Works of Creation*, published in 1691, and on its heels, *Miscellaneous Discourses concerning the Dissolution and Changes of The World*, published in 1692 and retitled *Three Physico-Theological Discourses* in 1693 and subsequent English editions. Both works were drafted decades earlier, as sermons given before Ray left Cambridge in 1662. Both were highly influential throughout the eighteenth century. One exerted unparalleled influence on the English scientific imagination and has been said "more than any other book [to have] initiated the true adventure of modern science." This is *The Wisdom of God* (henceforth *Wisdom*), which was expanded by the author in three subsequent editions until 1704 and then went into seventeen more editions (including translations in French and German) between 1709 and 1845. Its popularity is attested not only by the number of editions but also by the breadth of its influence: when the Scottish poet Robert Burns wrote his poems "To a Mouse" and "To a Louse," for instance, he had Ray's book in mind.[1] Indeed, *Wisdom* is a "poetic" book in the tradition of the biblical Psalms, aimed at showcasing divine creativity in the natural world rather than persuading atheists by means of logically compulsive proof. In this chapter I explore how Ray's *Wisdom of God* departed from preceding works to become a paradigmatic work of physico-theology: for Ray drew on contemporary natural history in unprecedented detail and with a fidelity to empirical method that did not characterize the writings of other Royal Society members, much less those of his Cambridge colleagues. Then I briefly contrast Ray's poetic celebration of Creation with the more argumentative "Newtonian" physico-theology that arose in the 1690s.

Ray's Scientific Humility in *Wisdom*

John Ray, Fellow of the Royal Society, botanist and ornithologist, and sometime Church of England divine, seldom fails to appear in English discussions of early modern natural theology. Although the term "physico-theology" is often not used, there is a tendency in this discourse to mark Ray out among English natural theologians, perhaps as paradigmatic of a certain type of natural theology.[2] Ray stringently advocated for the new sciences of experimentation and natural history and their ability to lend unprecedented insight into the divine, famously exclaiming in the preface, "Let it not suffice us to be book-learned, to read what others have written . . . but let us ourselves examine things as we have opportunity, and converse with Nature as well as Books. Let us endeavor to increase this Knowledge, and make new Discoveries," adding that this activity is "the business of a Sabbath-day."[3] He worked tirelessly to debunk old myths and to test received knowledge against direct observation in collaboration with a community of like-minded students of nature: this can be seen in his unflinching prosecution of accuracy even against his own former beliefs in successive editions of *Wisdom*.[4] Even in its first edition of 1691, Ray already shows a marked interest in natural historical and physiological detail, spending tracts of the text listing and describing natural phenomena rather than interpreting and editorializing. When he reaches a boundary of his own knowledge, or that of the scientific community, he acknowledges this, expressing puzzlement or incredulity on such topics as fetal formation, the structure of the ear, birds' beaks, the buoyancy of fish, and male nipples.[5] In each case, he is keen to learn, but not because the missing piece of information would shore up a logically compelling refutation of atheists.

For Ray was convinced from the start of God's existence, wisdom, and power and used his "reading" of nature to deepen theological insight rather than to establish the accuracy of these theological doctrines. This is suggested, first, in the work's origin as a sermon series structured as an exegesis of Psalm 104:24: "How manifold are thy works, O Lord? In Wisdom hast thou made them all." Ray explains in his preface, "The Holy Psalmist is very frequent in the Enumeration and Consideration of these Works, which may warrant me doing the like, and justific the denominating such a Discourse as this, rather Theological than Philosophical."[6] Second, although in his preface he briefly uses terms such as "argument," "demonstration," and "proof," his strategy is far from logically compelling, and he knew that. Instead he establishes an "interpretive community" of nature that cannot see nature as anything other than indicative of God's existence, wisdom, and power.[7] Unfit or confusing phenomena are for Ray evidence of human fallibility rather than of some divine error or shortcoming, much less of divine nonexistence. It is clear that *Wisdom* is not a refutation of atheists, but a celebration of the wonders of God's wisdom evident in the book of nature, meant to illuminate and expand on the revelation in scripture.[8]

Forerunners

Ray declares in his preface that he has drawn copiously on the work of others and that his main contribution, besides gathering disparate materials into one affordable book, is the addition of "Considerations new and untouch by others": numerous examples of purpose and beauty drawn from natural history and other sciences to lend more rhetorical force to the basic claim of God's wisdom and providence in creation.[9] He acknowledges a particular debt to Henry More (*An Antidote against Atheism*, 1653), Edward Stillingfleet (*Origines Sacrae*, 1662), Samuel Parker (*Tentamina physico-theologica de Deo*, 1665), Ralph Cudworth (*True Intellectual System of the Universe*, 1678), and Robert Boyle (*Disquisition about the Final Causes of Natural Things*, 1688), and he also regularly quotes his friend John Wilkins's *Of the Principles and Duties of Natural Religion* (1675).[10] Here I will briefly discuss how several of these works differed from *Wisdom* in terms of modern scientific content before giving more attention to Henry More, whose *Antidote against Atheism* lent the most material to Ray's *Wisdom*.

Though many of these works engage modern science to some degree, in general they pay too little attention to the new scientific method and its findings to be considered physico-theology. Although the three authors associated with Oxford—Parker, Boyle, and Wilkins—were members of the Royal Society, only Boyle practiced experimental science vocationally and none was engaged in natural history. Stillingfleet and Cudworth, both Cambridge men, stand at yet another remove from modern natural science, arguing by various decidedly nonempirical means: for instance, from an idea of God in men's minds, from universal consent, from the historicity of miracles and other accounts in scripture, and from the nature of being itself. These authors aim explicitly to refute atheists, and they spend their energy examining the nature of their own arguments—that is, what makes them compulsive and to what extent. Spirited attacks are launched at particular figures, including Aristotle (e.g., for the belief in the world's existence from eternity), Epicurus (e.g., for the assertion that the world came about by chance and the seclusion of divine from human affairs), and Descartes (e.g., for limiting God's role in the creation and preservation of the world). Stillingfleet does rehearse how the natural world—with its beauty, order, and usefulness to man—bears out the Psalmist's claims that the heavens declare the glory of God and that man is fearfully and wonderfully made. But Stillingfleet shows a devotion to the ancient physician Galen that Ray would not emulate, suggesting that Galen had written a "just Commentary on those words of the Psalmist" in the seventeen books of his *De usu partium*, which could therefore be summarized without emendation rather than taken piecemeal or verified.[11] Cudworth's most notable contribution to the prehistory of physico-theology is perhaps the notion of a "plastic nature," which Ray would adopt. "Plastic nature," as Cudworth summarizes, constitutes an attempt to steer a course between "fortuitous mechanism," which removes God from the workings of the world entirely, and a state of affairs in which "God must be supposed to do all things in the world immediately, and to form every Gnat and Fly, as it were with his own hands."[12] In formulating this intermediate creative force, Cudworth anticipates key challenges facing physico-theology,[13] but his *True Intellectual System* is not itself physico-theology.

Parker and Wilkins were great advocates of the new sciences, if not natural philosophers themselves: both were educated in divinity, and both eventually rose to the position of bishop, with Wilkins also presiding over colleges successively at Oxford and Cambridge. Their texts encompass efforts to demonstrate, like Boyle's *Usefulness of Experimental Naturall Philosophy*, that those sciences underscore rather than undermine orthodox theology. None of these works, however, constituted a marshaling of the new discoveries of experimentalism and natural history in order to read the book of nature alongside scripture for divine insights. Parker's book comprised a humanist rehabilitation of ancient learning and an attack on Cartesian mechanism (criticizing along the way recent works such as Cudworth's and Stillingfleet's for positing an innate idea of God).[14] For its preference for sense perception and lengthy arguments that the new natural philosophy countered age-old pagan error, the book gained Parker election to the Royal Society—but most of the text is spent refuting various atheistic arguments or tracing the history of paganism. The sections considering natural phenomena as evidence of divine providence (the latter half of chapter 2 and chapter 3 of book 1) constitute less than a fifth of the text, and even those chapters draw primarily on ancient sources such as Galen, though Parker does cite neurological work by his contemporary Thomas Willis.[15] Wilkins limits "physico-theological" material yet more aggressively, to just a few pages of his *Principles and Duties*,[16] which largely distills the arguments of Cudworth and Stillingfleet.[17]

Robert Boyle's contributions to natural theology are significant: he not only endowed the notable lecture series on the topic but also touched on natural theological matter repeatedly in his own works. Here I consider only his brief *Disquisition about Final Causes*, the work Ray cites as a source for his *Wisdom*. Like other works discussed thus far, *Disquisition* attacks both Epicurus and Descartes, who in different ways limited the interaction between the natural and the divine and thus impeded the way for physico-theology. Boyle's book, however, engages contemporary scientific findings with enthusiasm and omits refuting atheists with lengthy argumentation,[18] making it perhaps the closest in spirit to Ray's *Wisdom of God*. Interestingly, Boyle himself was uncertain how to classify the work.[19] Reasonably, modern readers tend to see it as standing outside the scope of physico-theology, or even natural theology,[20] devoted as it explicitly is to the prior question whether "there be any Final Causes of things Corporeal, knowable by Naturalists." As promised, Boyle parses what things can be said to have final causes and how far we can presume to argue from them.[21] Boyle's answer to this prior question is partly a posteriori, however, relying on lengthy description of the "excellent contrivance" of natural things—mostly organic, but also inorganic—to underscore his assertion (to a Christian audience) that God can reveal Godself in whatever ways God likes, including via evidence of design and purpose in the details of the natural world. It is therefore a duty of humans, Boyle

pronounces, to draw appropriate conclusions about God's wisdom and benevolence from the book of nature.[22] Several times during this discourse
Boyle issues a call for others to follow his suit, observing more than he has
observed of natural phenomena and deriving thence a deeper sense of those
divine attributes; John Ray would respond to this call.

But while Ray names Boyle with special reverence in *Wisdom of God*, it is
an author with a very different set of philosophical commitments who provided the main source for Ray's paradigmatic physico-theology.[23] This is
Henry More, in the second book of his *Antidote against Atheism,* published
in 1653 and shortly before Ray delivered the sermons that became *Wisdom of
God*. Ray organizes his material in the same way as More, moving from
matter and cosmology to plants, animals, and humankind. This is admittedly
a fairly traditional, one might say hexameral, organization; but Ray makes
clear that More is his immediate source, directly quoting him more than a
dozen times, much more than any other author. Ray adopts many of More's
teleological arguments wholesale, quoting him with little or no emendation
on heavenly motion, the fitness of the eye, the seeds of plants, and animals'
capacity for self-enjoyment, sometimes at a length of several pages. Given
Ray's acknowledged debt to More, it was perhaps ungenerous for Charles
Raven to refer to More as an "opponent of science," casting him as a backward
naysayer whose yoke Ray needed to throw off.[24] Nonetheless, in spite of their
similarities, there are crucial differences in method and aim between the
two works of natural theology. Unlike Ray, More deploys physico-theological
reasoning pragmatically, even reluctantly, toward the end of proving
Christianity.

It is a cue that More sees physico-theological reasoning as subservient to
some larger aim that he sandwiches his arguments "fetch'd from external nature" between books, respectively, on his doctrine of innate ideas and on
supernatural phenomena. Ray would jettison all of this material, focusing instead on what could be observed empirically. "Natural Philosophers, when
they endeavor to give an account of any of the Works of Nature by preconceived Principles of their own," Ray would explain, "are for the most part
grossly mistaken and confuted by experience."[25] More, by contrast, took a
high view of "preconceived principles" and was derisive of empirical science,
making it something of a paradox that his book would contain so much
physico-theological material. He saw empirical argument not as a necessary
component of his compelling proof of Christianity but as a concession to the
"weake and sunk minds of sensuall mortals." His strategy in book 2 of *An*

Antidote is to appear "in the shape of a meere *Naturalist*. . . . For hee that will lend his hand to help another fallen into a ditch, must himself though not fall, yet stoop." Nonetheless, having framed the book in that way, More evinces empirical sensibility: although he cites various sources ancient and modern, more often he simply describes natural structures and processes, explaining that he wishes to "contain my self within the compass of such objects as are familiarly and ordinarily before our eyes, that we may better take occasion from thence to return thanks to him who is the bountiful Author of all the supports of life."[26]

The question that then arises is how More was able to hit on such a strategy when his own philosophical commitments pulled him in the opposite direction. Where did More find the patience to gather so much data from the observation of nature, supplementing older arguments with detailed consideration of the intricate design of various physical phenomena to underscore the necessity of a cosmic designer? More's very suspicion, as a Platonist, of this more Aristotelian mode of argumentation may have prompted him to rethink how the natural world should be brought to bear on God's existence and nature: Is a posteriori reasoning really the appropriate way to arrive at theological conclusions? Or should the natural world be approached similarly to the text of Descartes or any other previous author, as a helpful but redundant reminder of eternal truths already accessible to healthy mental faculties? I suggest that More saw it this latter way, for throughout book 2 of *An Antidote* he speaks of natural phenomena as though they are the products of his own thought, or "excogitation," rather than as the works of a voluntarist God that must be observed in order to be known. More aims to show "that whereas the rude motions of matter a thousand to one might have best cast it otherwise, yet the productions of things are such as our own Reason cannot but approve to bee best, or as wee our selves would have design'd them."[27] This is true of all physical phenomena from heavenly and human bodies to insects and minerals: God has done exactly what we would have thought up ourselves.[28] This excogitative optimism is far from the epistemologically humble posture Francis Bacon had prescribed in 1620 in the plan for his *Instauratio Magna*, when he declared that "Man, the servant and interpreter of nature, does and understands only as much as he has observed," adding: "For God forbid that we give out [*edamus*] a fantastic dream for a pattern of the world."[29] More's exercise in patterning the natural world paradoxically forced him into an unusually intimate engagement with that world, because he needed to resolve every seeming case of unfitness if his logically compulsive argument for a wise creator was to stand. This logical rigor, aimed at over-throwing atheists, combined with More's scorn for empirical science, dis-tances his *Antidote* from Ray's *Wisdom* despite Ray's evident debt to More.

Richard Bentley and the "Newtonian" Natural Theology

Print natural theology experienced a turn in 1691, then, when *Wisdom* ap-peared with its litany of examples drawn from the new natural history il-lustrating God's ingenuity, creativity, and providence in the natural world. The 1690s also witnessed the beginning of the famous Boyle Lectures in natu-ral theology, a notable series whose authors likewise privileged new scien-tific discoveries in underscoring coherence between the natural world and the God of Christian scriptures. Here I briefly consider Richard Bentley's inau-gural Boyle Lectures, a series of seven sermons first published together as *The Folly and Unreasonableness of Atheism* in 1693. *Folly of Atheism* draws enthu-siastically on the new sciences, but there are features of the work that distance it from Ray's *Wisdom*. First, as is suggested by Bentley's title, a significant aim of *Folly of Atheism* is to refute atheism (Epicurean materialism in particular).[30] Second, in contrast with Ray's uncritical celebration of divine wisdom, Bentley views the book of nature as able to discredit the book of scripture rather than simply supplementing it.[31] In fact, Bentley's more logically rigorous mathemat-ical argumentation raises questions about whether any "Newtonian" natural theology can be such a celebration, or whether these works inevitably imply a kind and degree of certainty that distinguishes them altogether from Ray's "rather theological than philosophical" book. These features are interrelated: Bentley's more stringent apologetic goals may have seemed best served by the relative logical tightness on offer in Newtonian physics as opposed to the more empirical and detail-oriented natural history.[32]

Though conclusions drawn from organic and cosmological sciences may appear side by side in a work of natural theology, as they do in both Ray's and Bentley's texts, there was a recognition among contemporary natural philos-ophers that focusing on one or the other topos might lead to different kinds of natural theology, with one kind being superior. Boyle, and Ray after him, preferred organic subjects, with Boyle making this case very clearly in the second section of his *Disquisition about Final Causes*: "Whether . . . we may consider Final Causes in all sorts of bodies, or only in some peculiarly qual-ified ones." Boyle divides "bodies" into animate (including vegetables as well as beasts and humans) and inanimate (chiefly celestial bodies, but also terrestrial minerals). Among these, Boyle says that both animate and celestial bodies lend themselves to teleological inquiry, but that animate bodies are to be preferred. "For my part I am apt to think," he writes, "There is more of admirable Contrivance in a Mans Muscles, than in (what we yet know of) the Celestial Orbs; and that the Eye of a Fly is, (at least as far as appears to us,) a more curious piece of Workmanship, than the Body of the Sun."[33] It is noteworthy on what grounds Boyle prefers organic science: animal structure and function are more "admirable" and "curious" than cosmic motion.[34] These are the qualities Ray emphasized.[35] Evident too is Boyle's empirical openness to new knowledge: perhaps our understanding of cosmology will change to the point where the heavens appear an equally productive place to look for divine workmanship.

If for Boyle and Ray organic life excites more admiration than the more mechanical movements described by Newtonian physics, Bentley emphasizes the higher degree of mathematical certainty and logical intelligibility in the latter science. This advantage appears to have weighed with Newton himself, as his Scottish proponent David Gregory recorded: "In Mr. Newton's opinion a good design of a public speech . . . may be to shew that the most simple laws of nature are observed in the structure of a great part of the Universe, that the philosophy [e.g., natural theology] ought there to begin, and that Cosmical Qualities are as much easier as they are more Universal than particular ones, and the general contrivance simpler than that of Animals plants &c."[36] There is no contradiction between Newton and Boyle: Newton does not say that cosmic motion is more "admirable" than "animals plants &c.," but that it is simpler and more universal. As such, it is easier to draw compulsive conclusions about the necessity of God's existence and providence from cosmology and elementary physics, a task that occupies Bentley to a far greater extent than it does Ray.[37] Like Henry More, Bentley lays stress on the need for the cosmos to be a certain way (rather than any number of other ways) in order for reasonable people to conclude there is a providential God. "We ought to consider every thing as not yet in Being," he writes, "and then diligently examin, if it must needs have been at all, or in what other ways it might have been."[38] Ray, by contrast, was quick to capitalize when phenomena appeared well fitted to their purposes, but he gave humans no credit for being able to evaluate, from the ground up, what might be better or best: he believed, as had Bacon, that the only way to know what's best is to look at what is actually there, complicated as that reality appears.

Conclusion: More, Bentley, Ray

As a result, Ray was willing to leave numerous loose ends in his physico-theology, preferring to catalog and narrate rather than demonstrate and prove. Loose ends were altogether missing from More's strident apologetic against atheism, which required that the world be the best possible. With his combative, Newtonian approach, Bentley leaned in More's direction (though he claims for the world only "a meliority above what was necessary to be," rather than the optimality that More had claimed).[39] On the other hand, like Ray and unlike More, Bentley consciously addressed a Christian audience, structuring the "sermons" that would compose *Folly of Atheism* around exegesis of various biblical texts.[40] Perhaps Bentley, poised at the beginning of the explosion of physico-theology on the European continent and in England, illustrates the tension that would continue to face authors who wished to read the book of nature for ever-more-detailed evidence of God's wisdom and providence. While sincerely aiming to discredit various dangerous or "atheistic" ideas, Bentley shared Ray's conviction, taken from the Bible, that those ideas were folly and that a deeper understanding of the Creator's wisdom would develop as humans read each new page of the book of nature.

NOTES

1. Crawford, *The Bard*, 5253. Greyerz, "Early Modern Protestant Scientists and Virtuosos," 708, points out that Ray inspired germanophone "physico-theological poets" as well.

2. See Brooke and Cantor, *Reconstructing Nature*, 177; Thomson, *Before Darwin*, 73; Mandelbrote, "Uses of Natural Theology," 468; Ogilvie, "Natural History, Ethics, and Physico-Theology" 95. Gillespie, "Natural History," 38, calls Ray "the man who would define physico-theology and natural history in the familiar British pattern." In *The Oxford Handbook of Natural Theology*, Scott Mandelbrote, "Early Modern Natural Theologies," 86–87, places Ray under the heading "Physico-Theologians" but still refers to *The Wisdom of God* as a work of natural theology.

3. Ray, *Wisdom of God*, 124.

4. Zeitz, "Natural Theology," 130–31.

5. Calloway, *Natural Theology*, 101–4.

6. Ray, *Wisdom of God*, pref.

7. Zeitz, "Natural Theology," 127, points out that in Ray's work "Natural theology (and science) are imaged as cooperative enterprises participated in by a cooperative community into which the reader is being welcomed." The term "interpretive community" was famously used by Stanley Fish in discussing textual hermeneutics but can also be applied to the interpretation of nature. See Calloway, *Natural Theology*, 111–13.

8. Greyerz, "Early Modern Protestant Scientists," 708.

9. Ray, *Wisdom of God*, pref.

10. Ray does not cite Walter Charleton's *Darknes of Atheism Dispelled by the Light of Nature* (1652), but this work deserves mention as it is subtitled "a physico-theological-treatise" and has been called physico-theology by some modern scholars. A seventeenth-century text might use the adjective "physico-theological" without being a work of physico-theology; on this matter and on Charleton, see the essays by Peter Harrison and Scott Mandelbrote in this volume.

11. Stillingfleet, *Origines Sacrae*, 405.

12. Cudworth, *True Intellectual System*, pref.

13. Westfall, *Science and Religion*, 94. On Cudworth's sources, see Raven, *John Ray*, 456–57.

14. On Cartesianism as a target of early physico-theology, see Vermij, "Beginnings," 175–76. Parker's arguments have been very usefully summarized and placed in their (humanist) context by Levitin, "Rethinking English Physico-theology."

15. Levitin, "Rethinking English Physico-theology," 45–55. Levitin concludes that Parker's natural theology is "scholastic theology adapted to the norms of the new philosophy" (72). Parker would reiterate these arguments with some emendation in his 1678 *Disputationes*, 429–87. On Galen's influence on late seventeenth-century English physico-theology, see Ogilvie, "Natural History," 97–98.

16. Calloway, *Natural Theology*, 84–86.

17. Rivers, "'Galen's Muscles,'" 579.

18. Boyle, *Disquisition about Final Causes*, 32, makes clear that he does not view Descartes as an atheist, he urges instead that his physico-theological arguments should stand beside Descartes's argument for God's existence. He asserts the inadequacy of the Epicurean view without laboring to refute it, as this has been done by others.

19. Hunter, *Boyle*, 203, points out that Boyle was unsure whether the book should be classified as natural philosophy or theology.

20. Osler, "Whose Ends?," 161.

21. Boyle, *Disquisition about Final Causes*, pref.

22. Boyle, *Disquisition about Final Causes*, 33–37.

23. Raven, *John Ray*, 458.

24. Raven, *John Ray*, 25.

25. Ray, *Wisdom of God*, 28.

26. More, *Antidote*, pref., 74.

27. More, *Antidote*, pref.

28. E.g., More, *Antidote*, 49, 94–95.

29. Bacon, *Instauratio*, 45.

30. Calloway, *Natural Theology*, 126–31.

31. Calloway, *Natural Theology*, 122–24.

32. The Newtonian focus of Bentley's Boyle Lectures is well attested: see, for instance, Gascoigne, "Bentley," 222–23. On the "Newtonian type" of physico-theology, see John Brooke's essay in this volume.

33. Boyle, *Disquisition about Final Causes*, 2, 43–44.

34. Mandelbrote, "Uses of Natural Theology," has divided seventeenth-century English works of natural theology into those emphasizing wonder and those emphasizing natural law: my reading here concurs with his taxonomy regarding Ray and Bentley. Boyle, Michael Hunter has pointed out, is difficult to categorize in these terms; see Hunter, *Boyle Studies*, 45–46.

35. On Ray's stress on "wonder," see Mandelbrote, "Uses of Natural Theology," 468.

36. Memorandum by Gregory, Dec. 28, 1691; quoted in Jacob, *Newtonians*, 154. Jacob argues that the "public speech" referred to is the newly endowed Boyle Lectures.

37. Briefly, both men reject Aristotelian, Cartesian, and Epicurean hypotheses and consider cosmic motion as well as organic structures, but Bentley begins with the organic sciences and works up to cosmology in three lectures on "the origin and frame of the world," whereas Ray considers cosmology and inanimate bodies for roughly the first quarter of his book before spending the rest of the time on organic subjects.

38. Bentley, *Folly of Atheism*, 244.

39. Bentley, *Folly of Atheism*, 183.

40. Calloway, *Natural Theology*, 124–25.

CHAPTER NINE

Matters of Belief and Belief That Matters

German Physico-theology, Protestantism, and the Materialized Word of God in Nature

ANNE-CHARLOTT TREPP

German-language works of the seventeenth and eighteenth centuries in physico-theology tended not to focus on cosmology and big concepts of nature. Instead, they concentrated on singular objects of natural history, privileging specific plants and minerals. There were countless Melitto-, Litho-, Testaceo-, Rana-, Akrido-, Phyto- and Chorto-theologies. In an attempt to increase our understanding of this predominantly Protestant phenomenon, this chapter examines the correlations among Luther's ubiquitarian Christology, his theology of nature and of Creation, and the development of Protestant physico-theology in the seventeenth and eighteenth centuries.

The most obvious characteristic aspect of germanophone physico-theology is its objects of study: preferably small, often rather unspectacular natural creatures and objects.[1] Friedrich Christian Lesser, for example, wrote a *Lithotheologie* subtitled *Natürliche Historie und geistliche Betrachtung derer Steine* (1735), Friedrich Menz outlined a *Rana-Theologie* (1724) on frogs, and in his *Melitto-Theologia* (1767) Adam Gottlob Schirach highlighted the bees as God's creatures[2]—to name only a few of the works in question. Most germanophone authors of physico-theological texts were Protestants—at least as far as we can tell by today's state of the art. In order to understand this predominant confessional orientation, it may be worthwhile to go back to the origins of Protestantism and to investigate the nucleus of Lutheran natural theology and its view of Creation. It will be of particular importance to establish how Martin Luther as theologian dealt with the material world. I argue in this chapter that the physical and ubiquitous quality of Luther's Christology became a central and basic element in the formation of an apparently specifically Protestant approach to nature and its development into physico-theology. Connecting physico-theology with the foundational and central issues of Lutheran theology offers a new explanation for the fact that no orthodox Lutherans reformulated natural theology or engaged in physico-theology during the seventeenth and eighteenth centuries; rather, this was done—as paradoxical as this may seem at first—by representatives of an undogmatic transconfessional theology as well as a renewed piety that enlivened personal religious experience. It allows us to understand, secondly, why Immanuel Kant's philosophical refutation of the physico-theological proof of God's existence did not, as is generally assumed, signal the end of germanophone physico-theology, and, thirdly, why the affinities between physico-theology and Pietism proved to be an influential ingredient at the end of the eighteenth century, in the context of increasingly globalized research on nature.

Creatures as God's Materialized Word

Even though Creation does not seem to assume any meaning in Luther's initial central message on justification, the reformer did formulate a theology that connected tightly with nature and the Creation on account of his emphasis on God's presence within the Creation and the latter's sustenance through the word. Luther understood the creaturely world to be like an accumulation of God's words and, in so doing, adhered to the tradition highlighting the *Liber Naturae* as God's second book next to the Bible. Luther looked at all creatures as *verba creata*, as results of God's words of Creation, and believed that as such they would be enduring. God certainly had not created the world for only a short while (*"auff ein huy"*).[3] To be sure, God had rested after having created the world, but he did not end its *preservation*:[4] "Because [creatures] grow, multiply, are preserved and governed all in the same way from the beginning of the world to this day, it follows clearly that the Word endures to the present and is not dead."[5] Creatures are not only created by God's word, but they continuously *live* through it.[6] The word "always *speaks without end*."[7] Christ is the word become flesh, the Logos incarnated (following John 1). He is entirely present in the word, simultaneously in his divine and his human nature. The exchange between the two natures, the *communicatio idiomatum*, is radicalized by the reformer when he insists on the material (i.e., bodily) presence of Christ.[8] According to Luther, Christ is at the same time both present bodily and ubiquitous in the "bodily bread" (*Leibesbrod*) of the Eucharist. There Christ's body and blood are truly present. In Luther's refutation of the scholastic dogma of transubstantiation, the bread remains bread (without any transformation) in a bodily and material sense, and wine remains wine.[9]

How Christ was present during the Lord's Supper—physically or even in a ubiquitous manner—became the central issue debated by the reformers.

Huldrych Zwingli (1484–1531) and, to some extent, John Calvin (1509–64) and Martin Bucer (1491–1551) rejected Luther's understanding of the relationship between Christ's divine and human nature.[10] Luther's conception of the ubiquitous presence of Christ not only concerned the Lord's Supper but applied to the entire Creation and its creatures.[11] The unity of God with his entire Creation is assured by the incarnation of Christ—that is, by his "descent into human flesh"—and involves unity "for all *creatures*," even the lowest.[12] This explains Luther's interest in the nature of a cherry tree, the cherry's stone, or a grain, which is more evident than his interest in the divine cosmos. Because, through Christ's incarnation, God reveals himself in every individual being down to the most minuscule creature, everything—even the meanest creature—becomes God's living imprint. On the other hand, heaven as "God's palace" and the cosmological notions connected with it tend to lose their soteriological importance.[13]

The reception of Luther's "ubiquitarian Christology" (to use Jörg Baur's term) was prone to disruptions and resistances. The concept triggered massive criticism, especially within Protestantism. Immediately after Luther's death in 1546, Philip Melanchthon took the "ubiquitarian Christology" off the agenda and advocated a rational theism, at least concerning the theology of revelation, against the Christology based on real presence.[14] Nonetheless, Luther's radical view of the *communicatio idiomatum* did not disappear from the canons of confessional culture. Sermons and works of edification kept promoting it, and it thus becomes one of the causes of a lived piety based on personal experience.[15]

Physico-theology as a Way of Vital Knowledge of God

How men can attain a vital knowledge of, and admiration for, the omnipotence, wisdom, goodness and justice of the great God through an intensive observation of those normally neglected insects—this is the subtitle of Christian Friedrich Lesser's *Insecto-Theologia* of 1738: the observation of divine omnipotence and providence, active everywhere in God's creation, should remind the reader of his or her creaturely nature and, thus, lead to an enhanced "vital knowledge" of God.[16] The observation of creatures should "awake our gratitude" toward God, as Johann Albert Fabricius (1668–1736) wrote in the last book of his translation of William Derham's *Astrotheologie*.[17] By way of an in-depth study of certain natural occurrences and particular objects, men should be enabled to rise to the spiritual awareness of their maker. Physico-theological authors argued, in other words, that God's existence, his omnipotence, wisdom, and goodness, could be deduced from the admirable beauty, order, and purpose-fulness of nature. A central point in this argument was proving God's provi-dence that manifested itself in the steady and enduring presence of God's care even among the meanest creatures. This is how physico-theology carried on the practical natural theology of the preceding decades, reinterpreted in the context of the Lutheran dogma of ubiquity.[18] In order to understand ger-manophone physico-theology, it is thus necessary to inquire into the continu-ities and specific transformations of Protestant natural theology, without, however, neglecting the influence of English physico-theological works.

It is no coincidence that the group of germanophone Protestant physico-theologians was dominated by theologians and laymen who tended to be critical of the church and strove for a practical Christianity experienced in daily life.[19] This tradition was inaugurated by Johann Arndt's *Four Books on True Christianity* (*Vier Bücher, Vom wahrem Christhentumb*). This work of edification, first published as separate volumes in 1610, is unanimously con-sidered to be one of the most widely read and distributed books of German Protestantism.[20] It is a work of a mystical spiritualism of the "inner" word, aimed at reforming religious belief into a pious practice, an "active exercise" and "vital experience" (*lebendige Erfahrung*), and steering away from an over-emphasis on dogma. In his "Register of those ancient and modern writers, who have made it their concern to lead men to God by an examination [*Be-trachtung*] of nature and [its] creatures," that is, his preface to his translation of William Derham's *Astrotheologie* of 1732, Johann Albert Fabricius gave much space to the fourth book (the *Liber Naturae*) of Arndt's *True Christian-ity*.[21] For God's word could be received not only in the Bible and in the heart but also in nature.[22] Similar to God's presence in the soul in a mystical and spiritual sense, which is sustained by the immediacy of his word, God's word is also present, both spiritually and materially, in his creatures. It acts every-where as an "enlivening and sustaining force," "in all things." "The whole world [is] filled" with God's word. God exists even *within* creatures as "the life of all living things" (*aller Lebendigen Leben*).[23] By referring to Romans 11:36, Arndt implied that "From Him, *in Him*, through Him are all things"—thereby gradually varying from Luther "*Von ijm/vnd durch jn/vnd in jm/sind alle ding*"; unlike in Luther's theology of Creation, for Arndt God's word, as his spirit or "breath" (*Odem*), is immanent and entirely identical with the Cre-ation. Even taking into account the differences between them, which do not amount to a differentiation between an actual providence and one describing the wisdom and order of God's plan (*sapiential-ordinative Providenz*), both

Arndt and Luther belong to a mystical tradition.[24] Above all, they are both concerned with the actualization or, more specifically, with the permanent manifestation of the enlivening and sustaining word of God in nature.[25] References to Arndt's opus are very frequent in germanophone physico-theological writing, not only in the work of Fabricius and Lesser but likewise in Barthold Heinrich Brockes's *Irdisches Vergnügen in Gott* (1721–48) and in the Halle physician Friedrich Hoffmann's *Vernünfftigen Physikalischen Theologie* (1742).[26]

Among the Pietists in Halle and those persons who sympathize with their alternative religious attitudes, we can observe a close relationship between a "reasonable" reading of the book of nature and "a lively experience of Jesus Christ."[27] Especially during the early years of Pietism there was a fruitful, mutual exchange between the Enlightenment and the pious culture of Pietism.[28] Even though August Hermann Francke (1663–1727) judged the Bible to be central to his theology of conversion and spiritual rebirth, there was a remarkable presence at the Franckesche Stiftungen (Francke's Foundations) of the study of nature and natural philosophy on all different levels. It enjoyed a special presence in the various schools.[29] Striving for a lively knowledge of God (*lebendige Erkäntnüß Gottes*)[30] and "true godliness" (based essentially on Arndt's theology of personal experience) as an anticipation of the perfect heavenly godliness entailed introspection, as well as *external* experience and common sense (*Verstand*) "to enable the will to follow without coercion."[31] Akin to Luther's monistic conception, body and soul were looked at as a unity, which meant that the psychic and bodily striving for salvation were intrinsically connected.[32] At the same time, the inner renewal of man was considered recognizable only under the condition that it manifested itself in the outward "practice of piety" (*praxis pietatis*) for the enhancement of God's honor and neighborly love. The theologian and pastor Friedrich Christian Lesser, mentioned earlier, a former student at Halle, was one of the most popular German-speaking physico-theologians of the first half of the eighteenth century. He was accorded membership in the Deutsche Akademie der Naturforscher Leopoldina and in the Berlin Academy of Sciences in recognition of his works. For Lesser, the observation of nature was part of an "inner service to the praiseworthy God."[33] Addressing the group of alternative Christians and their vision of spiritual transformation of every single person, Lesser writes in the preface to his *Lithotheologie* that "we are natural men before we become spiritual Christians. This is why natural knowledge gained from the book of the world must precede Christian experience on the basis of the Holy Bible."[34]

Johann Hieronymus Chemnitz (1730–1800), also a member of the Leopoldina, who likewise studied theology in Halle, wrote in this spirit in his *Kleine Beyträge zur Testaceotheologie* in 1760: "If God's properties are displayed so clearly in the scorned snail and shell, if he remembers it, attends to it, without leaving or missing it, how should I not, being his child and servant, find consolation in him?" Thus he "drew conclusion after conclusion" in accord with "the redeemers' divine logic."[35] From 1778 to 1795 Chemnitz published the *Neues systematisches Conchylien-Cabinet*, the leading German Conch edition of the second half of the eighteenth century.[36]

As in the case of a majority of germanophone physico-theologians, Chemnitz's and Lesser's cognitive interest regarding nature focused on the smallest creatures, because God's creative power, his wisdom and enduring providence, manifested itself so wonderfully in them. On account of the perfection visible in them in every detail, Lesser typically understood these small beings, drawing, among others, on Christian Wolff's Natural Philosophy, mechanistically as "small machines."[37] Simultaneously, like other contemporary naturalists, he recognized a clear-cut limit to the "mechanization" of animals.[38] For, as he argued, the steady and perfect movements and actions of animals could be explained only in reference to God's permanent influence, "because the Lord of Nature [continually] sustains everything in its essence, strength, and order."[39] Nature was able to function perfectly and permanently on the basis of "natural causes" and "general laws" only because of God's *concursus,* that is, his continuous sustenance.[40] Lesser made it clear that he was certain "that God's hand is present in all movements of natural things. . . . Even though the power of natural causes acts in them, God brings his own into all movements."[41] A similar concept can be found in *Hydrotheologie* by Fabricius, where he described not only water circulation but also the functioning of human and animal bodies as well as plants as "Maschina hydraulico-pneumatica." All operations and processes complemented each other as perfectly as "clockwork." Their permanent and meaningful operation was caused alone by God, the "dearest creator" and "preserver of our life."[42] Similarly the Berlin preacher Johann Friedrich Wilhelm Herbst wrote still in 1792 of God's "beneficial omnipresence" in nature and in his creatures in his *Naturgeschichte der Krabben und Krebse* (1782–1804) and in his *Natursystem aller bekannten in- und ausländischen Insecten* (1788–1806).[43]

The sources presented here that display the perpetual and sustaining presence of God and his *concursus* in Creation could be best understood as a reflection of the contemporary adaptation, and also transformation, of Luther's conception of the ubiquity of Christ's real presence. Thus, it can arguably be claimed that the nucleus of germanophone Protestant natural theology consists of the perception of creatures as the materialized living word of God in nature. Given this theological concept, authors could coherently connect the theological notion of Christ's real presence with the mechanistic models of nature of the advancing natural philosophy.

Accepting the perception of the incarnation in nature of the divine word (*Logos*) as an explanation for the functionality and persistence of all actions in nature may result in new questions. We may ask, for example, whether, and to what extent, the representation of a bodily real presence brought about an increased appreciation of bodies and matter in Creation and nature, or whether it at least resulted in an independent appreciation.[44] We may also ask whether ubiquity understood in this sense was able to become a basic element for a theologically as well as scientifically relevant relationship to the things in this world. For, as God revealed himself through the materialized word in every individual creature, individual things immanent to the world, even the lowest in nature's hierarchy, gained a new dignity and transcendence not least in their bodily presence and materiality.[45]

The place where this new appreciation of individual objects became especially apparent was the collections of *naturalia*. Within the urban bourgeois world, such collections mushroomed from the late seventeenth century onward at a considerable rate.[46] Every naturalist and physico-theologian owned one, even if it was small. The objects presented in them likewise shared a religious significance and could serve as objects of exchange. In the first half of the eighteenth century, the authoritative German guide to collecting was the merchant Caspar Friedrich Neickel's *Museographia* (1727). Neickel can be associated with Pietism, and he claimed that the edifying usefulness of a collection came "close to a heavenly pleasure in the observation of nature."[47] As in no other place, the examination of a collection of natural objects joined external and internal sensations and thus greatly intensified their religious significance:

> Much balsamic and aromatic matter intensifies the pleasure of smell. The odor must allow a recognition of whether the matter is sweet or bitter, pleasurable or repulsive. Our ear hears the ringing and bluster within the shells of marine snails. . . . Sensation tells us whether a thing is soft or rough, cold as ice or warm, graspable or not. In sum, in the observation of nature all outward senses are a source of enjoyment and usefulness. But the inner man, his heart [*Gemüth*] and soul, senses most, . . . for it is the center where all the outward sensations are directed.[48]

Only the presence of a collection of objects in their specific materiality and authenticity allows a "deeper reading" (*Einsehen*) of the book of nature.[49] In order to appreciate the objects fully and to be able to recognize God in his Creation, it was necessary that the external observation be joined by the "inner eyes," the "eyes of the heart."[50] On a similar note, Chemnitz, the future editor of the *Neues systematisches Conchylien-Cabinet,* emphasized the senses and the role of objects. In regard to his collection of *naturalia* and especially the collection of conches, he reported how he talked to his visitors about "charming creatures of the sea" in order to let the beautiful creatures radiate their creator's beauty to the spectators' eyes, so they might see, feel, and find "the visible vestiges of the creator's existence painted in front of their eyes."[51] The edifying contemplation of, simultaneously, material transcendence and the authenticity of things transformed natural objects into media of a lively experience of God.

Considering the special affinity of the "new godly people" or individuals who pursue practical religious experience for "things," it is probably no accident that a famous cabinet of *naturalia* of an "eminent scientific quality" took shape precisely at the Franckesche Stiftungen in Halle.[52] Initially, the collection was installed in a room of the Paedagogium; later it was transferred for reasons of space to the orphanage built in 1701.[53] There, once a week, the pupils were instructed about "the nature and qualities, as well as usefulness and use" of the collected items (*Realia*).[54] While the instructional purpose of the collection has been emphasized, it is less well known that this was probably the first collection anywhere that was largely arranged and listed according to the recently published *Systema Naturae* of Carl von Linné (or Linnaeus).[55] The artist and naturalist Gottfried August Gründler fundamentally rearranged the collection between 1736 and 1741 in its spatial and systematic organization.[56] In the course of this reorganization Gründler coedited the first German edition of the *Systema Naturae* jointly with the professor of mathematics Johann Joachim Lange.[57] Most of the approximately forty-seven hundred objects were cataloged in a "highly detailed system" and conserved, as well as exhibited, in a total of sixteen cupboards.[58] The systematic order of the objects corresponded to the naturalistic, painted illustrations on the cupboards, which still convey a lively effect today. This order also highlights the special value that Pietism ascribed to the immediacy of the sensory perception of natural objects.[59]

Global Physico-theology: The Perception of Physico-theology, Natural History, and the Pietist Mission around 1800

Even the Tamils, whom the missionary Johann Peter Rottler (1749–1836) addressed on his journey to Ceylon in 1788, were encouraged by him to observe individual objects in nature and to recognize God's glory in them. Rottler later recalled his own time in Ceylon, "I was lying among many flowers and had the Linnean system next to me while I examined these wonderful works of the great Creator. How agreeable and enjoyable were the hours I passed there! Whatever I was not able to deal with on the road I took care of at the house of rest."[60]

This close systematic perception of nature, physico-theology, and Pietist religious practice was continuously maintained in the context of the global activities of Francke's Foundations.[61] This is why the missionaries of the Danish-Halle Mission active at southeast Indian Tranquebar, now Tharangambadi, from 1706 shared a well-documented interest in collecting, systematizing, and describing the local flora and animal world.[62] For Rottler and Christoph Samuel John (1747–1813), missionaries who arrived at Tranquebar during the 1770s, research on Indian nature was central to their mission.[63] As partners in a European-wide network of communication they participated in the Enlightenment project of a worldwide observation and classification of nature. These missionaries were the authors of scientific observations and treatises in correspondences and learned periodicals, and they communicated on a regular basis with natural philosophers and scientific associations in Europe and India. Their correspondents included the naturalist John Reinhold Forster (1729–98) and William Roxburgh (1751–1815), the first director of the botanical gardens of the British East India Company in Calcutta.

Many of these missionaries had been pupils or teachers of the Franckesche Stiftungen. They consciously tried to contribute objects for "the augmentation . . . of the cabinet of *naturalia*" at Halle.[64] In doing so they received important instructions from the missionary physician and former student of Linnaeus Johann Gerhard König (1728–85).[65] It is an established fact that the committee in charge of the mission at the center of Halle Pietism in Copenhagen had opted for König on account of his Linnaean expertise as a naturalist and because of the committee's own interests, especially concerning the Indian flora.[66] König comprehensively instructed the missionaries John and

Rottler in natural history and in the collection of natural objects, a knowledge which they, in turn, knew how to use in their missionary work: "When I deal with certain aspects of natural history, I not only teach the European and Malabaric children entrusted to me always how to strive for the lively and practical recognition of Jesus Christ on the basis of the Holy Writ, but also how to collect conches, fish, crabs, herbs, insects, and to help grow trees, flowers, kitchen herbs and plants."[67] The children and adolescents in their care were constantly reminded to collect natural objects and to recognize the existence of God and his omnipotence and goodness from the observation of these objects. In addition, disciplines of natural science were integrated in the curricula of the missionary schools from the end of the 1770s onward.[68] The missionary John claimed in 1792 that "God's revelation in nature was the only *general* revelation," and for that reason he established physico-theology as the basis of his missionary practice.[69] In referring to Linnaeus, Heinrich Sander's *On Nature and Religion* (*Ueber Natur und Religion*, 1779–80), and Johannes Florentinus Martinet's *Catechism of Nature* (*Katechismus der Natur*, 1777–79), he recommended to any newly arrived or prospective missionary delving into natural-historical and physico-theological literature, in order to enhance the success of missionary work among the Tamils.[70]

Conclusion

The starting point of this chapter was an investigation into the reasons for the special affinity of germanophone Protestants and physico-theologians for individual natural objects all the way to the smallest creatures. There were many ways of reading the "book of nature," to be sure, and also different approaches to any specific object. One of them, as I have argued in this essay, was based on Lutheran Christology and theology of Creation. Because, as Luther claimed, God revealed himself by way of Christ's incarnation as an enduring life-giving and life-sustaining word, any, and even the meanest, creature, and every natural object, could become a living witness to God's existence. As a result, it is possible to recognize the nucleus of germanophone Protestant natural theology in the notion of the "real presence," as well as bodily ubiquity, of God in his Creation, and its reflection (*Vergegenwärtigung*) in his creatures. This leads to the question whether the physically conceived "real presence" of God encapsulated an independent value system regarding the role of body and matter in nature, whose implications in theology and history of science are yet to be examined in their full dimensions.

From the end of the seventeenth century onward, Pietist circles witnessed the amalgamation of an interest in a rational cognition of nature inspired by the Enlightenment with the quest for a knowledge of God embedded in real-life and everyday experience offered by physico-theology. Things in nature and the world attracted a new interest. The representation of God's providence, recognized in the functionality, wisdom, and beauty of Creation, could now combine with systematic research on nature, offering an active (*lebendige*) experience of God. In this context the massively multiplying bourgeois *Kunstkammern* and, especially, cabinets of *naturalia* acquired a special role. They highlighted objects in their concrete materiality, as a kind of authentic witnesses of the "material side" of the word of Creation. The exhibited objects became media of religious edification on the one hand; on the other, they were subjected to systematization and order according to the newest scientific standards, as was the case with the cabinet of *naturalia* of the Francke Foundations. As a result of the far-reaching Pietist reforms and missionary programs, the physico-theological approach to nature was exported to the world outside Europe and thus "globalized." However, the physico-theology of the German-speaking world lived on beyond the second half of the eighteenth century and did not restrict itself to Pietist circles.[71]

It was nevertheless the missionaries and missionary physicians connected with Halle Pietism who brought physico-theology and the binary Linnaean system to India and encouraged its spread there through their instruction of the Tamil population. This is how the missionary Christoph Samuel John could claim: "I hold science and its propagation to be a part of religion and missionary work and their enhancement a duty."[72]

NOTES

1. Brooke, *Science and Religion*, 268–91; von Greyerz, "Early Modern Protestant Scientists"; Krafft, "*Gott*," 84–86; Trepp, *Von der Glückseligkeit*, 330–31.
2. Lesser, *Lithotheologie*; Schirach, *Melitto-Theologia*.
3. Luther, *Werke*, 24:25, 26.
4. Luther, *Werke*, 42:57, 15–20.
5. Luther, *Werke*, 42:57, 27–29: "Quod Verbum in principio fuerit. Quia autem crescunt, multiplicantur, *conservantur* et reguntur adhuc omnia eodem modo quo a principio mundi, Manifeste sequitur verbum adhuc durare nec esse mortuum."

6. Luther, *Werke*, 24:322, 6–323, 2; Beutel, *Anfang*, 95–112; Beutel, "Wort Gottes"; Bayer, *Schöpfung*, 1–6, 64–74; Bayer, *Luthers Theologie*, 62–83; Luther, *Large Catechism*, pt. 2, art. 1.

7. Luther, *Werke*, 24:37, 29.

8. Steiger, "Die *communicatio*." See also Frettlöh, "'Gott'."

9. Cf. the adaptation by historian of science Funkenstein, *Theology*, 70–72. See also the cultural historian Roper, *Doktor*, 65–68.

10. Steiger, "Die *communicatio*," 17.

11. See Luther, *Werke*, 26:336, 11–18. For Luther's Christology, see Baur, "Ubiquität" (2007); Baur, "Ubiquität" (2002).

12. Luther, *Werke*, 12:556, 21. See also Mumme, "Geist," 14–15.

13. Baur, "Ubiquität" (2007), 218–19.

14. Baur, "Ubiquität" (2007), 223–25.

15. Steiger, "Die *communicatio*," 16; Baur, "Ubiquität" (2007), 218; Baur, "Ubiquität" (2002), 238; Baur, *Luther*, 164–67, 185–89.

16. Trepp, *Von der Glückseligkeit*, 332–33; see also Christian Wolff's preface to Nieuwentijt, *Erkänntnüß* and Johann Albert Fabricius's foreword addressed to Barthold Heinrich Brockes in his German translation of Derham, *Astrotheologie* (1732).

17. Cf. Trepp, *Von der Glückseligkeit*, 212.

18. On physico-theology more generally, cf. Trepp, *Von der Glückseligkeit*, 306–466.

19. See Trepp, *Von der Glückseligkeit*, esp. 306–15, which critically discusses the place of German physico-theology vis-à-vis the range of Lutheran-orthodox positions; cf. Philipp, *Aufklärung*, 42–47, and Michel, *Physikotheologie*, 148–49.

20. Brecht, "Aufkommen der Frömmigkeitsbewegung," 134–50, esp. 139; 149–50.

21. For a contextualization of Fabricius's "Register" ("Scribenten"), see Trepp, *Von der Glückseligkeit*, 313–15. Like Barthold Heinrich Brockes, Fabricius pursued hermetic philosophy of nature and iatrochemical medicine; see Kemper, *Gottesebenbildlichkeit*, 1:285, 2:328–29. Kemper (*Gottesebenbildlichkeit*, 1:323) referred in this context to an "enlightened hermeticism." Fabricius dismissed the theological dogmatic controversies of the Lutheran Orthodoxy and frequented Pietist circles; see Trepp, *Glückseligkeit*, 311–15.

22. Arndt, *Christhentumb*, bk. 4 (*Liber Naturae*). See also Geyer, *Weisheit*, pts. 1–2.

23. Arndt, *Christentumb*, bk. 4, hex. 6, 190. See also Geyer, *Weisheit*, 237–41; Trepp, "'Adam's Knowledge,'" 43–45; Trepp, *Von der Glückseligkeit*, 37–60; Neumann, *Natura Sagax*.

24. For Luther, see Hamm, "Wie mystisch war der Glaube Luthers?"

25. Trepp, "Natural Order and Divine Salvation." For different conceptions of providence, see Bernhardt, "Handeln Gottes."

26. Brockes, *Irdisches Vergnügen*; Hoffmann, *Vernünfftige Physikalische Theologie*.

27. Trepp, *Von der Glückseligkeit*, 338–466; quoted from Lesser's letter to his son, Dec. 1738.

28. Hölscher, *Frömmigkeit*, 125–56; Schneider, *Hoffnung*, 112–26. For the latest discussions of the relations between Enlightenment and religion, see Grote, "Review-Essay."

29. Müller-Bahlke, "Naturwissenschaft," 362; Whitmer, *Orphanage*, 20–32.

30. Francke, "Unterricht," 115.

31. Francke, "Unterricht," 116.

32. Geyer-Kordesch, *Pietismus, Medizin und Aufklärung im 18. Jahrhundert*, 62–95; Helm, *Krankheit*, 11–56.

33. Lesser, *Schriften*, pref.

34. Lesser, *Lithotheologie*, xii.

35. Chemnitz, *Testaceotheologie*, 53–54.

36. In 1778, after the former editor of the *Neues systematisches Conchylien-Cabinet*, Friedrich Martini, had died, Chemnitz continued his work and in 1795 had published eight of the

eleven volumes. For several decades he was an active participant in the correspondence network centered at Halle. Cf. Ruhland, *Konkurrenz*, 232–33, 283–86.

37. Lesser, *Lithotheologie*, 628; Lesser, *Insecto-Theologia*, 2–3, 9.

38. Funkenstein, *Theology and the Scientific Imagination*, 70–77; Westfall, *Science*, 26–48, 193–205; Shapin, *Scientific Revolution*, 123–65.

39. Lesser, *Insecto-Theologia*, 159.

40. Lesser, *Lithotheologie*, 206; see also Lesser, *Insecto-Theologia*, 159.

41. Lesser, *Lithotheologie*, 234.

42. Fabricius, *Hydrotheologie*, 226, 229. For Fabricius's concept of God's permanent and formative presence in nature, see Krolzik, *Säkularisierung*, 179–80; Krolzig, "Wasser," 197–98.

43. Herbst, *Betrachtungen*, 53–54. Herbst regarded his studies of nature as a confirmation of the mystical and spiritual aspects of his theology of rebirth. He belonged to Pietist circles in Berlin of the late eighteenth century (Brecht, "*Die deutschen Spiritualisten*," 242–43); among others, he corresponded with the Halle-trained theologian and missionary in Tranquebar, Christoph Samuel John. On John's Pietist religion, see Trepp, "Missionierung," 248. Both were also members of the Gesellschaft Naturforschender Freunde zu Berlin. On Pietist identity in Halle in the late eighteenth century, see Albrecht-Birkner, "'Ich verspreche Ihnen,'" 192–97.

44. For Luther's revaluation of corporeality, see Roper, *Doktor*, 67–77.

45. On the sensory aspects of Luther's experience of grace, see Kaufmann, "Heilsaneignung," 24–30. Cf., by contrast, Karant-Nunn, "Gedanken." See also P. Hahn, "Sensory Culture."

46. See Daston and Park, *Wonder*, 265–76; Laube, *Reliquie*; Valter, *Studien*.

47. Neickel, *Museographia*, 260; cf. Trepp, *Von der Glückseligkeit*, 405–10.

48. Neickel, *Museographia*, 447–48.

49. Lesser, *Lithotheologie*, xiv.

50. Lesser, "Nachricht," 557; Lesser, *Testaceo-Theologia*, 6.

51. Chemnitz, *Testaceotheologie*, 59–62. For Chemnitz's *Testaceotheologie*, cf. Noak's concept of "theology of sensual perception" in Noak, "Schule," 502.

52. Ruhland, "Objekt," 72.

53. Francke, *Ordnung*, 3; concerning the collection, see Laube, *Reliquie*, 347–66; Müller-Bahlke, *Wunderkammer* (2012); Rieke-Müller, "Welt."

54. See Francke, *Ordnung*. On the tradition initiated by Comenius, see Peschke, "Reformideen."

55. For the religious basis of Linnaeus's research on nature and search for a system, see, e.g., Trepp, "'Adam's Knowledge.'"; Lindroth, "Linnaeus," 30–31; Jahn and Senglaub, *Linné*, 52–54.

56. Müller-Bahlke, *Wunderkammer* (2012), 29–37; Stelter, "Möglichkeiten."

57. Rieke-Müller, "Welt," 64; Ruhland, "Objekt," 76.

58. Ruhland, "Objekt," 77.

59. Laube, *Reliquie*, 331–35 and, on the "material self-recognition" of Halle Pietism, 381–83.

60. Rottler, "Reise," 87.

61. See Hommel, "Für solche [Theologen]," 183–92; Laube, *Reliquie*, 352–56.

62. The foundation of the Danish-Halle Mission in 1706 was encouraged by the Danish-Norwegian crown on account of its affinity for Pietism.

63. Hoppe, "Naturgeschichte"; Hommel, "Forschungen"; Nehring, *Orientalismus*, 75–81.

64. John's letter to Schulze, Apr. 4, 1787. In AFSt/M 1 C 29b: 17.

65. Hoppe, "Naturgeschichte," 153; Ruhland, *Konkurrenz*, 200.

66. Ruhland, *Konkurrenz*, 199–213.

67. John, "Auszug," 899.

68. Liebau, "Erziehung," 453; Ruhland, *Konkurrenz*, 234.

69. Nehring, *Orientalismus*, 75–81; Hommel, "Forschungen," 171; Trepp, "Missionierung," 246–52.

70. Liebau, *Encounters*, 94–95.

71. See Ruppel, "Pflanzen."

72. John, *Memoria*, as cited in Liebau, "Erziehung," 437; Ruhland, *Konkurrenz*, 233–34.

CHAPTER TEN

Pascal's Rejection of Natural Theology

The Case of the Port-Royal Edition of the *Pensées*

MARTINE PÉCHARMAN

I cannot forgive Descartes; in all his philosophy, he would like to do without God, but he could not help allowing him a *chiquenaude* to set the world in motion; after that he has no more to do with God.

Pascal[1]

Pascal's famous statement on Descartes, reported in Marguerite Périer's *Mémoire sur Pascal et sa famille*, is prime evidence for Pascal's opinion that Descartes's *Principia philosophiae*, while striving to ground physics in metaphysics, and thus to explain natural phenomena from a prior demonstration of God's existence, actually reduces God to a subaltern function. Pascal interprets Descartes's causal hierarchy of God and the laws of nature as detrimental to the role of God: beyond his initial *chiquenaude*, the Cartesian God is no longer required to explain natural phenomena; matter and motion are by themselves sufficient principles for all the complexities of nature.

This criticism, despite its inadequacy—Descartes's natural philosophy does not dispense with God once a first impulse is given to the whole system of matter and motion—highlights a significant point, the absence of any reference to God's design in Descartes's explication of physical phenomena. Pascal's statement, thus, seems to condemn the *Principia philosophiae* as dealing with a theologically blank world. From this, we might expect that Pascal, against Descartes's strictly mechanist explication of nature (which Pascal no doubt connects with the attribution to Descartes of an implicit atheism), intended to defend the opposite view that all natural phenomena display God's purposiveness. We might suppose, in other words, that Pascal's apologetic project integrated natural theology as an essential tool in the attack on atheism.

This supposition would be false, however. Indeed, in an intellectual context featuring multiple proofs of the existence of God, including many arguments from the divine design of the world, Pascal's apologetic project is characterized by a preliminary rejection of the whole range of proofs traditionally used by philosophers and theologians. In addition to a general rejection of all kinds of traditional proofs for God's existence, Pascal's critique focuses particularly on natural theology, so that the specific proofs of God's "existence, wisdom, and omnipotence" based on natural phenomena (the very motto of physico-theology) are, as it were, denied from the outset. It seems to me essential to detail this point by referring to the so-called Port-Royal edition of Pascal's *Pensées*. Published in 1670 (a pre-edition was available in 1669), it reigned alone for more than a century, until the new edition by the Marquis de Condorcet in 1776. From the late seventeenth century to the late eighteenth century, Pascal's arguments in the *Pensées* were known solely through the mediation of the Port-Royal edition, which had transformed a selection of fragments from his original papers into a continuous and somewhat arbitrary series of chapters. By focusing on this material, my aim is not merely to provide insight into Pascal's critical judgment on the proofs of God from nature. I also argue that Pascal's opinion on natural theology in the late 1650s runs somewhat counter to the view held by Port-Royal in the late 1660s. Interestingly, the uncompromising rejection of natural apologetics in Pascal's fragments caused difficulties for the Port-Royal edition. The "Messieurs" of Port-Royal were reluctant to endorse Pascal's explicit denunciation of physical proofs: Pascal's strong stance against natural theology had to be mitigated.

Filleau's *Discours*

In Pascal's repudiation of the diverse kinds of philosophical and theological proofs of God, the most vehement refusal is expressly that of "metaphysical proofs" (including, thus, the Cartesian proofs in the *Meditations*). A passage in chapter 20 of the Port-Royal edition—the chapter demonstrating that our knowledge of God must be acquired through Jesus Christ—asserts:

> The Metaphysical Proofs of God are so far off from human Reasoning, and so intangl'd, that they seldom work upon any; and if that should convince any one, it would be but for the moment that they beheld this demonstration, but an hour after they would be afraid of being cozen'd: *Quod curiositate cognoverint, superbia amiserunt*.[2]
>
> Moreover, this kind of Proofs, can only carry us to a speculative knowledge of God; and to know him in this manner, is not to know him at all.[3]

Yet the exclusion of proofs in the *Pensées* extended far beyond the dismissal of sophisticated "metaphysical proofs." It is noteworthy, from this viewpoint, that the insistence on Pascal's systematic critique of all usual proofs—in other words, on his radical originality in the contemporary context—constituted a leitmotif in the two prefaces successively penned for the Port-Royal edition.

A first draft of the preface, written by Nicolas Jean Filleau de la Chaise, a member of the Port-Royal committee preparing the edition of the *Pensées* after Pascal's death in 1662, did not get the agreement of Pascal's family, the Périers, so it was not adopted as the preface of the 1670 edition. After a separate publication in 1672, Filleau's *Discours sur les Pensées de M. Pascal* was eventually added to the Port-Royal edition starting from the 1678 edition.[4] Now, this *Discours* gives a significant space to the issue of Pascal's distancing his apologetic project from natural theology. Filleau first portrays the different contemporaneous trends concerning the proofs of God's existence. He lists the kinds of arguments ordinarily anticipated by readers of apologetic books in the context of the late 1660s and concludes that none of them can be found in Pascal's *Pensées*:

> Those that find nothing convincing without Geometrical Demonstrations, expect also Proofs of the Existence of God, and of the Immortality of the Soul, which may lead them from one Principle to another, as their Demonstrations do. Others require those common Reasons that prove but very little, or which only satisfie those that are already perswaded. Others desire Metaphysical Reasons, which for the most part are only refin'd Notions, that are not capable of making any great impression on the Understanding, and whereof 'tis always suspicious. . . . There be others that look for nothing but for that which is called common Places, and I can't tell what kind of Eloquence, and sound of Words void of Truth, which only dazzle the Sight and never reach the Heart.
>
> It is most certain none of all these shall find what they seek after in these Fragments.[5]

Filleau emphasizes that Pascal's aim was not so much to prove the existence of God as to urge humankind to seek to know God from the "sentiments" that, notwithstanding Adam's Fall, subsist from the first creation of human nature. In this view, Filleau is particularly keen on situating Pascal's specific plan by comparison with a natural theology project:

> If God has left Tokens and Marks of himself in all his Works, as it can no way be doubted, we shall sooner find them in our selves than in exteriour things that don't speak to us, and of which we have only a slight superficial Knowledge,

being not able to know the Ground and Nature of them: And if it be inconceivable that he has not imprinted in his Creatures what they owe to him for the Being he has given them, it will be much more probable Man shall find this important Lesson in his heart than in inanimate things, which fulfil the Will of God without knowing it, and for whom their very Being differs but little from nothing.[6]

Filleau's description assumes that for Pascal, marks of God are found everywhere. Actually, if "marks of God" means that God is manifest in nature, this is not at all Pascal's opinion. It is interesting, however, that on this supposition, Filleau outlines a justification of Pascal's rejection of natural theology. He contends, on behalf of Pascal, that the rejection of physical demonstrations of the existence of God is grounded in differences both epistemic and ontological within the works of God, according to whether one is considering humankind or the external world as God's creature. Filleau, indeed, stresses two points concerning the kind of proofs accepted by Pascal in opposition to natural theology:

1. Because all divine works bear signs or signatures of God, human self-knowledge is a safer way to discover these signs than the knowledge of outward things: the direct internal way must prevail.
2. Footsteps of God in human hearts bear witness to God's design in creation (to be honored by his creatures as their creator) better than footsteps in things without any sentiment of their ontological condition, the purely material things.

The conclusion is that it would be useless for humans to look for signs of God's existence and will "in the dead works of nature." Both the knowledge that God is and the knowledge of the duties that God asks from his creatures are to be sought within human hearts: "This is properly what was Monsieur *Pascall*'s Design; he would bring Men home to their Heart, and would make them begin rightly to know themselves: All other ways, though good, yet he thought was not so suitable and fit as this to their Nature."[7] In this regard, Filleau insists, the apologetic content of the fragments is quite conformable to Pascal's oral presentation of his whole project at Port-Royal a few years before his death. At the beginning of the general scheme planned, Pascal placed the rejection of the ordinary proofs of God: the proofs "drawn from the Works of God," the proofs from natural theology, which are "disproportion'd to the Natural State of the Heart of Man," had to be eliminated together with the metaphysical proofs.[8]

What kind of proofs, thus, are left in Pascal's *Pensées* after this general rejection? Filleau categorizes them as only "moral and historical proofs, and certain natural notions, and things of experience."[9] Strikingly, the *Discours* attributes to Pascal's *Pensées* not only proofs taken from the Holy Scripture but also a kind of proofs left undefined, proofs in connection with human experience. To be sure, Pascal's apologetic project favors the use of positive theology over that of rational and natural theology. Instead of geometric, metaphysical, teleological proofs, the *Pensées* contains proofs by prophecies, proofs by miracles, proofs by figures and types, proofs by Jesus Christ. But Filleau, when he describes the *Pensées* as dismissing proofs drawn from the geometric or metaphysical order of reason to privilege history, is eager to characterize Pascal's method in terms borrowed from part 4 of *La Logique ou L'Art de penser* by Antoine Arnauld and Pierre Nicole (1662), specifically from the last chapters about the rational rules guiding our belief of events, as well as from Nicole's *Traité de la foy humaine* (1664), part 2, chapter 5, and from Nicole's preface to Arnauld's *Nouveaux Elemens de geometrie* (1667). The epistemological model of Pascal's moral and historical proofs of God resides for Filleau in the certainty of factual truths, a certainty that is not inferior to that of geometric demonstrations, although it is irreducible to the scientific method of geometry:

> That there is a City called *Rome*; That there has been a *Mahomet*; That 'tis true *London* was burnt, are things would be hard to Demonstrate, nevertheless it were a madness to doubt of, or to fear hazarding ones Life upon the Truth of them, were there any thing to be got by it. The way whereby we attain these sorts of assurances, are no less certain, than if we were Geometricians, and should no less incourage us to Act, and 'tis only hereupon that we ground almost all we do.
>
> Monsieur *Pascall* undertook to shew, that the Christian Religion was as evidently true as anything that is undoubtedly believ'd amongst Men.[10]

Pascal's position, however, was more complex. In contrast with Port-Royal, his aim was not to extend the rules of logic, the rules of certainty, to the credibility of events or facts. His unpublished "Preface to the Treatise on Vacuum" shows that Pascal was interested, instead, in the division of knowledge into two sciences: historical sciences and rational sciences. The main reason why Pascal forbade all kinds of proofs of God derived from something other than the Holy Scripture was that, in theology, truth can be known only through authority, that is to say, through the biblical books: "To give a perfect certainty about those matters which are the most incomprehensible to reason, it suffices to show them in the sacred books; as to show the uncertainty of the most probable things, it is enough to show that they are not included therein."[11] It is fascinating, thus, that the first analysis of Pascal's *Pensées* devised as a preface for the Port-Royal edition sketches an apologetic project that can be related to Port-Royal rather than to Pascal. To this end, Filleau lessens Pascal's harsh exclusion of natural theology; he states that there are indubitable marks of God in the physical world. And he defines according to the Port-Royal model of factual truths the proofs (moral, historical, experiential) that Pascal opposed to a vain attempt to deduce faith from reason.

The 1670 Preface

It was crucial for Port-Royal to obviate the threat of a critical reception of Pascal's apologetic project as insufficient or incomplete. The preface penned by Étienne Périer to replace Filleau's in the 1670 edition of the *Pensées* also highlights Pascal's denial of all kinds of proofs of God other than moral proofs. The claim is that the proofs not found in the *Pensées* are proofs that Pascal refuses to consider as proofs of religious truths. In contrast with Filleau, however, when Périer mentions the whole range of standard proofs in Christian apologetics rejected in Pascal's *Pensées* as useless, he does not suggest that Pascal specifically condemns the proofs of natural theology. Among the proofs of God discarded in the *Pensées*, Périer includes proofs taken from "Nature," but these do not refer to nature as testifying about a divine finality:

> It appears . . . necessary, to undeceive some Persons, that happily may expect herein to find Proofs and Geometrical Demonstrations of the Existence of God, of the Immortality of the Soul, and several other Articles of the Christian Faith, this was not Monsieur Pascall's Aim. He designed not to discover these Truths of Christian Religion by such kind of Demonstrations, grounded on evident Principles, able to convince the most obdurate Persons; nor by Metaphysical Disputations, which for the most part rather divert than persuade the Mind; nor by common places, drawn from the divers Effects of Nature; but by Moral Proofs, which more touch the Heart than the Understanding; that is, he endeavoured more to affect the Heart, than to convince or persuade the Judgment.

Périer's concern is to stress Pascal's disavowal of natural theology insofar as it represents a rational theology, the use of reason on matters that cannot be submitted to reasoning, rather than insofar as it deals with the works of God as marks of God's "existence, wisdom, and omnipotence." Unlike the parallel drafted by Filleau between Pascal's project and natural theology, Périer's preface does not linger especially on the proofs of God from nature. Instead, it underscores the uselessness of all proofs "by natural reasons." Quoting a fragment by Pascal not integrated in the Port-Royal edition, Périer adduces as a paradigmatic case our failure to convince when we attempt to demonstrate God's existence from the dependence of mathematical truths on the truth of God's substance:

> *I will not here undertake to prove by Natural Reasons, the Existence of God, or the Trinity, or the Immortality of the Soul, nor other things of this Nature; not only because I should not think my self able to find in Nature sufficient to convince obstinate Atheists; but also because this Knowledge without Jesus Christ is useless and barren. Though a Man should be persuaded that the proportion of Numbers are Truths Immaterial, Eternal, and depending of a first Truth wherein they subsist, and which is called God, yet I should not think such a one much advanced in his Salvation.*[12]

This excerpt shows that in Pascal's view the proof of God (e.g., in Augustine's *De libero arbitrio*) from the immutability of the rules of numerical proportions, which requires the immutability of a first substantial truth in which they are inherent, no longer holds. In addition to omitting this passage quoted by Périer, the Port-Royal edition incorporates the passage that follows it (with slight changes) into chapter 20 after the fragment already mentioned on intricate "metaphysical proofs":

> The Christians Divinity consists not barely in knowing a God that is Author of Geometrical Truths, and of the order of the Elements, this belongs to the Heathens. It consists not barely neither in knowing a God that exercises his Providence over the Bodies and Goods of Men, to bless with a long and happy Life those which Adore him; this is the Portion of Jews. But the God of *Abraham*, the God of *Jacob*, is a God of Consolation; it is a God that fills the Heart and Soul that enjoys him . . . , which makes them uncapable of any other End but himself.[13]

In the original fragment, Pascal continues instead:

> All who seek God without Jesus Christ, *and who stop in nature*, either find no light to satisfy them, or come to form for themselves a means of knowing

God and serving Him without a mediator. Thereby they fall either into atheism, or into deism, two things which the Christian religion abhors almost
equally.[14]

When integrating this passage into chapter 20 a few paragraphs later, Port-
Royal makes a significant change. The words I have italicized on p. 147,
which denounce the path followed in natural theology, disappear:

> So that all such as seek God without Jesus Christ, do not find any light that
> satisfies, or can be any way profitable to them: For either they attain not to
> know there is a God, or if they do, it is to no advantage to them, because they
> imagine means of having Communion without a Mediator, with this God
> whom they have known without a Mediator. So that they fall either into Athe-
> ism, or into Deism, which are two extreams Christian Religion abhors both
> alike.[15]

Rephrasing Pascal's Position

The latent conflict between the apologetic project in the *Pensées* and natural
theology is regularly abated in the Port-Royal edition in a variety of ways. It
seems that Port-Royal was rather embarrassed by Pascal's attack on proofs
from the natural world. The disagreement is particularly striking in chapter 20. Indeed, before incorporating Pascal's fragment against the employ-
ment of abstruse "metaphysical proofs," Port-Royal transcribes, as the source
of the first paragraphs in chapter 20, a fragment expressing Pascal's opinion
on the use of physical proofs. In Pascal's original papers, this long fragment
forms the draft of a "Preface to the second part" of his apology for Christian
religion and announces that Pascal will "speak of those who have treated of
this matter":

> I admire the boldness with which these persons undertake to speak of God.
> In addressing their argument to unbelievers, their first chapter is to prove Di-
> vinity from the works of nature. I should not be astonished at their enter-
> prise, if they were addressing their arguments to the faithful; for it is certain
> that those who have the living faith in their heart see at once that all existence
> is none other than the work of the God whom they adore. But for those in
> whom this light is extinguished, and in whom we purpose to rekindle it, per-
> sons destitute of faith and grace, who, seeking with all their light whatever they
> see in nature that can bring them to this knowledge, find only obscurity and

darkness; to tell them that they have only to look at the smallest things which surround them, and they will see God openly, to give them, as a complete proof of this great and important matter, the course of the moon and planets, and to claim to have concluded the proof with such an argument, is to give them ground for believing that the proofs of our religion are very weak. And I see by reason and experience that nothing is more liable to arouse their contempt.

It is not after this manner that Scripture speaks, which has a better knowledge of the things that are of God. It says, on the contrary, that God is a hidden God, and that, since the corruption of nature, He has left men in a darkness from which they can escape only through Jesus Christ, without whom all communication with God is cut off. *Nemo novit Patrem, nisi Filius, et cui Filius voluerit revelare.*[16]

Now, from the 1669 pre-edition to the 1670 edition of the *Pensées*, Port-Royal considerably modifies Pascal's phrasing in this draft. In the 1669 pre-edition, only the first two sentences still constitute a faithful copy from Pascal's manuscript. Immediately afterward, Port-Royal adds substantial comments to Pascal's fragment, which I have italicized:

I admire the boldness with which some persons undertake to speak of God, in addressing their discourses to the impious. Their first chapter is to prove Divinity by the works of nature. *I do not call in question the solidity of these proofs; but I have little faith in the usefulness and the result expected from them; and although they seem to me to be conformable enough to reason, they do not seem to me to be sufficiently conformable and proportioned to the disposition of the spirit of those for whom they are intended.*[17]

For it must be observ'd, this Discourse is not directed to those that have a lively Faith, and that presently see, that all the World is nothing else but the Workmanship of that God whom they Adore: *It is to such the whole Fabrick of Nature speaks the praise of its Creator, and that the Heavens shew forth his Handy-works.* But for those in whom this light is gone out, and in whom one would willingly kindle it; those Persons, destitute of Faith and Charity, that[18] only see darkness and obscurity in all the Works of Nature, *it seems not to be the best way of instructing them,* to give them for Proofs only of this great and important Subject, the course of the Moon, and Planets, *or common reasonings,*[19] *against which they have ever had an aversion; the obstinacy of their understanding has made them deaf to this Voice of Nature, sounding continually in their Ears,* and experience shews,[20] that *very far from gaining them by this means, there's nothing on the contrary more like to hinder them, and to deprive them of all hope of knowing the Truth, than to think to convince them only by this sort of Arguments, and to tell them that they should plainly see the Truth in them.*

It is not in this manner that Scripture speaks, that knows the things of God better than we do. *It tells us indeed, that the Beauty of the Creatures teaches him who made them; but it doth not say, that they work this same effect in all the World. It warns us on the contrary, that when they do it, it is not by themselves, but by the light that God sheds forth at the same time in the Minds of those to whom he discovers himself by this means: Quod notum est Dei, manifestum est in illis, Deus enim illis manifestavit.*[21] It tells us *in general,* that God is a God hid, *Vere tu es Deus absconditus;*[22] and that since the Corruption of Nature, he has left Mankind in a State of darkness, from which they cannot be freed but by Jesus Christ, without whom we are deny'd all Communion with God; *Nemo novit Patrem, nisi Filius, et cui voluerit Filius revelare.*

For Port-Royal, however, to reproduce literally at the outset of chapter 20 the first two sentences from Pascal's draft was still too much. This beginning is reworded in the 1670 edition, and the first paragraph undergoes some other modifications:

> Most of those that undertake to prove the Divinity to the prophane, for the most part, do begin by the Works of Nature, and they very seldom succeed. I do not call in question the solidity of these Proofs consecrated by the Holy Scriptures, they are agreeable to Reason; yet sometimes they are not conformable enough, and sufficiently proportion'd to the Disposition of the Spirit of those for whom they are intended.[23]

Obviously, Port-Royal was keen to smooth Pascal's harsh judgment on proofs from nature. The recomposition of this original fragment allowed the Port-Royal editors to emphasize mainly that the postlapsarian condition makes humans unreceptive to these proofs. Moreover, Port-Royal's own conviction that the God demonstrable from nature could be accepted as the God of the Bible resulted in the decision not to publish another fragment that expressly refuses natural theology on the very authority of the Holy Scripture:

> It is an astounding fact that no canonical author has ever made use of nature to prove God. They all strive to make us believe in Him. David, Solomon, &c.,

have never said, "There is no void, therefore there is a God." They must have had more cleverness than the most clever people who came after them, who have all made use of it.[24]

So, what should Pascal's apologetic project have looked like in order to be faithfully reproduced in the Port-Royal edition of the *Pensées*? Port-Royal wanted it to be devised along the lines of Pierre Nicole's essay on the "natural proofs" of God's existence, also published in 1670. Nicole highlights there that the attacks of atheists cannot be countered if one maintains a sharp divorce between the scriptural proofs and the metaphysical or physical proofs. The latter must be amended so that they are no longer perplexing arguments:

> Some invented subtle and metaphysical arguments for the existence of God and the immortality of the soul, and others more popular and more sensible arguments by calling men back to a consideration of the world's order as to a great book ever exposed to their sight. . . .
>
> These are abstract and metaphysical ones. . . . But there are also some that are more sensible, more fit to most minds, and that are such that we would have to do violence to resist them, and these are the ones that I intend to bring together in this discourse.[25]

Nicole's essay accordingly puts at the forefront an argument summarizing the physico-theological approach. The steady courses of the vast bodies moving above us, the orderly progress of nature, the admirable connection between all parts in the world, the variety of inanimate things, the stunning arrangement of animal bodies, and so on are for our reason, Nicole writes, as many "wonders," necessarily the effects of "some cause" possessing in itself all the perfections contemplated in nature. And this cause is God the Creator.

The *Pensées* sounds a discordant voice. For Pascal, if physico-theology may get some acknowledgment, it cannot be as an apologetic project. Physico-theology fails to convince unbelievers; it can only strengthen believers in their belief. As a letter to Gilberte Périer from Blaise and Jacqueline Pascal on April 1, 1648, made unambiguous a decade before the *Pensées*, to understand that the natural world is the work of God who has "represented the invisible things in the visible" already presupposes the possession of faith.[26] The defense of religious truths must therefore aim at proving the existence not of

God the Creator but of God the Redeemer, and doing so from biblical authority and not the evidence from nature.

NOTES

1. Pascal, *Œuvres complètes*, 1:1105 (my translation). This comment is reported by Pascal's
niece. A *chiquenaude* is a fillip, a flick of the finger.

2. *Monsieur Pascall's Thoughts* (Walker translation, 1688), 134. Unless otherwise noted, all
English quotations are from this translation (dedicated to Robert Boyle); Walker worked from
the edition printed in Amsterdam in 1684 by Abraham Wolfgang. On this passage, see *Pensées*, Sellier ed., fragment no. 222 (Pascal is quoting Augustine, *De verbis Domini* 55.1–2).

3. *Monsieur Pascall's Thoughts*, 134. Added by Port-Royal.

4. *Discours sur les Pensées de M. Pascal* (published with Filleau's *Discours sur les Preuves
des Livres de Moyse* and his *Traité où l'on montre qu'il y a des demonstrations d'une autre espece et aussi certaines que celles de geometrie, et qu'on en peut donner de telles pour la religion
chrestienne*). Walker's translations of Filleau's three essays are annexed to his 1688 edition of
Monsieur Pascall's Thoughts. All English quotations from Filleau de la Chaise are from this
translation.

5. Filleau de la Chaise, *Discours sur les Pensées de M. Pascal*, 271–72.

6. Filleau de la Chaise, *Discours sur les Pensées de M. Pascal*, 273.

7. Filleau de la Chaise, *Discours sur les Pensées de M. Pascal*, 274.

8. Filleau de la Chaise, *Discours sur les Pensées de M. Pascal*, 275.

9. Filleau de la Chaise, *Discours sur les Pensées de M. Pascal*, 276.

10. Filleau de la Chaise, *Discours sur les Pensées de M. Pascal*, 276.

11. Pascal, *Œuvres complètes*, 2:778 (my translation).

12. The italics are Périer's. See *Pensées*, Sellier fragment no. 690.

13. *Monsieur Pascall's Thoughts*, 134–35. Walker's translation omits "the God of Christians"
present in the French following "the God of Jacob" in the passage.

14. *Pensées*, Sellier fragment no. 690 (Trotter translation, 1910, p. 183, under fragment
no. 556 slightly modified; my emphasis in italics).

15. *Monsieur Pascall's Thoughts*, 136.

16. See *Pensées*, Sellier fragment no. 644 (quoting Matt. 11:27). Trotter translation, p. 90,
under fragment #242.

17. *Pensées de M. Pascal sur la religion* (1669), 150 (my translation). This pre-edition, also
called pre-original edition, was sent in a few copies by Guillaume Desprez to the Approbators
during summer 1669; the original edition was published on January 2, 1670, with some variants (see McKenna, *Entre Descartes et Gassendi*; Pérouse, *L'Invention des Pensées de Pascal*).
Since there are no variants in the 1670 edition after this first paragraph of chapter 20, I quote
in what follows the Walker translation, pp. 133–34. Italics indicate Port-Royal additions to Pascal's fragment.

18. Port-Royal suppresses "seeking with all their light whatever they see in nature that can
bring them to this knowledge."

19. I change Walker's translation ("common notions"). Port-Royal suppresses Pascal's
statement that this kind of discourse would be received as a weak proof of the Christian
religion.

20. Port-Royal suppresses Pascal's mention of "reason" together with "experience" to in-
troduce the statement (also removed) that this kind of discourse is liable to be despised.

21. Rom. 1:19.

22. Isa. 45:15.

23. *Monsieur Pascall's Thoughts*, 132.

24. *Pensées*, Sellier fragment no. 702 (for the proof from the *horror vacui* in nature: Grotius, *De veritate doctrinae christianae*). Trotter translation, p. 91, under fragment no. 243 modified.

25. Nicole, "Discours contenant en abregé les preuves naturelles de l'existence de Dieu," 120–21. I am quoting the translation of this passage in Armogathe, "Proofs of the Existence of God," 307–8. For a different reading of this passage in Nicole, see Carraud, *Pascal et la philosophie*, 347–54.

26. Pascal, *Œuvres complètes*, 2:582 (Rom. 1:20).

PART FOUR

ENGAGEMENT WITH
THE NEW SCIENCE

Physico-theology or Biblical Physics?

The Biblical Focus of the Early Physico-theologians

RIENK VERMIJ

In much of the older historiography, physico-theology is defined or treated as a form of natural theology. That is, physico-theological works are discussed as contributions to a long tradition of arguments from design or of Christian apologetics and studied foremost as part of the history of theology and philosophy. This approach has been quite influential, partly because much work on physico-theology is written by theologians or church historians.[1] However, it is one-sided at best. As early as 1971, Bots pointed out that what distinguished eighteenth-century physico-theology from earlier similar arguments was that it was as much about the empirical investigation of nature as it was about theology.[2] I would even go further and claim that the study of nature was the dominant part. After all, among the major physico-theological authors, there are few professional theologians and many active investigators of nature.

Physico-theology was not in the first place a discourse of theologians who had to come to terms with the new sciences. Rather, it was a way for laypeople who had embraced the "new philosophy" to adjust the new understanding of nature to the existing religious worldview and social order. Although philosophical and theological arguments may have been important to that view, the world was understood primarily in biblical terms. Consequently, advocates of the new sciences had to bring their work into agreement with the Bible and the biblical worldview, including biblical miracles. This approach, at least, holds true for three prominent authors who are generally regarded as among the pioneers of physico-theology: John Ray (1627–1705), Bernard Nieuwentijt (1654–1718), and Johann Jakob Scheuchzer (1672–1733). The argument from design is certainly prominent in their work, but it is not their main message. By focusing on their defense of the traditional biblical world view, I hope to show that physico-theology is not just about the argument from design and should not be studied from a merely philosophical or theological perspective.

From Philosophy to Physico-theology

The new view of nature that had originated in the seventeenth century had sought its legitimation originally in philosophical arguments. In particular, Descartes's views had found wide following. Descartes claimed that all natural phenomena should be reduced to some "laws of nature," operating in a mechanical, predictable way. These laws had their ultimate foundation in God's immutability and eternity. The various questions this posed concerning the relation between God and nature became subject to a debate over what has been aptly described as a "secular philosophy."[3]

Cartesianism thus legitimated an understanding of nature that had no recourse to the Bible, or to any supernatural elements, yet was still theologically acceptable. However, from the beginning there was also serious opposition. These debates focused especially on the interpretation of specific biblical texts (although one can easily see that much more was at stake). The question was, Should the Bible be taken at face value as a description of reality, or should it be read in the light of the new scientific knowledge? A main complaint of conservative theologians against the Cartesians was that they made the understanding of the Bible dependent on mere philosophical hypotheses. So when Cartesians or others argued that the theory of Copernicus was not contradicted by the Bible, their opponents protested that the interpretation of texts such as Joshua 10:12 should not be dependent upon the whims of secular scientists. God's word should be explained following the traditional authorities (for Catholics) or should simply explain itself (for Protestants).[4]

The philosopher Baruch Spinoza brought matters to a head when he applied Descartes's principles not just to natural phenomena but also to the biblical miracles. It became obvious that it was hard to account both for the existence of unchanging laws to explain the phenomena of nature and for a personal and providential God who had performed the acts related in the Bible.[5] So by the closing decades of the seventeenth century, the philosophical legitimation of the new philosophy increasingly came to be seen as a failure. Because very few people wanted to do away with the sciences as such, people started to look for an alternative justification, one that was not in conflict with the traditional understanding of the Bible as the infallible word of God.

The alternative that gradually took shape in the years around 1700 became what we call "physico-theology." As such it should be clearly distinguished from the earlier "secular theology" that likewise tried to harmonize science and religion. Physico-theology was a reaction to traditional natural theology, not a form of it.[6] Philosophical consistency was not a main objective. Authors were tinkering with all kinds of traditional ideas: elements of natural theology, like the argument from design, were tried, but so were theological traditions like voluntarism and new forms of natural philosophy that were alternatives to the Cartesian or Spinozist approaches, such as experimental philosophy or Newtonianism. Other elements were taken from traditional exegesis or the study of biblical antiquities. As a consequence, any attempt to describe physico-theology by a well-defined set of principles is doomed to fail. There is no master plan.

If we regard physico-theology above all as a way to bring the new sciences in harmony with traditional religious views, it stands to reason that both traditional religion and the new sciences are its constituent parts.[7] Seventeenth-century natural theology had regarded arguments based on empirical research as second rate, fit only for simple people for whom rational arguments were beyond reach.[8] The eighteenth century, however, was wary of rationalism and hailed the empirical study of nature as the one and only way to knowledge. Consequently, authors would use it for the defense of the Bible and the refutation of Spinoza and other "atheists" as well.

John Ray

Ray's scientific credentials are evident. Although ordained as a priest of the Church of England, he spent most of his time on natural investigations. He was an early member of the Royal Society and published several books of a purely botanical nature. Religion was definitely important throughout his career, but it was only at the end of his life that he published on the subject—partly, as he claimed himself, to make up for the little time he had spent on his profession. His best-known book in this respect is *The Wisdom of God* (1691), often seen as the first real work of physico-theology. However, here I want to focus on his second book in the physico-theological genre.[9]

In 1692, Ray published *Miscellaneous Discourses concerning the Dissolution and Changes of the World*. A year later, a much-expanded version saw the light. Ray had added whole new sections on the Creation and the biblical Deluge. The new title summarized the content: *Three physico-theological Discourses*

Concerning I. The primitive chaos and Creation of the World. II. the general deluge, its Causes and Effects. III. the dissolution of the world, and Future Conflagration. The remainder of the title mentioned, as the 1692 version had done, the various scientific subjects dealt with in the book, such as the origin of mountains, fossils, and earthquakes.

The book saw many reprints, as well as a Dutch and a German translation. The titles of these versions omitted the neutral "three physico-theological discourses," replacing these words by titles that highlighted the book's focus on the divine plan of history. An English translation of the Dutch title was "The world from its beginning to its end." The German choice was "Remarkable clover-leaf of the beginning, change, and downfall of the world."

The text of *Miscellaneous Discourses* went back to a sermon on the dissolution of the world that Ray had presented much earlier. The reason to go into print in 1692 must have been that by that time the topic had become the subject of considerable debate. Thomas Burnet (ca. 1635?–1715) in *The Sacred Theory of the Earth* (first [Latin] edition 1681–89) had presented an elaborate explanation of how the Deluge and the final Conflagration could have been and could be effected by natural means. Burnet's work was highly controversial, especially because of what people felt to be his rather cavalier exegesis of the relevant biblical sentences. Burnet personified the exegete who went astray because he based his argument on philosophical hypotheses rather than on the plain text. Ray fits in a long list of authors (one might mention John Woodward [1665–1728] and William Whiston [1667–1752]) who, in the wake of Burnet's book, came up with their alternative views on the Deluge and the history of the world. Although Ray in his book abstained from polemics, it would seem that he published his discourses in order to present a more biblical counterweight to Burnet's physical description of the earth's history.[10]

Much attention has been given in the literature to Ray's position on fossils in the controversy. The paratext may have strengthened that impression, as the book contains three engraved plates with depictions of fossils, done in a plain factual style. (In addition, there is a plate with images of ancient coins.) But it should be emphasized that this is a book not on fossils but on the truth of the biblical narrative. The Dutch publisher brought its outward appearance more in line with the content (and probably his readers' expectations) by including four other engravings, done by the renowned engraver Jan Luiken, representing paradise, the Deluge, the earthquake, and the Conflagration.[11]

Ray discussed various possible causes of the future Conflagration: earthquakes, an eruption of the subterranean fire, and so on. In all cases, the possibility was assessed partly by natural probablility but, most importantly, also by agreement with the literal text of scripture. As to the question whether the dissolution of the world would be effected by natural or supernatural means, Ray refused to follow just a naturalistic path. The instrument would be natural—to wit, fire. The necessary ingredients, namely the fuel, would also be natural and everywhere plentiful. However, the fire had still to be lit and then diffused instantly and equally throughout the whole world. "And this must be the Work of God, extraordinary and miraculous."[12]

This was not just an ad hoc argument to save the status of the Bible but a consistent view on the workings of nature. Ray applied the same principle to the explanation of natural phenomena of his own time. This becomes clear from his views on earthquakes, a topic that Ray dealt withat some length, especially because of their possible role in the Conflagration. He discussed their natural causes, based on the then-current theories (which basically went back to Aristotle). But then, shortly before the *Three physico-theological Discourses* went to press, there came the startling news that the English colony of Port Royal, on Jamaica, had been completely destroyed by an earthquake. Ray thereupon included in his book a description of this disaster, as well as some further considerations.

Ray emphasized that this earthquake was not just a natural event but also a special providence of God. "The People of this Plantation being generally so debauched in their Lives,"[13] the earthquake might justly be regarded as a divine judgment. Ray did not doubt that there were times that wickedness was rampant and a general corruption of manners predominated. At such times, "God usually sends some sweeping Judgment, either utterly destroying such a People who have filled up the measure of their Iniquity, or at least grievously afflicts and diminishes them. . . . And we shall find it noted by Historians, That before any great publick Calamity, or utter Excision of a Nation, the People were become universally vicious and corrupt in their Manners, and without all Fear of God, or sense of Goodness."[14] This had happened in the biblical Deluge and in the destruction of Sodom and Gomorrah, and likewise happened in Port Royal. Just as Ray argued in the case of the final Conflagration, the earthquake of Port Royal was natural in the sense that the instruments for the destruction of Port Royal were naturally present in the earth. However, that it erupted at a moment of such great wickedness there "was the Finger of God, and effected perchance by the Ministry of an Angel."[15]

As an investigator of the natural world, Ray accepted natural causality and natural laws. As a Christian, he felt these had certain limits. God was not just

the great architect of the world, who had contrived everything in the best and wisest way. He was also still the vengeful God of the Old Testament, who punished sinners by extraordinary means. The whole book is written to uphold the biblical narrative. The biblical worldview should not be replaced by a naturalistic one.

Bernard Nieuwentijt

Bernard Nieuwentijt was a physician and minor Dutch regent. Although he did not really make a name for himself in science, at the local level he was an active propagator of experiments. He owned an air pump and other apparatus used for experiments by a college that met at his house. Some of the experiments he described in his later work were deemed worthy to be repeated by the Royal Society.[16]

Jan Luiken's "The Conflagration" (*left*) and "The Deluge" (*right*) from John Ray, *De werelt van haar begin tot haar einde* (Rotterdam, 1694). The publisher of the Dutch translation of John Ray's *Three physico-theological Discourses* included some new engravings, made by the famous artist Jan Luiken, which highlighted the book's focus on the Bible. Courtesy of the Rijksmuseum, Amsterdam. Object numbers RP-P-1896-19368-1000 (*left*) and RP-P-1896-A-19368-998 (*right*).

Nieuwentijt was the son of a Reformed minister. He appears to have had a somewhat unruly youth and was actually evicted from Leiden University. (He thereupon obtained his degree at Utrecht.) During this period, he appears to have been an avid Cartesian and, it would seem, sympathetic to Spinoza's ideas. Later in life, he returned to a more traditional religious belief and regarded his juvenile flirtation with Spinozism as a dangerous folly. For the rest of his life, the refutation of Spinoza's ideas would remain almost an obsession. At the same time, he needed to legitimate his scientific work in a way that would not give occasion to impiety or irreligion.[17]

In the 1690s Nieuwentijt appears to have been associated with a circle around the Amsterdam merchant Adriaan Verwer. People visiting this circle were of various confessional backgrounds (Verwer himself was a Mennonite), but they shared an interest in the new science and a rejection of Spinozism. Initially, Verwer and Nieuwentijt tried to use Spinoza's "geometrical method" against him to prove religious truths, but later Nieuwentijt rejected the geometrical method altogether and instead turned to experimental philosophy. After the second edition of Newton's *Principia* came out in 1713, Verwer and his friends quickly accepted Newton's work as a superior form of natural philosophy that was in agreement with scripture. In the following years, they played an important role in the propagation of Newton's ideas in the Dutch Republic.[18]

Nieuwentijt's crusade against Spinoza culminated in two hefty volumes: *Het regt gebruik der werelt beschouwingen* (The right use of contemplating the works of the Creator) in 1715, and *Gronden van zekerheid* (Foundations of certitude), published posthumously in 1720. The latter is basically a book on epistemology wherein Nieuwentijt refutes Spinoza's "geometrical method." This book was never translated. *Het regt gebruik*, on the other hand, saw not just many editions but also several (albeit somewhat truncated) translations. The book's fame was partly due to its being a detailed and lavishly illustrated overview of the scientific—in particular, the experimental—knowledge of the time. But at the heart of it was the religious message.

There are two main thrusts in the argument. The first is the well-known argument from design. Unbiased and experimental investigation demonstrates God's wisdom, power, and goodness. The world is not there just by accident or chance but has been contrived with great wisdom. Nieuwentijt fills hundreds of pages with examples of this design, on the basis of scientifically accurate descriptions. An example that he himself finds particularly strong ("capable not just to reassure an irresolute mind, but even to convince an obstinate atheist," as he formulates it) is the fact that, of all the muscles of the intestines, only those of the rectum receive nerves from the spinal marrow and can be contracted at will, so that people can hold back their excrement when decency requires it.[19]

The world is not governed by chance, but it is also not governed by necessity. There are no immutable, necessary laws, as the Spinozists claimed. The great variety of things in the world demonstrates that God has complete freedom to create and perform whatever he wants. Miracles in which God acts directly are perfectly possible and actual; Nieuwentijt gives some concrete examples, both from the Bible and from more modern times. The thunder is God's voice by which he calls sinners to repentance.[20]

The second main thrust is what concerns us most here. It is Nieuwentijt's argument that the study of nature can demonstrate that the Bible is God's word and should be interpreted in a literal way. In his view, a careful reading of the Bible shows that it speaks of many things that have been discovered only by the efforts of seventeenth-century scientists. Not only does the Bible know that the earth is a globe (Isa. 40:22 speaks of "the circle of the earth"); it also knows that the poles are flattened (Jer. 6:22: "Behold, a people cometh from the north country, and a great nation shall be raised from the sides of the earth"). Likewise, Nieuwentijt feels that Isaiah 42:5 can be translated as "He that created the air and its stretching into all directions," thereby attesting to knowledge of the "spring of the air," whereas Job 38:31 ("Can you bind the sweet influences of the Pleiades, or loose the bands of Orion") refers to Newton's theory of planetary attraction. All this demonstrates that the Bible possesses more than human knowledge.[21]

In some cases, Nieuwentijt directly engages with exegetical or religious problems, mostly to counter atheist views. A whole chapter is devoted to "the possibility of the resurrection"—that is, the resurrection of the flesh at the end of days. Nieuwentijt answers here the well-known question that posed, With what body would a cannibal who had lived his whole life on human flesh be resurrected? Apparently, the freethinkers of his day liked to bring such arguments forward.[22] Also, Nieuwentijt includes a long digression on the basin of King Solomon according to 1 Kings 7:23: "a molten sea, ten cubits from the one brim to the other: it was round all about . . . and a line of thirty cubits did compass it round about." Spinozists used this text to argue that the Bible's knowledge of geometry was less than perfect. Nieuwentijt counters this by suggesting that the Hebrew text really says that the basin was sexagonal.[23] Interesting too is his comment on Luke 12:27 (see also Matt. 6: 28–29): "Consider the lilies how they grow: they toil not, they spin not; and yet I say unto you, that Solomon in all his glory was not arrayed like one of these." Puzzled by this text, Nieuwentijt decided to use a microscope to investigate "wherein this sublime glory of the lilies might consist."[24]

Apart from his ongoing attacks on Spinoza, Nieuwentijt typically abstains from polemics. Burnet is mentioned only in passing. When in *Foundations of Certitude* Nieuwentijt argues for the existence of spiritual beings, he does so without referring to the highly controversial refutation of witchcraft by Balthasar Bekker (1634–98).[25] Likewise, Nieuwentijt refuses to take sides in the long-standing debate on the Copernican system in the Netherlands. He emphatically rejects that the Bible ever speaks "according to the erroneous opinion of the people," a point of view associated with Spinoza, and one that "has given occasion to many miserable people to entertain irreverent thoughts about the divinity of this Word," but he leaves open the possibility that the biblical sentences on the motion of the sun should be understood in a figurative sense.[26] It is clear that he himself does not support the Copernican theory. His book has a whole chapter "on the unknown," wherein he argues that the question of the motion of the earth is still open.[27] Still, as long as they treat the Bible with due respect, he will not call supporters of Copernicanism heretics. In the Dutch Republic, where the Copernican system was still a bone of contention among the various factions in the church, his cautious stance may have helped to make his book acceptable for a general audience.

It should be added that the English and French translations focused on the argument from design and skipped most of Nieuwentijt's discussion of biblical texts.[28] For Nieuwentijt himself, however, the defense of the Bible was really at the heart of his project. In bringing the study of nature into agreement with religion, the argument from design is only a first step. Bringing the sciences into agreement with the Bible is what really matters.

Johann Jakob Scheuchzer

Scheuchzer, like Nieuwentijt, was a physician. He got his main scientific training at the University of Altdorf (near Nuremberg) under Johann Christoph Sturm (1605–1703), the propagator of an "eclectic" philosophy wherein a large place was allotted to experiments. After taking his degree at Utrecht in 1694, Scheuchzer settled in his native Zurich as a city physician, and in the following decades, he played an important part in the intellectual life of the city. He published a large array of books, especially on the natural history of Switzerland. These books cover many fields in often painstaking detail.

Scheuchzer was much given to his scientific work. As natural investigation was still suspect to many people, he on occasion felt compelled to take on its defense against the objections of the orthodox. At the same time, he had to face the question how to bring his scientific activity in agreement with his religious views. He was certainly aware of the various pitfalls and felt the need to take a stance against atheism as well. Several of his early lectures at Zurich, for a local learned society, are concerned with the relation of science and religion. In 1696 he gave two lectures on the use of mathematics in theology, and in 1702 he gave three lectures on Spinoza.[29] Polemics against Spinoza, the Epicureans, and other "atheists" turn up regularly in his work.[30] On the other hand, he had a great admiration for Newton. He referred to the latter as early as 1711 and in later work spoke of him as the wisest and most accurate philosopher that had ever lived.[31]

As in the case of Nieuwentijt, Scheuchzer's hostility to the Spinozists went hand in hand with a defense of the authority of the Bible. So, in the introduction to the second, expanded edition (1711) of his textbook on natural philosophy, he explained the relation between physics and scripture. Physics is based on experience and reason, not on authority. The Bible does speak about certain natural facts, such as the Creation, but it is not a complete textbook on natural knowledge and cannot be used to decide controversies in physics. Still, "it is uncontrovertible that any time that [the Bible], in the representation of spiritual and celestial things, speaks about the nature and properties of natural things, this saying has to be taken as divine and infallible."[32]

Most of the time, Scheuchzer would skip abstract philosophical questions and focus instead on concrete questions, either in physics or in biblical exegesis. In 1713 he was granted permission to give public lectures at the Carolinum, the institute for higher education at Zurich. Two of his three weekly hours were spent on explaining in the vernacular the physico-mathematical texts of the Holy Scripture. From this work resulted some of his best-known publications, in particular his *Kupfer-Bibel* ("Bible in copper," that is, engravings), most often referred to as *Physica sacra*, a massive discussion, verse by verse, of the "physical" elements in the Bible, illustrated with hundreds of engravings.[33]

The biblical Deluge takes a prominent place in the book. More than thirty of the book's full-title engravings are devoted to the subject.[34] Scheuchzer discussed the physical explanations of the Deluge that had been put forward by Burnet, Whiston, and Dethlev Cluver (1645–1708). His ideas on the relation between natural and supernatural causes appear to be very similar to Ray's. Because the Bible is silent about the way in which the waters above and below the earth caused the Deluge, physicists are permitted to present their ideas on those questions, provided these ideas agree with the matter itself and with the constitution of the earth. However, any means that can be conceived for bringing about such a Deluge should not be regarded "as if they were the work and motion of nature." They are divine miracles.[35]

Like Ray's *Discourses*, Scheuchzer's *Physica sacra* is nowadays known mostly for what it says about fossils, and in this case, too, that does not do justice to the book. The Deluge is certainly not the only topic dealt with. Scheuchzer refers to natural history and also to many historians and antiquarians. In many cases, the physical elements that he discusses are rather what we would call "material culture." As an example, his discussion of the "molten sea" of King Solomon might serve. Scheuchzer spends about a dozen pages of text on the subject, discussing six authors. He also includes seven large plates of earlier reconstructions of the basin. The question of the ratio between its circumference and diameter, which had worried Nieuwentijt, is only a small part of Scheuchzer's treatment, although it is duly discussed. Although Scheuchzer was hesitant as to the correct solution, he had no problem stating that it was impudent to accuse scripture of error, as Spinoza had done.[36]

Scheuchzer admitted that the literal interpretation of the Bible does not always apply, but he insisted that one cannot deny the miraculous or supernatural character of many of the events related in the Bible. He did use modern scientific insights for his explanation of the Bible, but he refuted naturalizing exegesis. The Spirit of God in Genesis 1:2 is not a powerful wind, an elementary light, or a magnet but really is the Spirit of God.[37] Unlike traditional exegetes, he did not believe that Joshua 10:12–14, the miracle of the sun standing still, refuted the motion of the earth. Instead, he felt that the Bible supported the system of Copernicus. However, he did disagree with interpreters, such as Benedict Pereyra (1536–1610), Jean Le Clerc (1657–1736), and most of all Spinoza, who wanted to explain the miracle as a mirage or some other natural effect. Against the Jesuit Pereyra, Scheuchzer explained that he had no problem keeping divine miracles within the path of reason as long as the only miracles under consideration were those of the Catholic tradition, but "this is not at all appropriate for those that are included in the Holy Pages." In other words, a real miracle had happened, even if it had been the earth and not the sun that had stood still. Scheuchzer hoped to prove this from nonbiblical sources, especially Chinese astronomical records.[38]

Conclusion

It is definitely one-sided to claim that these physico-theological authors propagated a benevolent God, the all-wise architect of the universe, instead of the vengeful and actively interfering God of the Old Testament. Both elements are present in their work, but it is the defense of the biblical God that is their main motivation. This element has often been overlooked in modern historiography, partly, it seems, because of the idea that physico-theology was a form of natural theology.

Still, it is true that in the course of the eighteenth century the biblical elements faded into the background. In the case of the translations of Nieuwentijt's book, they were even actively suppressed. Later physico-theological works focused nearly exclusively on the argument from design. One may well wonder why this was the case.

Legitimation of scientific research was only one element in the rise of physico-theology. Another was the changing relation between church and state. By the end of the seventeenth century, there was increasing criticism of the confessional church model for political and purely theological reasons. People were as committed to religion as ever, but the religous wars and strife of the seventeenth century had left many people wary about ecclesiastical pretensions. The monopoly of the confessional churches was increasingly under attack. In this situation, disciplining the believers by threatening them with hell and brimstone became less convincing. Hence, in the course of time God's benevolence came to be emphasized over his ire.

In authors such as Ray, Nieuwentijt, and Scheuchzer, we see the beginnings of this shift. What they have in common is not just a strong commitment to biblical revelation but also an irenic mindset that made them unwilling to fight fellow Christians. They would denounce notorious atheists but as far as possible stayed out of interconfessional or intraconfessional debates. They were orthodox and devoted members of their respective churches but preferred reasonable arguments to doctrines and anathemas. They offered thereby a model for a new, peaceful way of Christian instruction and belief. Piety based on natural knowledge appeared to be an attractive alternative to the intolerance and dogmatism of orthodox theologians. That such piety was centered on the Bible went for people at the time without saying.

NOTES

1. Examples of theological treatments include Hirsch, *Geschichte*, 1:170–74; Zöckler, *Geschichte*, 336–464. See also Dillenberger, *Protestant Thought*.
2. Cf. Bots, *Tussen Descartes en Darwin*, 14–15.
3. Henry, "Metaphysics."
4. Cf. Vermij, *The Calvinist Copernicans*, 241–331.
5. On Spinoza, see Israel, *Radical Enlightenment*, 218–57.
6. See also Vermij, "Defining the Supernatural"; Barth, *Atheismus und Orthodoxie*, 249–54. A different view is given by Harrison: "Physico-theological approaches are thus exemplifications of this 'secular theology'"(chapter 2 in this volume).
7. Cf. Calloway, chapter 8 in this volume.
8. Cf. chapter 4 by Mandelbrote on Fénelon and chapter 8 by Calloway on More, both in this volume. Most studies on physico-theology see the period around 1700 as a watershed and

see a relation to the new forms of natural philosophy—in particular, experimental philosophy; e.g., Bots, *Tussen Descartes en Darwin*, 134–47.

9. See on *Wisdom of God* and Ray in general: Calloway, *Natural Theology*, 95–137; and Calloway, chapter 8 in this volume.

10. On Burnet, cf. Vermij, "The Flood." On the theological controversies, see esp. Mandelbrote, "Newton and Burnet." See also Boscani Leoni, chapter 16, and Mandelbrote, chapter 4, both in this volume.

11. Cf. Calloway, *Natural Theology*, 108–11, who emphasizes that Ray under all circumstances kept to scripture.

12. Ray, *Three physico-theological Discourses*, 3rd ed. (1713), 388–89 (quotation on 389).

13. Ray, *Three physico-theological Discourses*, 3rd ed. (1713), 269.

14. Ray, *Three physico-theological Discourses*, 3rd ed. (1713), 270–71.

15. Ray, *Three physico-theological Discourses*, 3rd ed. (1713), 271. Cf. Mandelbrote, chapter 4 in this volume, on Ray's appeal to special providence.

16. London, Royal Society, Journal book, Feb. 6, 1717 [= 1718]. Cf. Nieuwentijt, *Het regt gebruik*, 106–7.

17. On Nieuwentijt, see Vermij, *Secularisering*; Bots, *Tussen Descartes en Darwin*, 16–48; Ducheyne, "Curing Pansophia."

18. Vermij, "The Formation of the Newtonian Philosophy"; Mojet, "Early Modern Mathematics."

19. Nieuwentijt, *Het regt gebruik*, 65. Cf. Nieuwentijt, *Religious philosopher*, 1:73. This translation is much shortened; see Vermij, "Translating, Adapting, Mutilating." All translations are mine.

20. Nieuwentijt, *Het regt gebruik*, 522–26, 387.

21. Nieuwentijt, *Het regt gebruik*, 473–77, 346–48, 764–66.

22. Nieuwentijt, *Het regt gebruik*, 855–92. (Cf. Nieuwentijt, *Religious Philosopher*, 549–70.)

23. Nieuwentijt, *Het regt gebruik*, 597–602. (Cf. Spinoza, *Tractatus theologico-politicus*, ch. 2.)

24. Nieuwentijt, *Het regt gebruik*, 602.

25. Nieuwentijt, *Gronden van zekerheid*, 403–4.

26. Nieuwentijt, *Het regt gebruik*, 912.

27. Nieuwentijt, *Het regt gebruik*, 893–916. (Cf. Nieuwentijt, *Religious Philosopher*, 571–85.)

28. Vermij, "Translating, Adapting, Mutilating."

29. Kempe, *Wissenschaft, Theologie, Aufklärung*, 166–68.

30. E.g., Scheuchzer, *Geestelyke natuurkunde*, 1:2, 6–7, 587.

31. Scheuchzer, *Physica oder Natur-Wissenschaft*, 2:41, 96–100, 227–29; Scheuchzer, *Geestelyke natuurkunde*, 1:106.

32. Scheuchzer, *Physica oder Natur-Wissenschaft*, 5.

33. Kempe, *Wissenschaft, Theologie, Aufklärung*, 179–80. For reasons of accessibility, I used the Dutch version of Scheuchzer's work: *Geestelyke natuurkunde*.

34. Scheuchzer, *Geestelyke natuurkunde*, vols. 1–2, engravings 34–65.

35. Scheuchzer, *Geestelyke natuurkunde*, 1:89–90 (quotation on 90).

36. Scheuchzer, *Geestelyke natuurkunde*, 3–4:700–12, engravings 447–53. For Spinoza, see 4:704.

37. Scheuchzer, *Geestelyke natuurkunde*, 1:7.

38. Scheuchzer, *Geestelyke natuurkunde*, 3:583–89 (with engraving 371).

CHAPTER TWELVE

Maxima in minimis animalibus

Insects in Natural Theology and Physico-theology

BRIAN W. OGILVIE

If the study of insects in early modern Europe had a catchphrase, it was *maxima in minimis*: the idea that nature, or nature's God, shows the greatest power in the smallest things. The phrase appears in the Latin translation of Jan Swammerdam's *General History of Insects* (1685), in John Ray's *Wisdom of God Manifested in the Works of the Creation* (1691), and as the motto on the frontispiece of Friedrich Christian Lesser's *Insecto-Theologia* (1738), a book-length treatise on the physico-theology of insects.[1] It has its origins much earlier, in book 11 of Pliny the Elder's *Natural History*. After praising the intricate construction of insects, creatures "of immense subtlety,"[2] Pliny admonished his readers (in Philemon Holland's 1601 translation): "We make a wonder at the monstrous and mightie shoulders of Elephants. . . . Wee marveile at the strong and stiffe necks of Bulls. . . . We keepe a woondring at the ravenings of Tygres, and the shag manes of Lions: and yet in comparison of these Insects, *there is nothing wherein Nature and her whole power is more seene, neither sheweth she her might more than in the least creatures of all.*"[3]

To be sure, Pliny did not use the exact phrase; his Latin read, "cum rerum natura nusquam magis quam in minimis tota sit" (for nature is nowhere more wholly present than in the smallest things). His phrase served to admonish readers not to pass over or despise things that might seem of small account. In that sense, it is an echo of Heraclitus's reminder to his guests that there are gods even in the kitchen, and of Aristotle's remark in *On the Parts of Animals* that even humble things are worthy of a philosopher's attention (where Aristotle recounts Heraclitus's remark in support of his own endeavor).[4] Yet, in its superlative form, the Plinian *nusquam magis tota quam in minimis*, as well as its pithier form, *maxima in minimis*,[5] had implications that go beyond Heraclitus's and Aristotle's apothegms. It implied that insects might be privileged objects of study for understanding nature and, by extension, nature's God.

In the sixteenth and early seventeenth centuries, those natural theologians who appealed to *maxima in minimis* did not necessarily apply the maxim to insects. There was no necessary connection between the detailed investigations of insects that began in the second half of the seventeenth century and the development of physico-theology. Rather, in the specific contexts of late seventeenth-century England and early eighteenth-century Germany, theologian-naturalists developed the connection. As I will argue, this development within physico-theology served to direct naturalists' attention to aspects of insect life that were not directly addressed by descriptive natural history. Hence, physico-theology was not just a byway in the history of natural history; it contributed to the formation of entomology as a discipline toward the end of the eighteenth century.

The Plinian Maxim in the Late Sixteenth Century

The Plinian maxim, though widely expressed in sixteenth- and early seventeenth-century works on insects, rarely appeared in the empirically driven argument for and from design that characterizes physico-theology.[6] Instead, it appears in what I consider to be a devotional form, often emblematic and metaphorical rather than demonstrative: authors (including artists) *assumed* that God's handiwork could easily be seen in insects, which offered object lessons for human beings, both moral and intellectual.

One example is the 1592 collection of engravings published by Jacob Hoefnagel based on the model books of his father, Joris Hoefnagel.[7] These engravings consisted of four parts, each with a title page and twelve additional engraved sheets (totaling fifty-two sheets). Most contained small naturalia, including frogs, mice, and snakes but, above all, flowers and insects. Each contained a motto (or sometimes a riddle), some drawn from scripture, others from other sources or created by one of the Hoefnagels. Two in particular, from the third set, underscore the possibilities, but also the limits, of natural theology.

In the third plate, a quotation from the ninety-second (or ninety-first) Psalm exhorts us to delight in nature, including a dragonfly, a caterpillar, a beetle, and an earwig: "For thou, Lord, hast made me glad through thy work: I will triumph in the works of thy hands." But in the ninth, we are warned against pursuing this delight too sedulously: "Let us not investigate divine things too closely with human reasoning, but led from the works, let us admire the workman." The final plate of the entire collection, meanwhile, draws the reader away from small things to the great things of the world: "O Lord and God, holy one of our ancestors, all created things extol your name: heavens, earth, straits, springs, and rivers; and whatever else this great machine of the world contains."[8]

Not surprisingly, Thomas Moffett cites Pliny in the preface to his *Theater of Insects*, largely completed by 1590 but first published in 1634, in compressed form: "Yet where is Nature more to be seen than in the smallest matters, where she is entirely all?"[9] But Moffett's examples, too, are largely metaphorical. The ant shows prudence, the bee, justice; both reveal temperance. In these cases, natural theology rests on commonplaces from biblical and classical antiquity rather than on careful observation of insect morphology, anatomy, or behavior.

Moffett's book was one of the two most important publications on insects before the 1660s. The other was Ulisse Aldrovandi's 1602 *De animalibus insectis libri septem*; Aldrovandi, however, does not mention the Plinian maxim. Both remained authoritative through the middle of the century. In the later seventeenth century, though, we can see the persistence of an emblematic natural theology side by side with a new physico-theological approach to insects.

Jan Swammerdam's Emblematic Insect

Jan Swammerdam's extensive observations of insect metamorphosis and anatomy are well known.[10] Swammerdam's two major works devoted to insects were published in 1669—his *General History of Insects* in Dutch, soon translated into Latin and French and (posthumously, in 1737–38) his expanded work, including detailed insect anatomies prepared with the aid of microscopes, which Herman Boerhaave published under the title *Bible of Nature*. Swammerdam did not choose the title, but it expresses his perspective; Boerhaave drew it from a letter from Swammerdam to his friend and patron Melchisedec Thévenot in which the Dutch anatomist regretted that their former friend Nicholas Steno had become a religious controversialist, turning away from studying nature's bible.[11]

These works would provide ample material for physico-theologians. And in his insistence that God followed an inexorable natural law, Swammerdam adumbrated a key aspect of physico-theology. In his own natural theology, however, Swammerdam was more traditional. In his 1675 *Ephemeri vita*, a lengthy book comprising a natural history of the mayfly interspersed with extensive religious meditations, Swammerdam offered a thoughtful religious justification for studying nature.

Since, therefore, such creatures are *Hosts of God*, and the wise Solomon has referred the sluggard to the small and lowest ant—yes, God through the Apostle says: *That none of his creatures is invisible to him, that they all appear naked and bare before his all-seeing eyes*—so is it the duty of all men, through all creatures, no matter how small, to climb to God the Creator himself, in order to worship his wisdom, goodness, and all-powerfulness in them. We cannot do that more suitably than to note how it is an example of our miseries, so that through the abasement and humiliation of the heart, we make ourselves worthy to possess the forever enduring goods, and their eternal world.[12]

Contemplating the mayfly's short, miserable life should discourage and diminish us. It ought to remind us of our insignificance and, in so doing, drive out the self like another Ishmael.[13] Swammerdam's natural theology thus remained firmly in the emblematic, exemplary tradition. I emphasize this point in order to underscore that there was no *necessary* connection between the detailed study of insects and a physico-theological outlook. In Edward Tyson's preface to the English translation of Swammerdam's *Ephemeri vita*, on the other hand, we see a marked difference. Tyson acknowledged the importance of natural theology, quoting Romans 1:20 (in Greek) but also, tellingly, a passage from Thomas Browne's *Religio medici*: "The Wisdom of God receives small honour from those vulgar heads, that rudely stare about, and with a gross Rusticity admire His Works; those highly Magnifie him, whose judicious inquiry into his Acts, and deliberate research into his Creatures, return the Duty of a devout and learned Admiration."[14]

Swammerdam's pessimistic meditations fit ill with this orientation, so Tyson omitted them, turning a thick octavo of more than four hundred pages into a slim quarto of sixty. The reason, Tyson explained, was that he aimed to improve natural philosophy by laying bare the creature's interior: "If we would understand how 'tis that Nature gives Life and Motion to these *Automata*, we must unloose the Case, and take asunder the several Wheels and Springs, and carefully observe how she joyns them all together."[15]

The English version did, however, include a brief nod to natural theology, in the form of a long prefatory poem by the physician and versifier Thomas Guidott. "On the History of the *Ephemeron*" reiterates the Plinian trope of *maxima in minimis*: "The Insect-world but lately known, / doth both his Skill and Glory too, declare. . . . / Nay, if we Great with Small compare, / We find these Little-Heraulds too, Proclaim / Jehovah's Mighty Name. . . . / And here in this Ephemeron we see / An Embleme both of Change, and of Mortality."[16]

Thus, in the front matter of the English adaptation of *Ephemeri vita*, we can see two forms of natural theology. Guidott's line "If we Great with Small compare" harks back to the general form of the Plinian maxim, in an emblematic form that echoes Swammerdam, most likely unknowingly: "An Embleme both of Change, and of Mortality." But Tyson's preface is physico-theological in its approach to nature as a *mechanism* whose "several wheels and springs" should be investigated in order to fulfill Browne's injunction to perform "deliberate research" into God's creatures. It is this impulse that would be pursued, either through personal observation or in the pages of books, by later physico-theologians who wrote on insects.

Insects in Physico-theology

An interesting precursor to the physico-theology of insects can be found in the 1613 work *On Divine Providence and the Immortality of the Soul* by the Flemish Jesuit Leonard Lessius. As I have noted elsewhere, Lessius's work shows intriguing parallels with physico-theology: it employs a wide range of design arguments in order to refute the "atheists" and "politicians" who deny divine providence.[17] In discussing how the "structure or making of living Creatures" was done "with reference to an end,"[18] though Lessius mentioned insects only in passing, his phrasing is significant: "Al parts or members in them are wonderfully faire, all most exactly framed, and all most perfectly agreing and fitting to the functions, for which they were made. Among so many kinds of which small living bodies, there is not one so base and vyle, which is not able to procure an astonishing admiration in whom behold them attentively. Yea by how much the creature is more base and abject, by so much the more the art of divyne Providence shineth in the fabricke and making of it."[19] Though Lessius did not use the phrase *maxima in minimis*, he clearly states its foundational principle. I have not yet had time to trace the history of this text's reception among natural theologians and physico-theologians; aside from Michael Buckley's work on the origins of modern atheism, it has received little scholarly attention.[20]

Whether they were aware of Lessius's work or not, the physico-theologians who wrote on insects pursued the path he indicated. In the four editions of his *Wisdom of God Manifested in the Works of the Creation* published during his lifetime, John Ray drew on his ongoing research on insects to expand and revise his physico-theological treatises.[21] William Derham's *Physico-Theology*, too, discussed insects extensively.[22] Indeed, many pages of Derham's discussion contained only a few lines of his Boyle Lectures, supported by extensive notes.

As Ray and Derham are relatively well known figures in the history of physico-theology, though, I will concentrate on two lesser-known figures, Friedrich Christian Lesser and Pierre Lyonet. Lesser's *Insecto-Theologia* was the first physico-theological text devoted entirely to insects. Published in German in 1738, the work was translated into French and published, with notes by Pierre Lyonet, in 1742; the French translation then served as a basis for an Italian translation published in 1751 and an English translation published in 1799, only three years before William Paley's *Natural Theology*.[23]

Born in Nordhausen (Thuringia) in 1692, Lesser studied first medicine, then theology, in Halle, where he drank deeply from the well of August Hermann Francke's Pietism.[24] After two years of study as a scholarship student in Leipzig, he spent six months in 1714 living with his uncle in Berlin, where he frequented the Royal Library. He then returned to Nordhausen, where he spent the rest of his life as pastor and administrator. He was a member of the Academia Naturae Curiosorum and was elected to the Prussian Academy of Sciences.

Lesser was not as committed and meticulous a student of insects as Swammerdam or Ray. Nonetheless, he was seriously devoted to learning about them and observing them firsthand. The introduction to his *Insecto-Theologia* (1738) listed what he considered to be the important literature published on insects from the first book on them, Ulisse Aldrovandi's *De animalibus insectis* of 1602, through the first installments of René-Antoine Ferchault de Réaumur's *Mémoires pour servir à l'histoire des insectes*, beginning in 1734, and the hot-off-the-presses *Biblia naturae* of Johannes Swammerdam, written in the 1670s but published only in 1737 and 1738. The fact that a pastor in Nordhausen knew of a book published in Leiden the year before his own testifies to his interest in the subject and the efficacy of his information network.

In Lesser's treatment of insects, we find what might be seen as a fulfillment of the promise made in the Jesuit Lessius's work more than a century earlier. By gathering under general headings many details of insects' diversity, numbers, classification, respiration, generation, metamorphosis, abode, behavior, food, defenses, care for their young, sagacity, and anatomy, Lesser offered a detailed catalog of the ways in which insects were ideally suited to the lives that they lived.

In his text, Lesser did not claim a privileged status for insects. He criticized Swammerdam's title *Bible of Nature* precisely because such a title should not be limited to insects but should properly apply to all of God's creatures; insects are but their last chapter.[25] And he himself wrote a *Lithotheologie* and a *Testaceo-Theologia* as well as the *Insecto-Theologia*. Nonetheless, his frontispiece cited the Plinian maxim. And he noted that God's creation was too manifold for any one finite human being to comprehend; for that reason, he had concentrated on insects. He noted as well that the recent invention of the microscope "enables us to penetrate into a sort of invisible region, and displays to our eyes a new world, composed of an infinite number of living beings."[26] In the case of insects, the microscope revealed the detailed structure of creatures like the cheese mite, at the very margins of human perception, further underscoring God's power and wisdom.[27]

To better understand the importance of detailed empirical investigation in the physico-theological treatment of insects, we can consider Lesser's text alongside Pierre Lyonet's notes and corrections to it. Lyonet, born in Maastricht in 1706 to a Swiss army chaplain of Huguenot ancestry, studied theology and law and then became a translator, foreign secretary, and master of cyphers for the States General of the Netherlands.[28] Lyonet took an early interest in insects and was gathering material for a history of them when the publisher of the French translation of Lesser's work asked him to review the translation. Lyonet did more: he corrected the translation but also, out of respect for both the truth and Lesser's original text, annotated the text with many corrections and additions.

These annotations, and the texts to which they refer, reveal the importance of detailed empirical study in the Pietist Lesser's and the Reformed Lyonet's physico-theologies. Both were concerned with regularity in nature as a sign of divine workmanship; both were opposed to granting any role to chance in natural processes. It is where Lyonet disagreed with Lesser, though, that we can see how the physico-theological treatment of insects, by gathering together examples from a broad range of species, could contribute to an understanding of the creatures that went beyond descriptive natural histories focused on particular species. Their discussions of instinct provide an instructive example. The debate over the sources of animal behavior in early modern Europe centered on whether animals exercised some form of judgment that was reasonable, or at least analogous to human reason, or whether they were driven by instinct as if they were mere machines.[29]

Lesser was firmly in the camp of instinct. He argued, "It is allowed that insects are destitute of reason; the wisdom therefore of their conduct, the justness of their precautions, and in a word every thing they do which is agreeable to reason, does not proceed from themselves. From whom then do they derive it? Who hath taught them the season and manner of propagating their species? Who hath directed them to lye with such compactness in their eggs without being the least uneasy? How do they know the precise moment when it is proper to issue from their eggs?"[30] Feeding habits, too, "demonstrate the great and incomprehensible wisdom of the Creator. It is certain that insects are devoid of reason; yet their whole economy seems to be the result of sound judgment."[31] After reviewing many, many examples of insects' apparent capacity to judge, Lesser concludes that "there is something in them all which, if it does not surpass the cunning and subtilty of the mind of man, at least very nearly approaches them. It cannot be the effect of chance, for there is an evident display of design, and a constant regularity, which demonstrates that an all powerful and all wise being directs them."[32]

Here Lyonet parted company with Lesser. Where Lesser claimed that insects act without reason, Lyonet responded,

> When we take a general view of the operations of insects, the great uniformity, which at once appears in the economy of each species, would make us believe, that they act merely by instinct. But, when we examine their proceedings in detail, and when we see, that they not only vary their operations, according to the necessity of the case; but that, when they are placed in difficult circumstances, in which, according to the ordinary course of things, they should not naturally find themselves, we observe, they do not fail to make the most of their resources, and that they can, with much industry, remedy accidents, and extricate themselves from very embarrassing situations, we are then tempted to allow them a portion of reason.[33]

Moreover, careful observation shows that insects of the same species can have diverse characters: some bold, others timid; some temperate, others voracious.[34]

Nonetheless, in granting reason to insects, Lyonet did not deny God's hand in creating them.

> Whether we suppose, that insects act from reason, or that they are constrained to act as they do, by a blind instinct, the glory of God is not the less conspicuous in either case. In the first, we cannot but admire the wisdom of the Creator, who has made machines which, without reason, can act as consequentially as if they were endowed with it; in the other, we must admire the same Wisdom, that could create so many different sorts of beings, of more limited knowledge than we are, but, nevertheless, sufficiently intelligent to provide for their own preservation, and that of their race. In the first case, God has exalted

the organic mechanism, to a degree of perfection, which matter alone does not seem capable of attaining; in the second, he has raised the brutes, to a pitch of perfection, superior to what we can conceive of organic mechanism.[35]

Either way—whether insects behave rationally or instinctually—God's hand is visible; the lowest creature reveals, in splendid fashion, his wisdom and goodness.

Conclusion: Physico-theology and the Origins of Entomology

In their attention to the number of insects, to their variety, and to their behavior—instinctual or not—and in their concern for the accuracy of their claims, physico-theologians did not merely repackage scientific knowledge in an apologetic form; they drew naturalists' attention to aspects of nature that the descriptive tradition in natural history tended to neglect. For example, Aldrovandi and Moffett provided detailed descriptions of individual species of dragonflies. So too did later writers on insects: Francis Willughby, John Ray, Carl Linnaeus, and Charles de Geer. But Lesser considered them under more general headings. Their sexual dimorphism, he explained, is part of a general pattern among insects in which the males are generally more beautiful than the females. The speed and accuracy with which they hunt their prey is part of a broader discussion of how insects' forms and behavior are ideally suited to their feeding habits.[36]

Physico-theological reflection was, of course, not the only source of such attention to aspects of behavior and comportment. The *Mémoires pour servir à l'histoire des insectes* of the French academician Réaumur also contains such generalizations, but Réaumur did not arrive at them by way of physico-theological considerations.[37] Still, where physico-theology was practiced, it inspired general considerations about insect behavior and the notion of an "economy of nature" that was emerging in this period. In that regard, I suggest, physico-theology was not merely dependent on natural history; in its pursuit of detailed evidence of providence and design, it made significant contributions to new ways of thinking about nature.

And even after the heyday of physico-theology, its echoes can be seen in the most important textbook of entomology in nineteenth-century Britain: William Kirby and William Spence's *Introduction to entomology*, first published in four volumes from 1815 to 1826 and reprinted many times throughout the century. In the introduction to the first volume, the authors note that they chose an epistolary form because they wanted to start with the "manners and economy" of insects, not with their morphology and classification. And this was, in good part, because one of the authors'

> first and favourite objects has been to direct the attention of their readers "from nature up to nature's God." For, when they reflected upon the fatal use which has too often been made of Natural History, and that from the very works and wonders of God, some philosophists, by an unaccountable perversion of intellect, have attempted to derive arguments either against his being and providence, or against the Religion revealed in the Holy Scriptures, they conceived they might render some service to the most important interests of mankind, by showing how every department of the science they recommend illustrates the great truths of Religion, and proves that the doctrines of the Word of God, instead of being contradicted, are triumphantly confirmed by his Works.[38]

As J. F. M. Clark points out, these words represent the views of only one of the authors, the High Church Tory clergyman Kirby, not the political economist Spence.[39] Nonetheless, they remained in the seventh edition of 1858, well after Kirby's death in 1850. They were repeated, too, in the 1828 German translation by Lorenz Oken.[40]

Ironically, the very success of physico-theologians in gathering detailed evidence of the suitability of insects—and other creatures—to their environments and ways of life would serve to undercut the physico-theological approach when that evidence was, instead, employed in support of evolutionary theories of life. Yet, from the Victorian enthusiasm for insect collecting to the butterfly houses, insect zoos, and insectariums of the twenty-first century, we can see the persistence of the ancient idea that nature, if not nature's God, is best revealed in its perceptually marginal forms.

NOTES

1. Swammerdam, *Historia insectorum generalis* (1685), 51; Ray, *Wisdom of God* (1691), 130, referring to God as *maximus in minimis*; Lesser, *Insecto-Theologia* (1738), frontispiece.

2. "Inmensae subtilitatis animalia." Philemon Holland's 1601 translation renders this as "those living creatures, which are the most subtill of all others that Nature hath brought forth": Pliny the Elder, *Historie of the world*, 310. John Bostock's 1855 version, on the other hand, reads *subtilitas* as difficulty of investigating: "insects, a subject replete with endless difficulties": Pliny the Elder, *Natural history*, 3:1. Holland's reading seems more likely to me, though both translators appear to have missed Pliny's use of oxymoron.

3. Pliny the Elder, *Historie of the world*, 311; Pliny *Naturalis Historia* 11.1, Perseus Digital Library: "sed turrigeros elephantorum miramur umeros taurorumque colla et truces in sublime iactus, tigrium rapinas, leonum iubas, *cum rerum natura nusquam magis quam in minimis tota sit*" (my emphasis).

4. Aristotle, *Parts of Animals* 1.5, 645a15. In *Against Celsus*, the third-century CE theologian Origen offered an alternative to the trope of *maxima in minimis*: "Ants gnaw seeds when storing them so that they do not germinate but may be kept for the entire year. This is not due to their own reasoning, but to the bountiful mother Nature who thus adorns even brutes, so that even to the smallest things she adds some of her genius [*ingenia*]." Quoted in Latin by William Derham, *Physico-Theology* (1714), 381–82n (my translation).

5. The maxim *Deus magnus in magnis, maximus in minimis* is often attributed to Augustine, but I have not been able to find the passage in the Augustinian corpus.

6. On forms of early modern natural theology, see the helpful synopsis by Katherine Calloway, *Natural Theology*, 8ff.; and Re Manning, *Oxford Handbook of Natural Theology*.

7. Hoefnagel and Hoefnagel, *Archetypa*, facsimile, translation, and commentary in Vignau-Wilberg, *Archetypa studiaque patris Georgii Hoefnagelii*.

8. Hoefnagel and Hoefnagel, *Archetypa*, pars 4, tabula 12: "O Domine atque Deus, nostrorum sancte parentum, / Collaudant nomen cuncta creata tuum. / Caelum, terra, fretum, fontes, et flumina: et in se / Haec ingens mundi machina quidquid habet."

9. Moffett, *Insectorum sive minimorum animalium theatrum*, praefatio, sig. a1v: "Tamen ubi Natura magis, quam in minimis tota sit?" My translation is from Topsell, *Four-footed beasts and serpents*, sig. [Ffff5]v.

10. See, e.g., Cobb, "Malpighi, Swammerdam and the Colourful Silkworm"; Lindeboom, *Het cabinet van Jan Swammerdam*; Ruestow, *The Microscope in the Dutch Republic*; Schierbeek, *Jan Swammerdam*.

11. Swammerdam, *Letters to Thévenot*, 81–82.

12. Swammerdam, *Ephemeri vita: of afbeeldingh van 's menschen leven*, sig. *3r–v: "Want alsoo dit soort van dieren *Heyrlegeren Gods* sijn, ende dat den wijsen Salomo de luyaarts tot de kleene ende de neerstige Mier heeft geweesen, jaa dat Godt door den Apostel segt: *Dat hem geen van sijne schepselen onsichtbaar sijn, die haar alle naackt ende bloot vor sijn alsiende oogen vertoonen.* Soo is het de schuldige plicht van alle menschen, om door alle schepselen, hoe geringh datse sijn, tot Godt den Maacker der selve op te klimmen, om sijne Wijsheydt, Goetheydt, ende Almachtigheydt daar in aan te bidden. Datwe niet bequamelijcker kunnen doen, als de selve tot voorbeelden onser ellenden aan te mercken, om door die verneederinge ende verootmoediginge des herten, ons tot de besittingh van de altijdt duurende goederen, ende haare eeuwige weelden, waardigh te maaken" (italics in original).

13. Swammerdam, *Ephemeri vita: of afbeeldingh van 's menschen leven*, sig. *4r.

14. Swammerdam, *Ephemeri vita: Or the Natural history and anatomy of the ephemeron*, sig. A2r–v.

15. Swammerdam, *Ephemeri vita: Or the Natural history and anatomy of the ephemeron*, sig. A2v.

16. Swammerdam, *Ephemeri vita: Or the Natural history and anatomy of the ephemeron*, sig. [A4]r–v.

17. Ogilvie, "Stoics, Neoplatonists."

18. Lessius, *Rawleigh his ghost*, 86ff.; cf. Lessius, *De providentia numinis*, 65ff.: "Sexta Ratio, ex structura animalium & plantarum in ordine ad finem."

19. Lessius, *Rawleigh his ghost*, 109; Lessius, *De providentia numinis*, 82: "Omitto insectorum & vermiculorum diversissimas formas. . . . Omnia ad miraculum pulchra, omnia suis numeris absoluta, & ad functiones sibi congruentes instructissima. Nihil in his adeo vile, quod non summam admirationem moveat attentius consideranti. Quinimo quo quid est minus & abjectius, eo magis in eius structura divinae providentiae elucet artificium."

20. Buckley, *Modern Atheism*; Buckley, *Denying and Disclosing God*. Cf. Ogilvie, "Stoics, Neoplatonists," 775–76.

21. Ogilvie, "Insects in Ray's Natural History and Natural Theology"; Zeitz, "Natural Theology."

22. Derham, *Physico-theology* (2nd ed.), esp. 368–403.

23. Lesser, *Insecto-Theologia* (1738); Lesser, *Théologie des insectes*; Lesser, *Insecto-Theology* (1799).

24. On Lesser, see Trepp, *Von der Glückseligkeit*, chap. 6, 373ff.

25. Lesser, *Insecto-Theologia* (1738), 28n.

26. Lesser, *Insecto-Theology*, 14; Lesser, *Insecto-Theologia* (1738), 19.

27. Lesser, *Insecto-Theology*, 174; Lesser, *Insecto-Theologia* (1738), 308. Space precludes a detailed consideration of microscopy in physico-theology; see Cobb, "Malpighi, Swammerdam and the Colourful Silkworm"; Fournier, "Book of Nature"; Ruestow, *The Microscope in the Dutch Republic*; Terrall, *Catching Nature in the Act*; and Wilson, *Invisible World*, for the broader context of microscopy.

28. Unless otherwise noted, biographical details are from Van Seters, *Pierre Lyonet*.

29. See Richards, *Darwin and Evolutionary Theories of Mind and Behavior*, 22–27, and Richards, "Sensationalist Tradition."

30. Lesser, *Insecto-Theology*, 62; Lesser, *Insecto-Theologia* (1738), 85.

31. Lesser, *Insecto-Theology*, 103; Lesser, *Insecto-Theologia* (1738), 170. The second clause ("yet their whole economy . . .") was added by the French translator: Lesser, *Théologie des insectes*, 1:277.

32. Lesser, *Insecto-Theology*, 133; Lesser, *Insecto-Theologia* (1738), 226–27. The French translator added the clause about chance: Lesser, *Théologie des insectes*, 1:344.

33. Lesser, *Insecto-Theology*, 334; Lesser, *Théologie des insectes*, 1:148n.

34. Lesser, *Insecto-Theology*, 299; Lesser, *Théologie des insectes*, 1:69n.

35. Lesser, *Insecto-Theology*, 374–75; Lesser, *Théologie des insectes*, 1:294n.

36. Lesser, *Insecto-Theologia* (1738), 105, 211n.

37. On Réaumur, see Terrall, *Catching Nature in the Act*.

38. Kirby and Spence, *Introduction to Entomology*, 1:xi.

39. Clark, "History from the Ground Up."

40. Kirby and Spence, *Einleitung in die Entomologie*, 1:ix.

CHAPTER THIRTEEN

What Abbé Pluche Owed to Early Modern Physico-theologians

NICOLAS BRUCKER

Le Spectacle de la nature (1732–50), one of the largest publishing successes of the century in France—and in Europe thanks to translations in English, German, Dutch, Italian, and Spanish—is a popular account of recent physical research in the frame of a handbook whose educative principles are based on religion. The abbé Noël-Antoine Pluche (1688–1761), who followed a career as an educator, successively in colleges (Reims, Laon) and in the homes of wealthy noble families (Rouen, Paris), relied on a devotional project for intellectual growth. The young *chevalier* of the dialogue acquires knowledge while preparing his mind to receive the holy mysteries. The former must help him to reach the second. Thanks to his will to conciliate science and religion, Pluche, in spite of his Jansenist beliefs, endeavored to achieve a totalizing work, so that it combines the useful, the beautiful, and the true. Disliking systematic thought and promoting the school of concrete realities through a sensitive approach, he intended to reform sinful humanity. Aside from this philosophical background, the encyclopedic ambition of the book satisfied the expectations of the French public: in a single work *Le Spectacle* offers a synthesis of a wide range of sciences.

I start with a quotation from Voltaire's correspondence, a sarcastic greeting to Abbé Pluche: "My regards to Mr Pluche, who so intrepidly goes on copying books in order to display *Le Spectacle de la nature*, and who has become the ignorants' quack."[1] This mocking statement reflects a view of Pluche's work as simple compilation, his project to make us read the book of nature as a lure, and his use of numerous sources as plagiarism. This is precisely what I endeavor to show in this chapter, not to diminish Pluche's work, but rather to measure both how much he borrowed from former physico-theologians and, seen from a different perspective, to what extent he contributed to the renewal of physico-theology in the middle of the eighteenth century, not only in France but also, through the numerous translations, in many parts of Europe as well.

An Encyclopedic Aim

We do not know much about Pluche: the only biographical material comes from a eulogy by Pluche's publisher, Robert Estienne.[2] Estienne reports that Pluche, after having been principal of two colleges (Reims and Laon), served in Rouen as a tutor of Lord William Stafford Howard's children. Did he start learning English at that time? This is pure speculation. In any case we are sure he knew English, at least enough to read the work of English scientists. The proof lies in the 574 titles listed in Pluche's library catalog.[3] Among them we find seven volumes of the *Philosophical Transactions* (London, 1716–34), eight volumes of the *Spectator* (London, 1733), and several books in English, such as *An Essay toward a natural History of the Bible*, by John Hutchinson (London, 1725) and *Physico-Theology: Or, A Demonstration of the Being and Attributes of God, from His Works of Creation*, by William Derham (London, 1713). The latter title occurs twice in the catalog: a Dutch edition is also mentioned (translated by J. Lufneu, Rotterdam, 1730).

The catalog helps us to know more about the reading that may have inspired the general aim of Pluche's *Spectacle de la nature*. Aside from Derham we find in the list of Pluche's books: *The Wisdom of God Manifested in the Works of Creation* by John Ray, in a French translation (Utrecht, 1723); various works of Herman Boerhaave, Isaac Newton, Robert Boyle, Antonie van Leeuwenhoek, Antonio Vallisneri, Réne-Antoine Ferchault de Réaumur, and the Comte de Buffon; Thomas Burnet's *Telluris theoria sacra*; a French translation of a work by Bernard Nieuwentijt (*L'Existence de Dieu démontrée par les merveilles de la nature*, Paris, 1725); and seventy-two volumes of the *Histoire et Mémoires de l'Académie royale des sciences* (1666–1750). Dennis Trinkle detailed the scientific information used by Pluche in the chapters of *Le Spectacle*,[4] though he did not specifically focus on the physico-theological method. He based his analysis on the numerous marginal notes that are spread throughout the first three volumes. These notes include references to works on the topics that Pluche treats, sending the reader to more precise and detailed information. In addition, they lend authority to Pluche's work, and the references can also serve as a quite complete bibliography of recent works in natural history.

I count fifty-six different references in the three first volumes, most of them including the author's name and the title of the work, and most of them corresponding to the titles mentioned in Pluche's library catalog (see the appendix at the end of this chapter). The list reflects the culture of a French scientist around 1730, which extends to works from many European countries and implies reading knowledge of several languages (at least Latin, Italian, English, and Dutch). However, considering the frequency of occurrences, the *Mémoires de l'Académie des sciences* is cited most often (twenty-eight times), far ahead of the *Philosophical Transactions* (cited fourteen times). The *Mémoires* is especially quoted for the articles of Réaumur. Leeuwenhoek is the scientific author quoted most often (sixteen times), followed by Derham (nine).

These notes do not tell us much about exactly what Pluche inherited from physico-theologians, because in referring to them he generally ignores the religious dimension of their work and focuses on the scientific content. However, they confirm Voltaire's statement that Pluche has indeed copied from a great number of books.

Now the question is: Did these borrowings affect the general structure of the work? Are the design and the composition of *Le Spectacle* taken from physico-theologians? To answer these questions, I have compared Pluche's plan with those of three famous physico-theological works: Ray, *The Wisdom of God Manifested in the Works of Creation* (1691); Derham, *Physico-Theology: Or, A Demonstration of the Being and Attributes of God from His Works of Creation* (London, 1713); and Nieuwentyt, *Het regt gebruik der werelt beschouwingen, ter overtuiginge van ongodisten en ongelovigen* (The right use of contemplating the world, for the conviction of atheists and infidels, Amsterdam, 1715).

Unlike these authors, Pluche does not limit himself to the natural realm. He embraces human moral and social life, arts and techniques, the usual sciences, and the everyday skills by which humans have mastered the Creation. Physico-theology is obviously not the only target of his project. His design is wider. Pluche aims at fashioning some sort of total book, which would condense in a definite form the infinite variety of divine creation. However, totality remains a pious wish: the author concentrates in fact on few subjects. Surprisingly, he ends with a historical apologetics. In the classical meaning of physico-theology—praising the Lord's glory through a close examination of the wonders of nature—only the first three volumes of his eight-volume work can be considered fully relevant.

Pluche's Plan

When we consider these three volumes, we notice a general movement from the smallest to the biggest, from insects to mountains. The guiding thread is eyesight, according to the title *Le Spectacle de la nature*. The human ocular faculty turns toward realities it usually ignores. Head down, exploring the secret life of the creatures that inhabit grass, then head up, considering the tallest trees, the pupils in the dialogue learn how to see, extending both their knowledge and their spiritual capacity. The reference to optical devices, such as microscopes or telescopes, stands for a metaphor of a general attitude toward nature and a new intelligence of God's will.

Compared with the three famous canonical physico-theological works just listed, the distribution of the contents in *Le Spectacle* appears quite original. However, Ray, Derham, and Nieuwentijt each adopted a completely different plan: the former starts with celestial bodies, then goes on with elements, plants, and animals, and ends with the human body; the second, after having considered the globe, studies humans and, only afterward, animals and plants; the third goes directly to the study of humankind, then deals with elements and celestial bodies. Obviously, the genre of the physico-theological essay does not include any standardized plan. Despite these differences, the categories remain the same: sensible and insensible creatures, the four elements, terraqueous or celestial bodies, and so on. Pluche, though keeping this frame as it is, insists on providing an epistemological background to his treatment of similar topics in the first three volumes; in the fourth volume, he recalls the history of both experimental and speculative physics and thus refers, in a reflexive manner, to the conditions that made recent discoveries possible.

Pluche's choice seems to result from a mix of all three plans followed by others. *Le Spectacle* begins with animal and vegetal life, then examines the earth and the sky, and finally deals with humans, considered as individuals but also as members of society and as participants in the different tasks and activities in which they engage.

A Fly, a Flea, or a Mite

The main difference between Pluche and the canonical physico-theologians concerns his decision to start with insects. Why does Pluche so closely examine the anatomy and the ethology of the smallest creatures? Even if he knows the most important essays of the seventeenth century, which founded early modern insectology (Francesco Redi, *Esperienze intorno alla generazione* degl'insetti, Florence, 1668; Jan Swammerdam, *Historia insectorum generalis*, Utrecht, 1669; Leeuwenhoek, *Arcana naturae detecta*, Delft, 1695), he feels himself closer to his contemporary and compatriot Réaumur. He personally owned Réaumur's *Mémoires pour servir à l'histoire des insectes* (Paris, 1734–42). But as the first volume of *Le Spectacle*, which deals with insects, was published in 1732, it cannot directly refer to the *Mémoires*. Pluche made use instead of the various memoirs published by Réaumur since 1728 in the *Mémoires de l'Académie des sciences*. Aside from works that deal with pure physics, Réaumur shows an interest in insects, at first from a rather practical perspective, considering them as noxious creatures and trying to find means of protection from them. In the introduction to his memoir on moths, the scientist gives an apology for this area of natural history: "Even if researches in natural history only aim at displaying the prodigious variety of beings in the universe and only help us to conceive a greater idea of the author of so many marvelous works, they should not be qualified as frivolous."[5]

This defense of the study of living creatures against the suspicion of frivolity proves that this branch of knowledge, compared to mathematics, physics, or astronomy, still lacks authority. Therefore one must consider *Le Spectacle de la nature* as an effort to promote insectology to the level of the major sciences.[6] When in 1734 Réaumur writes: "Some authors seem to wish the number of observations on insects increased, as now increases the number of demonstrations of God's existence,"[7] he may be thinking of Pluche, but in any case he likely has in mind Ray, Derham, and Nieuwentyt. The former, echoing Pliny, states: "There is a greater Depth of Art and Skill in the Structure of the meanest Insect, than thou art able for to fathom, or comprehend."[8] Invoking a famous maxim, Ray calls God *"Maximus in minimis"*[9] and defends natural philosophical activity through a comparison between a watchmaker and a naturalist. Why would the former, when framing a small watch, win the esteem of the public, whereas the second would be blamed for observing the shape, behavior, and life of a bee?

Derham calls insects "the little contemned branch of the animal world" and, further, "a despicable part of the Creation."[10] Nieuwentyt claims for the study of insects: "The glory of their Creator . . . does not appear less in the structure of a fly, a flea, or a mite, than in the making of the biggest elephant."[11] When he decides to begin his essay with insects, he adopts a definite position in a matter under debate at the time. His argument for doing so has a scientific basis: he intends to help the promotion of insectology. But it also includes a theological background: starting with the tiniest—and the most despised— part of the creation, he shows his ambition of embracing the totality of the universe. The physico-theological project cannot exist but in an exhaustive manner.

The Use and Design of *Le Spectacle*

Recalling what we have already found out, we may consider Pluche a follower of the famous English and Dutch promoters of early modern physico-theology. Nonetheless, Pluche's project presents many original features, notably in its literary form and in the mundane and pedagogical aim of the book.[12] In a quotation extracted from the very end of the third volume, we see that Pluche clearly wants to distinguish himself from his forerunners. In this long letter, the Prior of the dialogue—whom we may consider a mouthpiece for Pluche himself—explains the use and design *of nature*, as he says, and in so doing he also tells us the purpose *of the book* we have in our hand. The metaphor of the looking glass is extremely intriguing. "What then, shall we say, is the Use and Design of Nature? Shall we compare it to a Looking Glass, which is made to represent something more than the Glass itself;[13] or to an Ænigma, which, under remote Similitudes and Terms, conceals some Meaning, which we are glad to find out?"[14] The reflexive function of the metaphor points out the aim of the *Spectacle*: an epistemology of knowledge.[15] The target is no longer nature itself or the unveiling of God's will through its examination. This glass is not a glass one sees through but a mirror that, while reflecting nature, gives of the latter a brand-new conception and, reflecting the spectator at the same time, sends him back to his own interiority. That is why in the title *Spectacle de la nature*, the word "spectacle" may be considered of more importance than the word "nature." Or, to say it differently, the nature that is given to be known in Pluche's work is a modified nature, a nature that has become the thing (*la chose*) of scientists, an object to which humans devote study and experimentation (from a scientific point of view), but also a nature that has been redeemed (from the theological point of view).

To clear up his position, Pluche attacks, in the second part of the paragraph quoted, the traditional and clerical physico-theology. "The many large Volumes which have been written to prove the Existence of God, of which every reasonable Man is as thoroughly convinced as of his own; the many Sermons and Theological Lectures, which are founded in some Countries to establish this Truth, which common Sense will teach every Man; are so many Discourses, in some sort affronting to the Understanding of their Auditors and Readers, at best unprofitable and needless."[16] Under "Sermons and Theological Lectures" Pluche explicitly refers to the Boyle Lectures, established in 1691 in the will of Robert Boyle for the purpose of refuting atheism. Clearly Pluche does not want to be identified with physico-theologians, for at least two reasons that we can guess. From a literary angle, he has chosen a literary genre to reach a broad and "worldly" audience—the "gens du monde" who frequented the salons and other fashionable cultural venues. *Le Spectacle* is written as a set of dialogues, following on the great success of Fontenelle and his *Entretiens sur la pluralité des mondes*; it also includes letters at the end of almost each volume. From a philosophical angle, following Pascal (but not the Pascalins, as Martine Pécharman explains in this volume), he rejects the idea that one could prove the existence of God on the ground of natural display. In a nature marked by original sin, God cannot but be a hidden God.[17] Ann Blair formulates the contradiction in those terms: "A Jansenist natural theology seems an oxymoron."[18]

Pluche stands against a naive conception of finalism. The question is not to prove the existence of the divine watchmaker. As anyone would admit, this is no valuable purpose. The question is rather how to know more about Creation, and therefore how to better revere the Creator.[19] Pluche's conception of apologetics is nearer to that of Jacques Abbadie,[20] from whom he borrowed the attempt to prove Christianity on historical grounds. After the so-called "vulgar Theology,"[21] developed in the first seven volumes, Pluche devoted the eighth and last one to a learned theology, authorized by scriptures.

Science and Religion

Against the repetitive manner of Derham, who concludes at least five hundred times that creatures are the work of an intelligent being and a skillful artist, Pluche considers only the actual perfection of the natural realm before stating anything. What he targets, following Leibnizian philosophy, is a consideration of the whole. The theology of *Le Spectacle de la nature* considers God as a global creator: his design cannot be but general. One shall not clear up the hidden parts of the machine but rather understand the design of the Almighty. The question turns not on the *how*, but on the *why*. "The farther you advance in the Knowledge of your Riches, the more you are convinced that all Nature is one Whole; the several Parts of which are mutually aiding and assisting to each other, having been connected together by infinite Wisdom for our Use and Benefit."[22] In this way, Pluche shares the point of view of Ray, when he rejects the idea that everything in the world was created for the sake of humankind: "For it seems to me highly absurd and unreasonable, to think that Bodies of such vast Magnitude as the fix'd Stars, were only made to twinkle to us."[23]

Contrary to Derham, who considers that the display of nature necessarily leads the atheist to God, Pluche thinks that one must be convinced by other means. He is nearer to Ray's view, as expressed in the following lines: "For as all other Sciences, so Divinity, proves not, but supposes its Subjects, taking it for granted, that by natural Light Men are sufficiently convinc'd of the Being of a Deity . . . but these Proofs, taken from Effects and Operations, expos'd to every Man's View, not to be deny'd or question'd by any, are most effectual to convince all that deny or doubt of it; neither are they only convictive of the greatest and subtilest Adversaries, but intelligible also to the meanest Capacities."[24]

Pluche also shares with Ray the idea that one cannot judge the use and design of each component of the creation. "Our Understanding [is] too dark and infirm to discover and comprehend all the Ends and Uses to which the infinitely wise Creator did design them."[25] On many occasions, Pluche shows his reluctance for the new and the extraordinary. "We keep our Admiration for what is new or uncommon, and it is not so much the Marvellous as the Novelty in any Thing, that pleases and awakens the Attention most."[26] The Prior expresses his contempt for "profound Naturalists"[27] who claim to explain physical laws of the universe, such as the tide. Hypotheses on the tide divided natural philosophers at the time in two camps: Cartesians versus Newtonians. "How vain then must be the Attempt to measure the Sea, when we know neither the Extent nor Depth of it!"[28] We find the same sort of statement about those who debated the Flood. Those polemics expose the diversity of human opinions, diversity that is classically considered a proof of the universal error to which humans are condemned. It is no surprise that Pluche refuses to participate in any such debate: "The Uncertainty of these Inquiries is sufficient to discourage us from entering into them."[29]

Pluche's garden is a close, familiar, and secure world. It stands for a continuation of the domestic sphere. As one leaves the mansion for the terrace and then the garden, one progressively enters the fields of knowledge, step by step, through an experimental process that leads from more familiar objects to more abstract and difficult topics. The realm of possible knowledge has no apparent limit. Therefore, one might get lost, and the author regularly warns the reader not to go too far: he or she needs to be guided. Speculative knowledge is condemned, because it is considered potentially contradictory of God's views. Pluche's discourse combines a faith in the power of reason and a suspicion toward any speculative drift. The criterion for a right use of reason is usefulness, which is one of the mottoes of the European Enlightenment.

In opposition to speculative tendencies, Pluche's garden offers a down-to-earth wisdom, according to which each small effort at knowing must be rewarded. This principle paradoxically contributes to strengthening trust in a true and innocuous knowledge. Pluche's *Spectacle* aims at restoring the innocence of knowledge, to fulfill human participation in God's design. The learned person is like Adam before the Fall. Science may lead on the way to God. Pluche's physico-theology is therefore neither a long naive wondering nor a mere thanksgiving; it raises the reader to the principle of being human and opens a chance of redemption.

Pluche endeavors to check and confirm the compatibility between physical knowledge and divine will. He offers an authorized and Christianized natural history. He makes knowledge enter his garden and treats it as something to transmit in a pedagogical frame. *Le Spectacle de la nature* is in fact a *Spectacle* of knowledge, a general overview of available knowledge.

Conclusion: Toward Consensus

Voltaire's statement about Abbé Pluche copying books is definitely right, but in a way he probably did not imagine. In *Le Spectacle de la nature*, nature is all but natural. It is made out of hundreds of books. Pluche himself calls nature "the best and choicest Library."[30] Marginal notes are not designed to lead the reader to the sources. They are there to show the miraculous convergence of knowledge among learned men all over Europe, provided that they strictly follow the experimental method. By that means they avoid controversy: "For there [in nature] we neither find Errors nor different Opinions, nor Controversy, nor Prejudice, nor Contentions."[31] This consensual conception of knowledge makes of the garden a place of peace and restfulness.

Though Pluche took all his matter from authoritative natural philosophical sources, he imprinted on his theology a more personal stamp: some sort of gentle Jansenism, favorable to the "mondains" or worldly elites of his time. Contradictions lie at the heart of this natural theology: visibility (of phenomena) and invisibility (of God); optimism and original sin; confidence in and suspicion toward understanding.

Pluche is more complex than the mere popularizer that historiography has generally considered him to be. He played the role of a genial mediator, introducing to a large audience, in France and in Europe, the latest state of knowledge of the physical world. On the theological side, he clearly reinterpreted the work of his forerunners, trying to innovate in both form and content. That is why, inverting our primary question, we can now ask: What does physico-theology owe to Abbé Pluche?

APPENDIX: Marginal notes in Pluche, *Le Spectacle de la nature* (I–III)

Reference (author and title)	Frequency	Reference (author and title)	Frequency
Actes de Leipzig	1	Mariotte (*Du mouvement des eaux*)	1
Aldrovandi (*De Animalibus Insectis*)	1	Marsilli (*Histoire de la mer*)	2
Bergier (*Histoire des grands chemins de l'Empire*)	1	*Mémoires de l'Académie des sciences* (of which 8 are to Réaumur)	28
Bibliothèque universelle et historique	1	Montfaucon (*Monuments de la monarchie française*)	1
Bochart (*Hierozoicon*)	3		
Boissart (*Traité des monnaies*)	1	Mortimer (*The Whole Art of Husbandry*)	5
Borelli (*De Motu Animalium*)	2		
Bourguet	1	Nieuwentyt (*Existence de Dieu*)	3
Boyle (*De Gemmarum Origine*)	1		
		Philosophical Transactions of the Royal Society of London	14
Boyle (*De Mira Subtilitate . . . Effluviorum*)	1		
Burnet (*Théorie de la terre*)	1	Pomet (*Histoire générale des drogues*)	2
Careri	1	Puget (*Observations sur la structure des yeux de divers insectes*)	1
Derham (*Théologie physique*)	9		
Geoffroy le Jeune	1		
Goedart (*Metamorphosis*)	5	Rapini (*Hortorum Libri*)	1
Guillelmini	1	Ray (*Historia Plantarum*)	2
Hales (*Vegetable Staticks*)	4	Ray (*Synopsis Methodica Avium et Piscium*)	2
Hartsoecker (*Essai de dioptrique*)	2		
Hooke (*Micrographia*)	2	Robault (*Traité de Physique*)	3
Jonston (*Thaumaturgia Naturalis*)	1	Ruuscher (*Natuerlyke historie van de couchenille*)	1
Journal des savants	2	Ruysch (*Theatrum Universale Omnium Animalium*)	1
La Hontan (*Voyages*)	1		
La Quintinie	1	Savary (*Dictionnaire universel de commerce*)	10
Leeuwenhoek (*Arcana Naturae*)	10		
		Swammerdam (*Histoire générale des insectes*)	1
Leeuwenhoek (*Epistulae Physiologicae*)	6		
		Vallisneri (*Dialogo . . . insetti*)	1
Lémery (*Traité universel des drogues simples*)	4	Vallisneri (*Opere fisico-mediche*)	2
Liger (*Amusements de la campagne*)	1	Vallisneri (*Saggio d'istoria medica e naturale*)	3
Lister (*Exercitatio Anatomica; De Cochleis*)	5	Vanierius (*Praedum Rusticum*)	1
Malpighi	4	Wilkins (*Of the Principles and Duties of Natural Religion*)	1
Malpighi (*Anatome Plantarum*)	2	Willughby (*Ornithologiae Libri*)	5
Malpighi (*De Gallis*)	1	Woodward (*A Natural History of the Earth*)	2
Malpighi (*De Ovo incubato*)	2		

NOTES

1. "Mes compliments à Mr. Pluche, qui continue si intrépidement à copier des livres, pour étaler le *Spectacle de la nature*, et qui s'est fait *le charlatan des ignorants.*" *Remerciement sincère à un homme charitable* (May 10, 1750), in Voltaire, *Complete Works*, 32a:204–5.

2. "Éloge historique de Monsieur l'Abbé Pluche," in Pluche, *Concorde de la géographie.*

3. Pluche, *Catalogue des livres de feu M. l'Abbé Pluche*, 117–75.

4. Trinkle, "Pluche's *Le Spectacle de la nature.*"

5. Réaumur, "Histoire des teignes ou des insectes qui rongent les laines et les pelleteries," 140.

6. We fully share the thesis defended by Brian Ogilvie in this volume concerning the link between the development of physico-theology and the promotion of entomology as a science.

7. Réaumur, *Mémoires pour servir à l'histoire des insectes*, 1:4.

8. Ray, *Wisdom of God* (1743), 180.

9. On this maxim, see Ogilvie, chapter 12 in this volume.

10. Derham, *Physico-Theology* (1723), 359.

11. Nieuwentijt, *L'Existence de Dieu démontrée par les merveilles de la nature*, 401. We find the same comparison in Ray (*Wisdom of God*, 1743, 156). The French translation of Nieuwentijt Pluche may have used is slightly different from the Dutch original edition. I thank Rienk Vermij for this information.

12. On this point see Gipper, *Wunderbare Wissenschaft*, 169–258; Chassot, *Le dialogue scientifique*, 213–42.

13. "For now we see through a glass, darkly; but then face to face: now I know in part; but then shall I know even as also I am known." 1 Cor. 13:12. Reference suggested by Katherine Calloway.

14. *Spectacle* (1766), 3:303. All citations are to this English edition.

15. Sudduth, *The Reformed Objection to Natural Theology*, 223: "Natural theology is an *epistemically loaded* project." See also Brucker, "Noël-Antoine Pluche," 335–38.

16. *Spectacle*, 3:303–4.

17. Pascal, *Pensées*, 1991, 436–37 Laf. 781, Sel. 644. Pluche concludes volume 3 on the importance of the heart as opposed to reason, a typically Pascalian theme: "Every thing refers itself to the Heart of Man, and the grateful Heart refers all to God" (*Spectacle*, 3:351).

18. Blair, "Jansenist Natural Theologian," 93.

19. Sudduth shows that Reformed English thinkers, from 1640 to 1790, viewed natural theology as mainly operating in the context of the Christian life. Sudduth, *The Reformed Objection to Natural Theology.*

20. Abbadie, *Traité de la vérité de la religion chrétienne.*

21. *Spectacle*, 3:303.

22. *Spectacle*, 3:161.

23. Ray, *Wisdom of God* (1743), 177.

24. Ray, *Wisdom of God* (1743), pref.

25. Ray, *Wisdom of God* (1743), 57.

26. *Spectacle*, 3:114.

27. *Spectacle*, 3:122.

28. *Spectacle*, 3:334.

29. *Spectacle*, 3:122.

30. *Spectacle*, 3:115.

31. *Spectacle*, 3:115.

CHAPTER FOURTEEN

Antonio Vallisneri between Faith and Flood

BRENDAN DOOLEY

Author of more than 585 discrete publications (at last count), Antonio Vallisneri (1661–1730), for three decades the eminent naturalist at the University of Padua, was not shy about divulging his ideas; nor did he rely entirely on print. His correspondence, most recently estimated at some twelve thousand units, four times that of Royal Society secretary Henry Oldenburg, constituted a veritable record of the European learned world of his day, especially in the fields of natural history and biology. The size of his collections of coins, medals, miscellaneous antiquities, firearms, chinoiseries, medical instruments, laboratory equipment, human and animal body parts, biological objects including fossils, not to mention miscellaneous wonders, some of them incorporated posthumously in a natural history museum at the University of Padua, eludes precise quantification. No list as such survives, although something of the sort may have formed the basis of a portion of the autobiography edited by Gian Artico di Porcia and published in the first volume of the collected works. There we are given only a four-thousand-word summary largely made up of imprecise aggregates like "[Item] The series of fish, petrified or else seemingly embalmed inside the layers of a rock," or "[Item] The series of various marine plants called corallines, or other sorts discovered upon same."[1] How many specimens constituted each series is not for us to know. Here I focus on his theories of the earth and of the generation of living things, which were more closely connected to each other, and to physico-theology, than current scholarship has allowed.[2]

Debate was still raging among supporters of the theses of Thomas Burnet, John Woodward, and Gottfried Wilhelm Leibniz concerning the structure of the earth and the processes involved in forming the apparent topographies when Vallisneri intervened in 1721. Fossils had become essential elements even though Burnet ignored them. For a phenomenon known to Strabo and remarked upon by Leonardo da Vinci, serious attention came only as formerly acceptable world pictures, for a variety of reasons, began to lose credit. Already by 1644 in the *Principia philosophiae* Descartes offered an alternative Genesis story as much in contrast with the biblical one as with the increasingly fashionable Lucretian one.[3] As a translator (into Italian) of the *Meditationes* of Descartes, Vallisneri followed closely as intellectual rivalries gained added energy from confessional fervor.

Burnet's hypothesis, partly inspired by Descartes but containing many original elements, presented a darkly spectacular earthly abode once perfectly formed in layers extending out from a denser liquid center, but destroyed and inundated by the gushing tide that burst through the crust by divine fiat, leaving behind all manner of improbably articulated disruptions—in short, a jagged "ruin," that is, the world we have now.[4] The combination of innovation and biblical reference proved enormously influential even among those who remained skeptical or, at times, scandalized. There were growing appetites to be satisfied for proving the power of reason and the truth of Christianity—both of which were gathering their share of enemies as the century wore on. There was already some precedent for such a theory in the Danish naturalist Nicolas Steno (Niels Steensen), a former Padua professor and associate of the Accademia del Cimento, who suggested in his *De solido intra solidum naturaliter contento dissertationis prodromus* (1669) dedicated to the Grand Duke of Tuscany that "the way the present condition of any thing discloses the past condition of the same thing is above all other places clearly manifest in Tuscany," where "inequalities of surface observed in its appearance today contain within themselves plain tokens of different changes," conceivable in inverse order from last to first.[5] "Six distinct aspects of Tuscany we therefore recognize. Two when it was fluid," he goes on, without reference to any biblical Flood, "two when level and dry, two when it was broken."[6] Burnet added to this kind of account, among other features, drama and religion.

By the time he joined the debate, Vallisneri had acquired a considerable reputation in research concerning all forms of life. Following his studies in Bologna under Marcello Malpighi and Girolamo Sbaraglia in the 1680s, he began the extensive fact-finding and sample-collecting tour around the hillsides of his native Garfagnana region that would lay the foundations of his museum.[7] Several early articles in the sporadically published periodical *Galleria di Minerva* laid out his methodological and theoretical vision, concerned with the generation of living things and the structures of geological change. After various appointments at Padua beginning in 1700, he advanced to the first chair of theoretical medicine in 1711.[8] As cofounder and editor of the influential *Giornale de' letterati d'Italia*, launched the previous year on the model of the *Journal des sçavans* and numerous others then in vogue, he complemented the expertise of Scipione Maffei and Apostolo Zeno, humanities luminaries in their own right, by dealing chiefly with the naturalist side of things and providing articles and book reviews on related matters composed by himself and others.[9] His editorial line in the *Giornale* was the same as in his teaching, namely, a critical reflection on received wisdom of all kinds, with strong reliance on what has come to be defined as empirical evidence. Praise or blame he gave to authors on the basis of their conformity to such norms, as well as their agreement with conclusions he broadcast in his other writings—for which he did not hesitate to provide his own anonymous reviews in the pages of the *Giornale*.[10]

The geological proposition in all this activity can be read to a certain degree in the light of the biological one—as might be expected from the author of an *Academic Lecture on the Order of Progression and the Connection between All Created Things* (1721). In lessons and published writings Vallisneri laid out the principal arguments for his theory about reproduction based on three cardinal notions: similars reproduce, all creatures come from eggs of some kind ("ovism"), and fecundation simply vivifies preformed material already in existence from previous generations ("preformism" and "involucrism").[11] It is well to recall that the issue of *Whether Generation Is Effected by the Spermatic Worms or by Eggs* was included in the subtitle to the 1721 *History of the Generation of Humans and Animals*—published, not coincidentally, in the same year as the work on marine fossils.[12] Each of these cardinal notions in the author's mind across two decades underwent certain subtle modulations that will not concern us here. Suffice it to say that a good portion of the work was devoted to debunking any theory attributing a predominant role to the spermatozoa. Vallisneri's preformist position is simply stated: the entire human race, in his view, was contained initially in the ovary of Eve.[13] Indeed, scattered throughout his works and lessons are the names of the major figures in the controversy, from Robert Hooke to Richard Waller, from Théophile Bonet to Nicolas Hartsoeker, from Jan Swammerdam to Antoni van Leeuwenhoek. Handsome illustrations guide the reader or audience members (in the case of presumptive teaching visuals) through the maze of different explanations for the increasingly vast inverse infinitum that the microscopic testimonies seemed to reveal. In an expansive moment he suggests that the clay whereof Adam was made most naturally included the accompanying worms, which would have been passed on to the female and subsequent generations.

In making his design argument for preformism, Vallisneri, and still more ardently some of his contemporaries, hearkened back to a work published when he was about fourteen, namely, *De la recherche de la vérité* by Nicolas Malebranche (1638–1715), in which the abundance of wonders, among natural things great and small, by sheer number, variety, and workmanship, is said to inspire contemplation of divine perfection. "The great world is considered perfect because inhabited not only by the beauty of so many animals, large and small, noble and ignoble—to use the vulgar phrase," Vallisneri suggested in 1713, drawing an analogy with current urban life: "and if some empty open space is deserted even by the markets surely it is pointed out and discredited as being inhospitable, baleful, and entirely inglorious."[14] Inside and outside the body, the more, so to speak, the holier—and the more attuned to the fundamental axiom of plenitude.[15] But how to square the theological and philosophical points with the research program? "I am one of those common philosophers," Vallisneri protests, "who in physical matters does not rise much above the ground."[16] He adds, in an excess of humility, "I attempt to fly only in proportion to the measure of my short clay-covered wings." Nor does he "risk trying to travel so far back to where I find only holy darkness, and where all human understanding, however sublime, becomes confused and lost."[17] The distinction between natural knowledge idioms and biblical idioms, articulated a century before by Galileo in the *Letter to the Grand Duchess* (printed for the first time in Italy, in "Florence," i.e., Naples, in 1710),[18] here receives a new endorsement, "whereas I find more occasions to venerate those sacred words by maintaining a respectful silence rather than attempting to understand them completely."[19] The relation of such views to the history of physico-theology within the Galileian tradition obviously deserves further study.

In the *Academic Lecture* he appealed to "the famous Leibniz," with whom he disagreed on numerous points but not regarding the chain of orderly progression.[20] He directed particular attention to "the ultimate differences, not just in the viscera and the internal parts, but in the external parts of so many animals, so many plants, so many minerals, so many figured stones, of the new world as well as the old one." And in outlining the structure of "this inalterable order, and this great abundance of all types [*genera*] and species," he pointed out the fundamental interdependencies of all, whereas "each [had] an absolute need for the other, such that, when one was lost, many others would be dragged to ruin, and this very regular symmetry of the universe would perish instantly."[21] His basic outline commenced with "the purest earth, called terra vergine or primigenia, after which comes the whole vast series of all more or less simple earths, colored or uncolored, mineral or non-mineral, endowed with virtues or vices or neither, or celebrated for other potentialities, or not." Next came the crustaceans and various stonelike creatures, followed by the "many animals whose structure resembles ours, many of whom are even excellent imitators of our gestures and our customs, some of them lacking only speech," and finally a reference too vague to attribute with any certainty to the latest frustrations from university politics in an age of reform, "whereas on the contrary many humans have nothing but speech to tell them apart from beasts."[22]

But apart from the Great Chain of Being and basic morphology, what joined the three natural worlds—animal, vegetable, and mineral? According to one inherited tradition, the answer lay in the macrocosm/microcosm analogy. The viscera of the earth—inner canals bringing necessary fluids to and from a source of inner heat—in some way paralleled the viscera of a complex organism; and the analogy signaled a vaguely unexplainable connectedness even in the realm of causation. More and more under attack in the late seventeenth century because of mounting pressure from alternative materialist and mechanistic theories, the analogy still had powerful support from the last holdouts of the Paracelsian school of natural philosophy.[23] Likewise still influential were approaches that referred to the basic cosmic forces underlying any operations involving the elements (whatever these were understood to be) composing things and beings, although recent scholarship by William R. Newman and others has cautioned against ignoring the great divide between the two relevant communities of practice, astrological and alchemical.[24] "The chain between rocks and organisms" (to use Vallisneri's phrase) also involved the mode of reproduction: but here were numerous traps for the unwary naturalist. To consider that inert matter might organize itself into a live thing still appealed to minds schooled in the peripatetic theory. What then was the spark of life?[25] Neither the vitalist hypothesis of an *anima mundi*, eventually reinterpreted and repurposed by the Cambridge Platonists, nor the Aristotelian idea of vital heat or pneuma could possibly resolve all the practical and theological implications to everyone's satisfaction. According to the approach championed by William Harvey, Daniel Le Clerc, and even Francesco Maria Nigrisoli (who was targeted for particular demolition in the pages of the *Giornale de' letterati* by the editors' Venetian patrician ally Antonio Conti),[26] there had to be a vital spirit (variously termed *lux seminalis, aura seminalis,* or even "plastic nature," according to the different versions), but apart from the obvious clash with the ethos of conceptual conservatism, the admission of an immaterial origin could seem dangerously close to violating the principle that like makes like.[27] Equally speculative was the naturalistic hypothesis about both kinds of generation, considering that for instance metals grew from certain metal-producing seeds in the ground, as John Webster claimed,[28] whereas creatures on the biological side came from seeds of some other sort; but here we encounter resistance from the "ovist" position that all organisms originated not from a seminal agent but from an egg of some kind.

For any view of generation, fossils offered an interesting challenge. They would have to have been produced; but by what?—and equally puzzling, when? Vallisneri held to a few basic principles in this regard. The specimens were unequivocally organic—as Robert Hooke and others had insisted; therefore, the usual laws of reproduction applied and not the extravagant claims of some.[29] "What force is this indeed, so hugely amazing and powerful, not to say creative, which forms organic bodies without there being any egg, in nests—especially ones not their own, causing real fish, real seashells, real oysters, real snails, to leap forth like a juggler's trick," he exclaimed, reacting to one view of fossil origins.[30] Of course, if they were once organisms like any other on the earth, the discovery of some seemingly unidentifiable with any current species was a fault of ignorance, not the artifact of a unique process. The catalog of known species was admittedly far from complete even without endorsing John Ray's position regarding extinct species.[31] As to chronology, Vallisneri was somewhat coy, at least in print, as much was at stake here, and definitive answers were not forthcoming. Until the mid-seventeenth century any evidence of a much older planet than the one implied in a scripturally consolidated sequence of human history from Creation to the present was incorporated into existing categories with more or less success. But how to handle the utter absence of chronicle references to the scattering of marine fossils in the most unlikely places by an obviously tumultuous formation of mountains, rivers, and valleys? Isaac Vossius in his *De vera aetate mundi* (1659) added some fourteen hundred more years to the accepted current interpretations of world dating. That was far less than Isaac La Peyrère's adventurous Preadamite theory, which argued, with little chance of wide assent, for the earth as a far older planet than ever imagined, and for the Genesis story as an account of only one people among many who must once have stalked it.[32] Edmund Halley offered astronomical and geometric evidence to support a theory that came so close to cosmic eternalism as to have confused numerous later historians.[33] Whether Vallisneri was aware of Newton's assertion that "absolute, true, and mathematical time, of itself, and from its own nature, flows equably without relation to anything external" is debatable.[34] But he clearly felt that the standard view framing time in relation to biblical events was becoming obsolete.

As to the Flood, Vallisneri professed genuine consternation. "I could get lost in all the examples related to this question," he complained.[35] The topic was in the mouths of "thousands," whereas opinions and so-called observations seemed in equal supply. There was certainly no doubt about "the extraordinary external mutations of the earth," nor could a post-Flood time frame categorically be denied considering the action of "water receding from one place and advancing in various different ways into another, or having swallowed many parts of the continent, or changing in some other still unexplained way the external crust on which we stand, leaving us to bother our heads about how this had occurred."[36] But attributing the most profound surface changes to the Noahtic Flood was not only philosophically wrong; it was a monstrosity. The diluvialists, he said,

> expostulate that they have observed the stratifications of the mountains to be not all horizontal nor ordered so perfectly as one might expect had God created them thus, nor covered neatly by various inundations and then filled in, but instead they seem so slanted and bent in various directions, broken, lacerated, dislodged, and brutally composed, as to suggest that such could not have been done for any other reason than as a divine punishment carried out through some ferocious, stupendous, tumultuous, universal, and most powerful agency, as the Great Flood surely was, or at least shortly after that, to show succeeding generations the unmistakable signs of Divine justice.[37]

The final purpose would have been to bring to fruition a plan not of goodness and truth but of violence and retribution intended for "stunting or diminishing the tendencies of a too-rich and too-fruitful earth," basically to keep otherwise unruly humans "on a tighter rein, so their exuberance amid the overflowing felicity should not entice them to rebel against the supreme Benefactor" or, presumably, anyone else.[38] The vision behind this analysis and, by extension, that of the individuals who propounded it, he says, citing Burnet's *Telluris theoria sacra* (possibly thinking of bk. 2, chap. 11), was that humans necessarily, "distracted by toil, shamed by perpetual poverty, should run for succor only to Him, as the most miserable and despondent ordinarily adored Him, imploring his help so the fields sown and ploughed might respond to the sweat of the industrious farmer." Such a plan, he comments, would not have been of providence but of pettiness. Indeed, "I observe in this objection an undigested mixture of physics and morals," he adds—not physics and theology, because theology, in his view, it is not.[39] Any eventual political implications of such schemes in regard to kings and peoples, on the other hand, the Venetian Republic's naturalist left for others to decide.

Understandably enough, pagan writers such as Aristotle and Strabo did not take fossils to have been left behind by any biblical Flood; but neither did sixteenth-century sages such as Girolamo Fracastoro or Andrea Cesalpino, according to Vallisneri. In his interpretation of their statements, combined with a reading of observations by Jean Astruc in the *Mémoires de Trévoux*,[40] "natural inundations of the sea in the most remote and obscure times (God only knows when) left behind our mountains, as we said, around Pisa, Leghorn, Boutonnet, or other places while flowing out to cover and saturate other more distant areas." If indeed the waters subsequently receded somehow by being sucked down into a vast chasm within the earth, as in Leibniz's theory, he added, "the case would not be without the urging of Divine Providence, which foreseeing that all types of living things had to increase their numbers upon the earth over the centuries, had to allow for this surface to increase in order to sustain, accommodate, and nourish the greater numbers."[41] From these reflections he advanced to considerations about the relation between God and geology, attempting, inasmuch as possible, to derive the major principles from the evidence and not vice versa, as too commonly occurred. Only a relatively small amount of habitable terrain would have sufficed for a small population; but over time proportions could change dramatically. Here is where a special intervention might have worked to the benefit of the natural world, somehow swallowing up a good part of the sea water and "thus exposing more land and mountains so the present and future inhabitants might expand their confines and find enough space" for their needs.[42] "In other words, I considered that Almighty God, loathing overpopulation, desired the habitation always to be in proportion and adapted to the needs of the inhabitants; and that he did not lack the means to carry this out, either by having it done using nature's ordinary laws and those prescribed for this very well ordered machine without having recourse to miracles, or else by invoking his supreme omnipotence from time to time."[43] No matter which form of Christianity might benefit from such an analysis, further conclusions were entirely unwarranted.

His cautious approach stopped short of actually explaining a miracle, which after all, as Francesco Luzzini has pointed out, had become largely superfluous for his system.[44] "The Deluge occurred," Vallisneri pronounced; and by it "God punished . . . the treacherous ingratitude of human beings."[45] That said, "I cannot understand how this took place, if I do not resort to . . . his unpredictable will and endless omnipotence." If the work of the naturalist touched the limits of human reason in regard to the "miracle" in question, good sense and his own conceptual reticence suggested a change of direction back to more solid ground. "We cannot understand completely what we can daily see and touch with our hands, but we wish to know such a portentous prodigy," he mocked. Let the many would-be philosophers "long of robe and short of vision" resort to the laws of nature in order to discuss what occurs "despite nature."[46] He would be having none of it.

At this point worth noting is the structural resemblance here to certain arguments and expressions of the French abbé Noël-Antoine Pluche, twenty-seven years his junior but by 1720 already embarked on his *Spectacle de la nature*, which would begin to appear two years after the older scholar's death. "The message that Pluche diffused" to his "many audiences," writes Ann Blair, "was an unusual mixture of effusive admiration for a vast array of examples of the wondrous and the useful in nature with constant reminders against the excessive use of reason to seek deep explanations and build theoretical systems."[47] Scott Mandelbrote contrasts such an approach with one that insists instead on reaching for intellectually satisfying demonstrations of divine power.[48] In Vallisneri's publications we find a certain caution joined to a fideistic impulse, the latter reminiscent also of the *Lettere familiari contro l'ateismo*, which appeared in Venice in 1719, by Lorenzo Magalotti, erstwhile secretary of the Accademia del Cimento and chief author of its *Saggi di naturali esperienze*, a veritable manual on conceptual restraint applied to the possible explanatory power of experiments and observations conducted even under the most rigorous conditions.

Still, what Vallisneri published could at times run counter to what he said to friends, creating an interesting conundrum for the intellectual historian. In the works that entered the mainstream of academic debate, he allowed the evidence to suggest theological reflections that might serve as a buttress to faith. In private, his agile navigation between faith and experience seemed to vanish, replaced by glimpses into a more troubled vision. With particular regard to chronology, "My beloved Mr Louis, the Earth is far older than is believed," he revealed in 1718 to Louis Bourguet, the Swiss-based naturalist and a frequent correspondent, in another attempt to entice the latter away from diluvialist convictions similar to those which would eventually sour his relations with yet another friend, Johann Jakob Scheuchzer, whom he had once considered for a Padua position.[49]

> We can see how many changes occur on the Earth in just a few centuries: rivers shift, older mountains go down and new ones arise, there are seas and valleys now where dry land once was, or land and fields where once were water and seas. The great plain that surrounds the Po river was once a swamp . . . now there are cities and castles. . . . Earthquakes, volcanoes, the rains sometimes immense, the sea storms, the wind force and other can cause the strangest changes. And what if . . . the sea that surrounds Italy would once have been high up to the mountains?[50]

How to reconcile this with the sacred teachings? "Except for the faith we owe to the Holy Text . . . who assures us of the Deluge?" Then came the leap into comparative histories: "The Chinese question it, and so do a lot of testimonies that now . . . I have no time to show."[51] We too will leave it at that.

In the light of such statements, the evident self-censorship in the published work appears less jarring; and, in fact, abundant evidence attests to private or semiprivate deliberations, allowing an examination of how these impinged upon print publication. Motivation and sincerity are tricky to assess but not impossible. "When we resort to miracles, natural history provides everything," he avers, once again inferring Galileian doctrine, in this case regarding the naturalist's obligation to explain biblical passages when possible according to developing ideas about nature's laws.[52] But miracles could also serve a strategic purpose in Counter-Reformation discourse. "I indeed often use them in my treatise," he says. "But do you know why? To silence our priests: otherwise I imply that the events I speak of did not happen in the way Woodward and many scholars with him imagine."[53]

Well aware of the peculiarities of the current confessional situation from the standpoint of what could and could not be said, Vallisneri viewed the learned world nonetheless as being divided according to effective and less effective methodologies. Rather than attributing the less effective to areas of a particular confessional affiliation, he referred instead to the standard set by his predecessors at Padua and in the Tuscan grand duchy, most recently confirmed by the Fisicomatematica academy in Rome and the Investiganti in

Naples, where the analysis of observations attempted to strike a balance between vision and validation: perhaps the two most difficult themes in the natural philosophy of the time—and for us too.[54]

NOTES

1. Here and following, *Opere fisico-mediche stampate e manoscritte di Antonio Vallisneri,* 1:liii–lvi.

2. Rappaport, "Vallisneri and Europe." Themes in that article relevant to the present study were further developed in Rappaport, *When Geologists Were Historians,* 217–22. More recent bibliographical discussions are found in the chapters of Generali, *Antonio Vallisneri: La figura, il contesto, le immagini storiografiche,* esp. the editor's introduction, v–xxx, and in the present connection, chapters by Michael Cunningham and Ezio Vaccari.

3. Garber, *Descartes' Metaphysical Physics,* 118–19.

4. Burnet, *Sacred Theory of the Earth* (1684), vol. 1, bk. 1, chap. 3. In general, Pasini, *Thomas Burnet,* updated by, among others, Magruder, "Thomas Burnet."

5. Steno, *The Prodromus of Nicolaus Steno's Dissertation,* 262 (63 in the original). On which, see Kardel and Maquet, *Nicolaus Steno,* 763–826.

6. Steno, *The Prodromus of Nicolaus Steno's Dissertation,* 263 (69 in original).

7. Generali, *Antonio Vallisneri: Gli anni della formazione e le prime ricerche,* chap. 6. Especially rich in detail is Vallisneri's own autobiography, *Notizie della vita e degli studi del Kavalier Antonio Vallisneri,* in the current regard, 79–81.

8. Dooley, "Talking Science at the University of Padua in the Age of Antonio Vallisneri"; Dooley, "Scienza parlata, scienza scritta"; Dooley, "Learned Journals and Teaching at the University of Padua."

9. Generali, *Antonio Vallisneri:Ggli anni della formazione e le prime ricerche,* chap. 7; Dooley, *Science, Politics, and Society,* 98–129.

10. For instance, of his *Corpi marini,* in the *Giornale de' Letterati d'Italia* 37 (1725), 156–94; on which, see Lombardi, *Un nume del Settecento,* 191.

11. Savelli "L'opera biologica di A. V."; and the relevant portions of Bernardi, *Le metafisiche dell'embrione,* pts. 3 and 4.

12. Vallisneri, *Lezione Accademica intorno all'ordine della progressione,* 421–37.

13. Pinto-Correia, *The Ovary of Eve,* 22.

14. Vallisneri, *Nuove osservazioni, ed esperienze intorno all'ovaja scoperta,* 114.

15. Noteworthy is the discussion by Lovejoy, *Great Chain of Being,* chap. 4. For this and other modal theories, worth noting also is Knuuttila, *Reforging the Great Chain of Being,* esp. the chapters by Michael David Rohr and Jaakko Hintikka.

16. Vallisneri, *Nuove osservazioni, ed esperienze intorno all'ovaja scoperta,* 113.

17. Vallisneri, *Nuove osservazioni, ed esperienze intorno all'ovaja scoperta,* 113.

18. For the fortunes of the 1636 Strasbourg bilingual edition, I refer to the critical introduction to Galileo Galilei, *Lettere a Cristina di Lorena,* 15–28.

19. Vallisneri, *Nuove osservazioni, ed esperienze intorno all'ovaja scoperta,* 114.

20. Vallisneri, *Lezione Accademica intorno all'ordine della progressione,* 424.

21. Vallisneri, *Lezione Accademica intorno all'ordine della progressione,* 437.

22. Vallisneri, *Lezione Accademica intorno all'ordine della progressione,* 436.

23. Debus, *The French Paracelsians,* 1–14, 183–208.

24. Newman and Grafton, *Secrets of Nature,* 1–38.

25. Here and subsequently, apart from sources already mentioned, see Duchesneau, *Les modèles du vivant de Descartes à Leibniz*, chap. 7; which updates Roger, *Les sciences de la vie dans la pensée française*, 418–41.

26. *Giornale de' letterati* 12 (1712): 240–330.

27. Vallisneri articulates his views on these theories in *Istoria della generazione dell'uomo e degli animali*, sec. 1, vol. 12, pp. 300–301 (231–32 in original). Concerning the wider context, note the introductory essays by M. T. Monti, lxxvii–lxxxvii, and by François Duchesneau, clxv–clxvi. Concerning the Conti episode, Badaloni, *Antonio Conti*, 38–41.

28. Webster, *Metallographia*, 71; on which, Norris, "Mineral Exhalation Theory of Metallogenesis"; and concerning Isaac Newton's position, Dobbs, *Foundations of Newton's Alchemy*, 86.

29. Consider Drake, *Restless Genius*, 69–86.

30. Vallisneri, *De' corpi marini*, 15.

31. Ray, *Three Physico-Theological Discourses*, 3rd ed. (1713), 173. Concerning Ray's ideas of classification, see Ogilvie, "Insects in Ray's Natural History and Natural Theology."

32. Popkin, *La Peyrère*, chap. 3. In addition, Rossi, *I segni del tempo*, chaps. 1–2. Insights from here are promised a second life in Dooley, "The Experience of Time."

33. Concerning the disputes, Levitin, "Halley and the Eternity of the World Revisited."

34. Newton, Scholium to the Definitions in *Philosophiae Naturalis Principia Mathematica*, bk. 1, 6–12. Concerning the discussion of Newtonianism, Ferrone, *Scienza natura religion*, 237–316.

35. Vallisneri, *De' corpi marini*, 46.

36. Vallisneri, *De' corpi marini*, 46.

37. Vallisneri, *De' corpi marini*, 65.

38. Vallisneri, *De' corpi marini*, 65.

39. Vallisneri, *De' corpi marini*, 65.

40. Astruc, *Mémoires de Trévoux*, 8:512–25.

41. Vallisneri, *De' corpi marini*, 46.

42. Vallisneri, *De' corpi marini*, 46.

43. Vallisneri, *De' corpi marini*, 47.

44. Luzzini, *Il miracolo inutile*, 139–226. On this theme, useful reflections also in Tucci, "'Il parlare della Santa Scrittura e l'operare della natura.'"

45. Vallisneri, *De' corpi marini*, 24.

46. Vallisneri, *De' corpi marini*, 24.

47. Blair, "Jansenist Natural Theologian," 92.

48. Mandelbrote, "Uses of Natural Theology." Concerning the general structure of natural theological argument, see Brooke, *Science and Religion* (1991), 192–225 or (2014), 261–306.

49. Luzzini, "Flood Conceptions in Vallisneri's Thought," 86, citing Vallisneri, *Epistolario (1714–1729)*, 353.

50. Luzzini, "Flood Conceptions in Vallisneri's Thought," 86, citing Vallisneri, *Epistolario (1714–1729)*, 353.

51. Luzzini, "Flood Conceptions in Vallisneri's Thought," 86, citing Vallisneri, *Epistolario (1714–1729)*, 353.

52. Vallisneri, *Epistolario (1714–1729)*, 738.

53. Vallisneri, *Epistolario (1714–1729)*, 738.

54. For further illumination on these points, see von Greyerz, "Protestantism, Knowledge, and Science."

PART FIVE

AESTHETIC SENSIBILITIES

A Language for the Eye

Evidence within the Text and Evidence as Text in German Physico-theological Literature

BARBARA HUNFELD

Literature usually consists of an invitation to look at things from a different perspective. Thus, it not only conveys content but also directs our attention to its construction. This chapter uses the latter perspective. It addresses physico-theology not from the point of view of historical documentation but from the angle of literary studies and, considering this particular orientation, it does not offer a case study of yet another example of a national tradition of physico-theology but rather aims at the elucidation of a basic principle in the construction of physico-theological thought. As a result, the following argument is of a structural nature. Physico-theological thinking follows a basic pattern: it adheres to the conviction that any natural thing or object offers graspable evidence in the light of which the world becomes an apodictical witness of its own interpretation as creation. The phenomenon at hand and its interpretation thus seem to coincide within the framework of an idiosyncratic hermeneutic perspective. This hermeneutics suggests that the text it interprets is in fact self-explanatory, whereas references to ever new physico-theological examples, at the same time, indicate the need to *interprete* the world. In order to elucidate this particular context, the present contribution looks at the poetry of arguably the most outstanding representative of German physico-theology, Barthold Heinrich Brockes, as an ideal-typical model.

Seeing and reading refer to different concepts. German physico-theological authors of the first half of the eighteenth century tend to telescope them in a strange manner. This results in an evidentiary utopia, that is, in the idea of a written text that reveals itself to the eye without deciphering, in a language that can be seen. This is tantamount to the hermeneutical dream of immediate comprehension without any misunderstandings, tedious detours, and wrong ways. Simultaneously it represents the semiotic dream of a representation that includes the indisputable presence of its object, thus making possible the experience of that which the text represents.

The affinity between physico-theological and semiotic thought is obvious. They are both governed by the hope of establishing unequivocal facts, whether these concern the knowability of God in Creation or the relationship between signifier and signified. Physico-theology thus becomes literature's Promised Land. At the same time, literature, where nothing is firmly established, which is the home of all contradictions that are unbearable in real life, provides the inverse mirror image of physico-theology's evidentiary claim.

In literature, the physico-theological acquisition of knowledge is tantamount to an aesthetic problem. *Evidence* (Latin: *evidentia*) is a claim about content that calls for a formal resolution. The *aporiae* that result from this throw light not only on the state of literature at the threshold of modernity but also on physico-theology from the special perspective of literature, where it is precisely indistinct thought that proves productive.

If one looks at the opening sentences of William Derham's *Astrotheologie*, the great inspiration of German physico-theology, it becomes clear that this book begins by addressing a language that is not a language at all. It is the language of things, the language of the created world. And even though it is impossible to hear the voice of the things that speak here, what the things have to say is unmistakable. Here I follow the German translation of the English original by Johann Albert Fabricius (1668–1736) published in 1728: "King David says: the Heavens declare the Glory of God, and the Firmament sheweth . . . his Handywork. One day tells the other, and one night manifests and communicates it to another. This language of the heavens is so unmistakable, and their text so apparent and readable, that all and sundry, even barbarian people untrained in the sciences and languages, can understand them and read their message. This is no language whose voice cannot be heard. Its thread travels to all countries and its voice to the end of the world."[1] Regardless of whether a physico-theological text invokes empirical considerations, it usually begins with an invocation of the authority of scripture. David's psalm authorizes the physico-theological perspective, whose empirical analysis will in turn confirm the biblical text. Even the title page corroborates the main concern: "a demonstration of the being and attributes of God, from a survey of the heavens."[2]

This connection between Holy Scripture and the empirical study of the heavens established right at the start results in a fusion of reading and seeing. The title page and introduction do the same. Because the title page highlights the value of ocular observation, it confirms the intrinsic connection between seeing and reading. Language and writing are signs that require deciphering, interpretation, and reading. However, in that visible writing of the heavens, which can be seen by everybody, any anxiety of interpretation (*Deutungsnot*) is redeemed—as is demonstrated simply by the rationalistic notion of "distinctness."

Or could it be that this redemption actually fails? Derham is conscious of the hermeneutical problem inherent in any use of language. Listening to the speech of another always poses a risk of misunderstanding, which requires additional explanations in order to avoid problems of comprehension. Precisely where Derham establishes his decisive conception of the language of the heavens in his "narration" of God's honor calls for interpretation. Let me supply the words left out in the passage cited above: "The Psalmist saith[a], *The Heavens declare the Glory of God; and the Firmament sheweth*, publickly declareth, telleth forth, or preacheth his *Handy Work*, as the Hebrew Word signifies[a]: that Day unto Day uttereth Speech, and Night unto Night sheweth, or tells forth, Knowledge."[3]

Just like Derham, Barthold Heinrich Brockes (1680–1747), the most important author of German literary physico-theology, invokes the writing of the heavens and the language of the world.[4] Because Brockes's physico-theology is a literary project, it is interesting for us to examine how he conceives of the relationship between empirical seeing and reading of the signs. All ocular witnessing of the world must become a language. Brockes's *Irdisches Vergnügen in Gott* (Earthly Pleasure in God), a multivolume compendium of poems (1721–48), transfers the world into a book, in a way similar to what the philosopher Hans Blumenberg has termed "die Lesbarkeit der Welt" (the readability of the world).[5] This kind of readability involves the solution of two problems: first, everything that can be seen must become readable in a physico-theological manner; and, second, the poems must claim to represent everything that can be seen, including even the experience of seeing.

At the beginning of the *Irdisches Vergnügen* the intended solutions both threaten to fail.[6] The first poem in volume 1 focuses on the feeling of being overpowered, experienced by the observer of the heavens. The ocular survey of the universe, being the most comprehensive view of the world, examines vision and language—that is, the tools of the world's experience Brockes has in mind. The text operates close to an abyss for two reasons. First, it points to the Copernican shock, which is rendered acceptable by being transformed into the experience of the sublime. This, in turn, reestablishes the readability of the world. Second, and much more important, the project of the *Irdisches Vergnügen* appears threatened temporarily as vision and language are brought to the limits of their performance. The sheer unmeasurable abundance of the empirical that cannot be rendered in words appears as something that opposes itself to any literary classification and reading.

What cannot be circumscribed may possibly be seen but cannot be categorized visually or understood in reading. The aesthetic problem of representation also involves the task of examining the limits of cognition. Traditionally, seeing as a form of cognition and the reading of signs enjoy an ambivalent reputation. In 1 Corinthians 13:12, this concept is expressed as: "We now look through a glass within a dark word."[7] This reflects the complaint about the imperfection of cognition, at least as far as this world is concerned. The immediacy of cognition is a utopia whose fulfillment, according to 1 Corinthians, is limited to the reign of God: "When the completeness has come," piecemeal cognition by way of mirrors and dark words will have an end. Only then, beyond images and language, will humans be able to see "face to face" (1 Cor. 13:12). Until that time, signs such as the metonymic language of 1 Corinthians will reign. This language points to its own imperfection and, at the same time, looks at a future in which all language ends because there will be no more need for mediation. The metaphor of looking through a mirror and word encapsulates the hope for a future, still-distant overthrow of metaphors.

Seeing, as this makes clear, has its limits. But the signs that must assist vision also enjoy a mixed reputation; in the worst case they persist even when there is nothing any more to be seen because the remoteness of memory has replaced life's immediacy. "Stat rosa pristina nomine, nomina nuda tenemus" (of the former rose only the naked name remains); Bernard of Cluny's famous sentence from the first half of the twelfth century sees the signs as naked because they lack the beauty, size, and fullness to which they point. They are echoes of all that is not at our disposal.[8]

Literature offers an answer to this dilemma, not so much in solving what cannot be solved as in letting contradictions become productive. The concept of evidence is at the basis of such a contradiction. It suggests that there is no epistemological problem, nothing to compromise understanding, because one can see in the signified that to which the signifier points in its immediacy, as a presence "ante oculos" (before our eyes). From a semiotic perspective, this is a fantasy, a dream about a kind of seeing that is successful only within the world of signs, within the game of representations. That there is no way out of the maze of representations is shown by Brockes's *Irdisches*

Vergnügen in Gott on two different levels: where it establishes the nature of boundary experiences, as well as where the project is conventional.

In the first half of the eighteenth century, while Brockes guides the reader past the wonders of nature in ever-new poems, seeing has become popular again, thanks to the Enlightenment. In the rhetoric of rationalism and in the sensualist quest for the provenance of empiricism, vision and cognition are being telescoped into one procedure. Brockes's didactic poems translate the new perception of the world into a linguistic picture book, not unlike the images and didactic rhymes in contemporary primers. Instead of the alphabet, the reader is taught to read the letters of the book of nature. The topos of the book of nature fills the signs with meaning and dispels the distrust hitherto felt against them.[9]

Readers of sonnets of the seventeenth century were used to the fact that poems tell you how to see the world. Within this context, Brockes replaced the emblematic illustration of the world with empiricism and exchanged the topoi of the Baroque, as well as manneristic word games, with concrete description. The aim of his poems is not a general assessment but rather the individual physical properties of things. At the end, however, there is transcendence (just as in Baroque poems) whereby all wonders of the world are read in view of recognizing God.

This is an ever-recurrent pattern that demonstrates to readers not only how to pronounce their prayers but also how they would experience the world, if the sheer multitude left them some time to do so. On a linguistic level, the experience whose immediacy is invoked as a condition of physico-theological knowledge is prefigured. Brockes's texts thus make explicit the point that they are merely paraphrases and bad mimesis. However, the lack of signs has its own significance in that it is itself a sign pointing to the inimitable creative powers of the true, divine author.

Brockes's *Irdisches Vergnügen* adopts the universal claims of an encyclopedia in its attempt to transfer macrocosm and microcosm into the book. Largeness and smallness must be recognized and put in perspective. Most prominent in this respect are Brockes's poems about stars and flowers, sometimes about both within the same text. Here, the relatedness of things is a topic and likewise their abundance, and in both of these the readers can lose themselves. This is recognized not as a possible insufficiency of seeing but rather as an additional verification of the creator, as it is his overabundance that results in the disorientation of the observer.

Order in the world of language is to prevent this overabundance from becoming a problem. In the *Irdisches Vergnügen*, this arrangement first of all rests upon the physico-theological perspective that ushers in the topos of the book of nature; second, it is steadily confirmed through the repetitious pattern of the texts. Repetition not only provides structure to the individual poem and the sensations it evokes. It also supplies training in a mode of perception and interpretation that stabilizes the reader. Both procedures for creating order (the physico-theological perspective and repetition) are procedures of representation that favor reading rather than seeing as a basis for the access to reality

This principle can be explicated through the study of an illustration. The frontispiece of the eighth volume of the *Irdisches Vergnügen in Gott* of 1746 promises an all-encompassing order and compactness of the world described. There is no open horizon, no panoramic view, but rather a framed perspective that August Langen has described as a rationalistic defense against uncontrolled empiricism.[10] The space and the things within it are arranged along a central perspective, whose lines of construction could be extended beyond the image and behind the observer into the reality that is not yet part of Brockes's inventory of the world. However long this extension, it still defines seeing in the by now familiar way. In the same way, Brockes's poems in his multivolume work construct the pattern of an ever-unchanging perspective. What we observe is order as an effect of perspective representation. It engenders its own interpretative logic. Away from the center of the frontispiece's construction, but in its sensible middle, we notice a window in the façade of the mansion that reminds us of God's eye. From here, as from an aureole, the things seem to emerge, which allows for their perception as divine emanations.

Likewise, the theme of the frontispiece makes clear that the world is seen with the particular overlay described in these pages: it displays the garden as nature transformed. The geometry of the mansion's architecture is transferred to the garden, which in its symmetry represents a contemporary French garden. The central perspective corresponds to the all-encompassing sight of God, as well as to an access to reality reflecting Enlightenment thought. The caption claims that nature and art in the garden are relatives. We encounter a similar suggestion in the verses of the flower poem quoted in note 9, where the observer of the blossoms is able to read "about the creator's wonderful creatures . . . much on many leaves" (*von des Schöpfers Wunder-Wesen . . . Viel, auf vielen Blättern, lesen*). The double meaning of

Frontispiece of the eighth volume of Brockes's *Irdisches Vergnügen in Gott* (Hamburg: Herold, 1746) by Christian Fritsch from a template presumably by Johann Joachim Pfeiffer. Wiki Commons.

"leaves" brings together the leaves (*Blätter*) of the flower and the pages (*Blätter*) of the book. They converge in the idea of reading; both point to writing. In fact, the frontispiece does not really display nature and art as sisters; rather, nature appears to be a subject of art, a precondition for mastering an empirical approach. This does not, however, apply to the *Irdisches Vergnügen*. As God's art, nature is superior to human art. This only heightens the value of the kind of reading that intends to fuse the two.

According to the etymological dictionary of the Grimm brothers, the word "to read" (*lesen*) has several historical meanings.[11] First, it denotes collecting individual or scattered things; second, it means selecting, in the sense

of collecting something out of a larger batch; third, it stands for making order and arranging; and, fourth, it signifies reading in today's understanding.
Brockes's *Irdisches Vergnügen* represents reading in these multiple meanings: it
is an enormous collection of the things of the world, which are individual and
scattered, and which are transferred into the order of the text, as well as into
the order of a context of discussion. This is how they become readable. In the
oldest meaning of the word *lesen* denoting the collection of individual or scattered things, the Grimm brothers refer to its origin in "the observation of
walking in a track or following a line, which offers the walker a specific yield."[12]
This is precisely how Brockes's texts operate, by accumulating substantial gain
to conquer reality in this manner. The track/trace (*Spur*) is a frequent metaphor in the *Irdisches Vergnügen*. It describes the world as an index sign, as the
creator's trace, which one simply must follow to avoid getting lost.

However, the *Irdisches Vergnügen* cannot select and is unable to limit itself to just one of the meanings of *lesen* listed by the Grimm brothers. It also
tries to ascribe everything to the totality of the book, a hopeless undertaking.
Maybe this wasted effort has something to do with the fact that the theory
of signatures is no longer considered reliable in the eighteenth century. It is
true that Brockes's observations of nature initially follow the old track of the
traditional view of signatures, but the Enlightenment little by little leads to
the unfolding of an empiricism that gains independence in plain view of the
readers.

The synchronicity of seeing and reading described by Brockes's physico-
theology is part of the ambivalent history of representation. Aristotle's char-
acterization of art as mimesis was followed during many centuries by a reflection about the nature of representation. One important answer insisted
that art imitates just like a picture, which not only allows the viewer to recognize the original but actually suggests its presence. It would follow that
representation is the presence of the signified imagined by art and suggested
by representational techniques.

The art of the still life provides an example, as in the painting *Stilleben
mit Früchten* by Catharina Treu (1743–1811). It was long considered to be of
minor importance, a finger exercise of a master. However, by applying the
technique of optical deception, which appears to the spectator as an invitation to enter the picture, it reveals the outstanding illusionism of the mi-
metic aesthetic. This establishes a maximum of closeness to the original.
Brockes applies the same technique to his poems. When he describes his
"bowl with fruits" (*Eine Schüssel mit Früchten*), he creates a literary still life

Catharina Treu (1743–1811), *Stilleben mit Früchten* (Still life with fruit, undated). Courtesy of the Martin-von-Wagner-Museum of the University of Würzburg.

with words.[13] One after the other, his verses investigate the sensual presence of things in all its details, be it in the play of colors or in the "delicate velvet" of a peach's skin. It is owing to these special qualities of Brockes's literature that the contemporary public praised his representations as "poetic pictures" that renewed Horace's "ut pictura poesis" (poesis must follow the example of painting).[14]

Still lives, whether painted or written, do not enact mimesis simply in the display of form. What is decisive is rather the representation that arranges and interprets what has been found in nature so that the picture acquires the quality of an examination of the essence of displayed reality. Looking at the picture or at the text not only involves a "source of legitimate ocular lusts" (*Quell erlaubter Augen-Lüste*);[15] in addition, the presence and abundance of things become signs of a presence and abundance whose essence lies beyond the things. The representation transcends the things that are depicted. The aim is to create a representational perspective that results in the fusion of seeing and reading. The still life of Catharina Treu demonstrates the closeness of seeing and reading in this conception of representation. One notices a silver plate in the upper right-hand corner and, on it, an orange and a lemon. On the orange sits a fly. The painting depicts nature so faithfully that it manages to play a trick on nature itself, to use an old saying, insofar as the spectator could confuse image and reality. The fly that is attracted by the fruit encapsulates in this way the legend of Zeuxis, whose paintings of grapes reportedly attracted birds seeking to peck them. But the fly also has an additional meaning. What attracts the fly is the ripeness of the fruit that threatens to become putrefaction announcing the imminence of death. Thus the image can be understood as a warning against *vanitas*: all the abundance before our eyes vanishes. The affirmation of empiricism and its interpretation as a sign merge in a tilting movement (*Kippbewegung*). In this movement, the presence suggested by the representation is extinguished in a twofold manner. What can be seen in its immediacy will be lost. And because that which follows can only be deciphered and not be seen, even in the moment of seeing we fail to have immediacy.

The concept of evidence might offer to suppress this tilting movement in promising both "indirect perception in immediacy and immediate seeing in indirect perception" (*Mittelbarkeit in der Unmittelbarkeit, Unmittelbarkeit in der Mittelbarkeit*). Evidence within the text is evidence as text in a double sense: it is established by the text and offers evidence as the imagination of experience, which in turn is organized according to the principles of representation. This pairing is full of tension as it forcefully merges two poles that appear to be incompatible. However, in art and in the special language of literature, opposites can be evident as unity while the moment of literary evidence lasts.

The utopian attempt to suppress this tilting movement is necessarily in vain. If it succeeded, it would rob utopia of its productivity. Consequently, the tilting movement remains a constitutive element of the concept of evidence. It is a tilting movement not only insofar as the perceptions of seeing and reading might exchange roles; it can also be understood in the sense that something within the concept itself risks tilting. The utopia precisely points to the needs that it should fulfill. There is, in the first place, the need for an empirical experience. Although reading is necessary to prevent the spectator's confusion, the establishment of evidence requires the kind of presence that appears convincing. Hence the highlighting of ocular veracity (*Augenscheinlichkeit*) or of things being put before our eyes (*Vor-Augen-Stellen*), and even of a graspable (*mit Händen zu greifenden*) vividness, used time and again by German physico-theologians such as Brockes, Daniel Wilhelm Triller (1695–1782), or Fabricius.[16] In this context, Brockes's "poetic paintings" (*poetische Gemählde*) are considered to be exemplary. Triller, who contributes a panegyric to the second volume of the *Irdisches Vergnügen*, published in 1727, emphasizes that Brockes actually "depicts objects instead of only describing them" (using the expression *nach dem Leben geschildert*, "depicted from life").[17] It is with similar words that Fabricius praises Brockes, to whom he dedicates his 1728 translation of Derham's *Astro-Theology*. The German verb *schildern* points to a kind of painting and picturing that transcends mere describing as understood by hermeneutical theory and does not abide by the rules of language as a more indirect system of signs.

Whatever likeness is claimed for the image is threatened by a loss of vividness, even for Brockes, as the monotony of his framing of perception, the stillness and paralysis of the world in the sign, do not permit any real experience. As is well known, the letter can kill. This is why there is, secondly, a need for that kind of particular significance which reigns supreme in physico-theological literature, because its signs are considered a thing (*res*). In his previously mentioned panegyric, Triller cites Johann Arndt in hypostatizing the book of nature as God's book written "in perfect words." It is "a living book" with "living letters." The signs are no mere words but living things. Thus, Brockes can pronounce: "Your soul will soon comprehend this language . . . looking through your eye. . . . You can see this language, and apprehend and understand it by your sight."[18]

To return once more to the still life, painted as well as written: Brockes's poem about the "bowl with fruits" ("Schüssel mit Früchten") is too long to be used here as an illustration for the tilting of the evidence, so I use Treu's painting for that purpose instead. I do not mean the tilted chalice, a frequently invoked symbol for life in decline or even death. I consider more important two other moments of tilting that are visible next to the chalice. On the right the slices of bread balancing on the edge of the table threaten to fall down, and the cloth thrown on the table could also slide down, taking with it the pieces of bread. On the left there is the fruit basket with its contents (a "reading" in the meaning of a salvaging "collection"), which might tilt and let the fruit roll out, disassembling the collection.

The reader of Brockes's texts is confronted with similar movements. That which has been arranged painstakingly finally bursts the chains of representation because seeing and reading drown in the flood of things and even more in the flood of words. There is no sublime overpowering of the reader, which could turn into the experience of God, comparable to the overpowering of the observer in the first poem addressing the universe at the beginning of the *Irdisches Vergnügen*. Rather, the reader is overwhelmed because the signs acquire independence and eradicate any seeing and also any reading. The experience shows why evidence is necessary and also why it is an impossible proposition to provide such evidence, unless it is part of an aesthetic dream.

NOTES

Translated by Kaspar von Greyerz.

1. "König David saget: Die Himmel erzehlen die Ehre Gottes, und die Feste verkündigen, . . . und preisen seiner Hände Werck. Ein Tag sagt es dem andern, und eine Nacht thuts kund, oder erzehlets der andern. Diese Sprache der Himmel ist so deutlich, und ihre Schrift so erkenntlich und leserlich, daß alle und jede, auch die barbarischen Völcker, die in Wissenschaften und Sprachen sonst keine Geschicklichkeit haben, dennoch dieselben verstehen, und was sie anzeigen, lesen können. Es ist keine Sprache noch Rede, da man nicht ihre Stimme höre. Ihre Schnur gehet aus in alle Lande, und ihre Rede an der Welt Ende." Derham, *Astrotheologie* (1728), 216–17. Fabricius's translation differs from the original version in the fifth edition of Derham's *Astro-Theology* (1726) in that the "psalmist" of the original becomes "King David" (pp. 1–2).

2. In Fabricius's 1728 translation: "Anschauen des Himmels, und genauer[e] Betrachtung der Himmlischen Cörper, Zum augenscheinlichen Beweiß Daß ein GOTT . . . sey."

3. Derham, *Astro-Theology*, 5th ed. (1726), 1–2. Note [a] refers to the Hebrew words in need of interpretation. Emphasis is in the original.

4. On Brockes see, e.g., Buch, *Ut Pictura Poesis*, 64–96; Kemper, *Gottebenbildlichkeit und Naturnachahmung im Säkularisierungsprozeß*; Ketelsen, *Die Naturpoesie der norddeutschen Frühaufklärung*; Peters, *Die Kunst der Natur*; Preisendanz, "Naturwissenschaft als Provokation der Poesie"; Weimar, "Gottes und der Menschen Schrift"; Hunfeld, *Der Blick*, esp. 63–70.

5. Blumenberg, *Die Lesbarkeit*.

6. Cf. Hunfeld, *Der Blick*, 38–100.

7. Translator's note: This is a translation of verse 12 in the German Luther-Bible, which explicitly refers to a mirror, whereas the King James version does so only indirectly: "For now we see through a glass, darkly." "The word" is missing in the King James Version.

8. Bernard of Cluny, *De contemptu mundi* i.952.

9. As an example: "Da ich zwischen Blumen gehe / Und, mit tausendfacher Lust, / Tausendfache Farben sehe; / Wird das Herz in meiner Brust, / Nicht nur durch die bunte Pracht, / Und durch den Geruch gerühret; / Sondern mein vergnügter Geist, / Wird zu dem, der sie gemacht, / Voller Brunst empor geführet. / Von des Schöpfers Wunder-Wesen, / Lässet ihrer Farben-Zier, / In gefärbten Lettern mir, / Viel, auf vielen Blättern, lesen." Brockes, "Blumen-Betrachtung," in *Irdisches Vergnügen*, 2nd ed. (1740), 5:96.

10. See Langen, *Anschauungsformen in der deutschen Dichtung des 18. Jahrhunderts.*

11. Grimm and Grimm, *Deutsches Wörterbuch*, 12: cols. 774–86.

12. Grimm and Grimm, *Deutsches Wörterbuch*, 12: col. 774: "In der anschauung des gehens in einer spur oder einem striche nach, der dem gehenden eine bestimmte ausbeute gewährt."

13. Brockes, "Eine Schüssel mit Früchten [A bowl with fruit]," in *Auszug der vornehmsten Gedichte aus dem von Herrn Barthold Heinrich Brockes in fünf Teilen herausgegebenen Irdischen Vergnügen in Gott*, 371–77.

14. Horace, *De arte poetica* 361.

15. Brockes, "Eine Schüssel mit Früchten" [A bowl with fruit], in *Auszug der vornehmsten Gedichte aus dem von Herrn Barthold Heinrich Brockes in fünf Teilen herausgegebenen Irdischen Vergnügen in Gott*, 372.

16. To name three prominent cases in point next to Brockes, see Daniel Wilhelm Triller's panegyric contributed to the second volume of Brockes's *Irdisches Vergnügen* (1727), the title page of the German edition of Derham, *Astrotheologie* (Fabricius trans.), and Fabricius's "Scribenten" in the second edition of Derham's *Astrotheologie* of 1732.

17. Daniel Wilhelm Triller, "Zufällige Poetische Gedanken . . . ," in Brockes, *Irdisches Vergnügen*, 1st ed. (1727), vol. 2, n.p.

18. Brockes, "Grosse Buchstaben," in *Irdisches Vergnügen*, 2nd ed. (1740), 5:322: "Deine Seele wird die Sprache . . . durch dein Auge, bald verstehen. . . . [Di]ese Sprache kanst du sehen, Und durch dein Gesicht vernehmen und begreiffen."

CHAPTER SIXTEEN

A Hybrid Physico-theology

The Case of the Swiss Confederation

SIMONA BOSCANI LEONI

To analyze the development of physico-theology in the Old Swiss Confederation, I focus on two protagonists of this movement during the eighteenth century: first, on the naturalist and physician Johann Jakob Scheuchzer (1677–1733) and, second, on Élie Bertrand (1713–97), who was a French-speaking Reformed theologian and about a generation younger than Scheuchzer. Both authors allow us to follow the development of physico-theology in the French- and German-speaking areas of Switzerland, and they represent an interesting example of the translation (in a literal and in a figurative sense) of physico-theological texts and topics from Britain to Switzerland.

In particular, I want to highlight the mutual influence between British and Swiss physico-theologians through the analysis of one of the movement's most sensitive debates beginning at the end of the seventeenth century—that is, the dispute about the connection of Noah's Flood with the existence of mountains, and about the origin of the earth. This was a very complex debate because one of the central issues at stake was the reliability of the scriptural account in Genesis. From its early stages, this controversy attracted the interest of Swiss scholars, especially because of its discussion of the providential role of mountains. The debate on mountains as part of the divine Creation and on the diversity and usefulness of the alpine nature played a very important role in the development of natural science in the Old Swiss Confederation. Beginning with the Renaissance, or even earlier, the Alps became one of the constitutive elements of the cultural "identity" of the country.[1] From the end of the seventeenth to the second half of the eighteenth century, the discussion of the providential role of mountains played a prominent role in the scientific work of Swiss savants such as Scheuchzer and Bertrand. Both naturalists played a paradigmatic role in the reception and transformation of the British debate on a local level, Scheuchzer in Zurich and Bertrand in the Pays de Vaud, which was part of the early modern territory of Berne. Zurich and Berne were the most powerful Reformed dominions of the Old Swiss Confederation, which was a conglomerate of rural and urban, Catholic and Protestant communes and their territories. Scheuchzer fostered tight contacts with the Royal Society and with several British scholars, and he was the first naturalist who went on scientific expeditions through the Alps. Bertrand represents a very interesting example of the reception of Scheuchzer's geological and paleontological works in the French speaking part of the Old Confederation. Both authors demonstrate that for Swiss scholars the study of nature was impossible without considering the Alps.

My essay is divided into two parts. First, I focus on the arguments about the origin of the earth exchanged in Britain during the second half of the seventeenth century and on the dissemination of explanations for the existence of mountains in the work of several British physico-theologians. Then, I consider the reception of these arguments and explanations in Scheuchzer's and Bertrand's works on natural history. Far from a one-way circulation from A to B, that is, from Britain to Switzerland, the circulation of views and ideas is in this case more complex. This is largely because many arguments invoked by British physico-theologians can be encountered in several works of Swiss scholars of the Renaissance, especially in treatises written by Conrad Gessner.

The Debate on Mountains and the Creation of the Earth in Seventeenth-Century England

A central British protagonist of the controversy about the interpretation of Genesis and the origin of the earth was without doubt Thomas Burnet (1635–1715), a Scottish theologian, master of the Charterhouse, and chaplain to King William III. The publication of his *Telluris theoria sacra* in 1681 marks the beginning of a protracted and heated debate. The English version of the book, *The Sacred Theory of the Earth*, appeared three years later, and between 1689 and 1691, a new edition with two additional parts was published both in Latin and in English. In the first book of his *Sacred Theory*, Burnet describes the Flood; in the second, paradise; in the third, the old world's final conflagration; and in the last, the new heavens and the new earth. Burnet assumed in the first two books that the terrestrial globe known to himself and his contemporaries was not identical with the original world. He argued that the prediluvian world was perfectly smooth like an egg; only after the Flood did the mountains appear. As a result, Burnet describes them as a disfigurement of the earth's original crust and interprets them as a reminder of God's punishment for humankind. The irregularities of the earth and the lack of symmetry offended his sense of proportions. The earth, Burnet claims, is a heap of ruins: "We must therefore be impartial where the Truth requires it, and describe the Earth as it is really in itself . . . yet if we consider the whole surface of it, or the whole Exteriour Region, 'tis as a broken and confus'd heap of bodies, plac'd in no order to one another, nor with any corrispondency or regularity of parts." By reading on, we soon find what kind of opinion Burnet had about mountains: "But as we justly admire its [a mountain's] greatness, so we cannot at all admire its beauty or elegancy, for 'tis as deformed as it is great. And there appearing nothing of Order, or any regular Design in its Parts, it seems reasonable to believe that it was not the Work of Nature, according to her first Intention, or according to the first Model that was drawn in Measure and Proportion . . . , but a secondary Work, and the best that could be made of broken Materials."[2] Among the three major irregularities of the earth that offended Burnet's sensibility, we can find—after the channels formed by the sea and subterranean caverns—the mountains and the rocks that he had seen on his Alpine travels: "These Alps we are speaking of are the greatest range of Mountains in *Europe*, and 'tis prodigious to see and to consider of what extent these heaps of Stones and rubbish are."[3]

In her classic work, *Mountain Gloom and Mountain Glory*, Marjorie Hope Nicolson drew attention to the fact that Burnet's book appeared six years before Isaac Newton's *Principia Mathematica* (1687), and she remarked: "Much more than the *Principia* which was widely acclaimed, Burnet's *Telluris theoria sacra* provoked reply, defense, attack."[4] One of the most controversial points of Burnet's theory was his argument that the earth, being deformed and composed of broken materials, given its irregularities, did not correspond to God's original model. This caused a chain of reaction. Hence, we can find statements in defense of the beauty and of the usefulness of mountains as a part of God's creation and God's providence in various contemporary physico-theological works, among which are passages from the works of John Ray (1627–1705) and William Derham (1657–1735).

In the second edition of his *Miscellaneous Discourses concerning the Dissolution and Changes of the World*, which was published in 1693 under the title *Three physico-theological Discourses,* John Ray several times expresses his opinion about "mountains and their usefulness" and even inserts a chapter entitled "A Discourse concerning the Use of the Mountains," in which he writes:

But here it may be objected, That the present Earth looks like a heap of Rubbish and Ruines; And that there are no greater examples of confusion in Nature than Mountains singly or jointly considered. . . . To which I answer, That
the present face of the Earth, with all its Mountains and Hills, its Promonto-
ries and Rocks, as rude and deformed as they appear, seem to me a very beautiful and pleasant object. . . . 2. They are useful to Mankind in affording them
convenient places for habitation . . . serving as Skreens to keep off the cold and
nipping blasts of the Northern and Easterly Winds. . . . 3. A Land so distinguished into Mountains, Valleys and Plains is also most convenient for the
entertainment of the various sorts of Animals, which God hath created. . . . 4.
The Mountains are most proper for the putting forth of Plants; yielding the
greatest variety, and the most luxuriant sorts of Vegetables for the maintenance of animals . . . , and for medical Uses.

He also mentions the importance of metals, minerals, and fossils, stressing
the role of the mountains for the generation and maintenance of rivers and
fountains and as boundaries and bulwarks against incursions of enemies.[5]

Ray added this chapter about the usefulness of mountains to the second
edition of his best-selling work *The Wisdom of God Manifested in the Works
of the Creation* (the first edition appeared in 1691), in which he openly criticized Burnet's theory.[6] William Derham, in turn, was inspired by Ray's works.
In his *Physico-Theology: Or, A Demonstration of the Being and Attributes of
God, from His Works of Creation* (the first edition appeared in 1713), he repeatedly refers to Ray and expresses similar views. In the fourth chapter of his
Physico-Theology, entitled "Of the Mountains and Valleys," Derham writes
that "my survey now leads me to shew, that the Mountains are so far from being a Blunder of Chance, a work without Design, that they are a noble, useful yea, a necessary Part of our Globe."[7] On the following page, he specifies
the various positive aspects of hills and mountains. The air is "colder, more
subtile," and healthier; mountains and hills are "Skreens to keep of the cold
and nipping Blasts of the Northern and Easterly Winds" (a direct quotation
from Ray's *Wisdom of God*). A few lines below, we read: "Another benefit of
the Hills is, that they serve for the Production of great Varieties of Herbs and
Trees. . . . Mountains do especially abound with different Species of V e g e -
tables, because of the great Diversity of Soils that are found there. . . . Mountains serve for the harbour for various animals," To support his arguments,
Derham quotes Theophrastus and John Ray. Mountains and hills are also
important for the generation of spring water and for the generation of minerals

and metals; they are "Boundaries and Bulwarks for several Nations of the Earth," as Ray argued earlier on.[8]

In concluding this chapter, Derham observes: "The Hills are a grand Agent in so noble and necessary a Work: . . . the admirable Tools of Nature, contrived and ordered by the infinite Creator . . . and so dispense this great Blessing to all Parts of the Earth."[9]

The Perception of Mountains from the Renaissance until the Eighteenth Century

The discussion of mountains in Burnet's works and in those of other physico-theological authors cannot be understood without a necessarily brief contextualization of the perception of mountains since the Renaissance. Many historians have interpreted the Renaissance as an initial positive phase of the perception of mountains, followed by a second phase, which began in the eighteenth century in the wake of literary works such as *Die Alpen* (1732), a poem by the Bernese Albrecht von Haller, and *Julie ou La nouvelle Héloïse* by the Genevan Jean-Jacques Rousseau (1761). While antiquity and the Middle Ages depicted an ambiguous image of mountains, which was at times negative (mountains as *locus horribilis*, inhabited by dragons), at times positive (as *locus amoenus*, the dwelling place of the gods), in the Renaissance there was a turning point. Humanists and travelers made contact with these giants of nature: one even began climbing mountains and developed a scientific interest in them.[10]

From this perspective, the role played by the Swiss Renaissance scholar, physician, and naturalist Conrad Gessner (1516–65) was essential. He advocated a positive perception of mountains (specifically the Swiss Alps) as a place where one could experience an "aesthetic pleasure" and discover the usefulness of nature (for botany and medicine). In a famous letter to his friend Jakob Vogel (d. 1562) of Glarus, later published as an introduction to his treatise *Libellus de lacte, et operibus lactariis, philologus pariter ac medicus* (1541), he wrote:

> I have resolved, my learned friend Avenius [Jakob Vogel], for as long as God
> grants me life, to climb at least one mountain in the season when flowers are
> in bloom: I want to *herbalize*, exercise my body and refresh the mind. How
> lovely, is it not, to admire the awesome mountain and be able to lift up one's
> head as far as the clouds! How sad, those people who spend all day long cooped
> up at home, instead of enjoying the wonders of this world; they are like dor-

mice, sleeping through the winter in their little lair, forgetting that humankind were created to admire the existence of a superior power, Almighty God. There are several other reasons that draw me irresistibly to the mountain. In particular, one is that our mountains, as is commonly accepted, are taller and richer in plants than all other regions; and this prompted a keen desire in me to study them more closely.[11]

Gessner also expressed his enthusiasm for mountains in other texts. A good example is his description of the ascent of the Frakmont (more commonly known as Pilatus, a peak overlooking Lucerne), which the scholar climbed on August 28, 1555 (the account of that hike is included in his *De raris et admirandis herbis*).[12] Similar fervor animates the dedicatory epistle to Christoph Pfäfferlin, which was subsequently published by Benedikt Marti (also known under the Latin name Aretius, 1522–74) in his account of the ascent of the Stockhorn and the Niesen.[13]

There was nothing exceptional in the Zurich scholars' interest in the Alpine landscape of that period; their curiosity is part of a general movement discernible in the writings of several Swiss and European humanists. Some examples of this positive perception of mountains may be found in the description of climbing the Pilatus by Joachim von Watt (known under the name Vadianus, 1481–1551), published in his commentary on Pomponius Mela's geographic work.[14] Likewise, in the poem *Stockornias* (1536) by the Biel pastor Johannes Müller (also known as Johannes Rhellicanus, 1478/88–1542), we find a depiction of the ascent of the Stockhorn, near Bern.[15] Our review would not be complete without mentioning the poetic composition devoted to an excursion on the Ütliberg (near Zurich), written by Theodor Ambühl (also known as Collinus, 1535–1604).[16]

In his famous *De Alpibus commentarius* (Zurich, 1574) another Zurich humanist, Josias Simler (1530–76), presented a similarly positive image of the Alps. It was at that time that mountains gradually began to be seen as a typical component of the Swiss landscape. In *De Alpibus*, the Alpine range was considered as a whole, presented from a historical, geographic, and naturalistic standpoint.

Although the publications mentioned so far belong to different literary genres, noticeably all of them evince a new way of relating to mountain landscapes with a lively interest and a scientific motivation: not only Gessner but Marti and Müller, too, were animated by a desire to explore nature and to discover new local vegetable and animal species.

We can find other examples of scholars advancing this trend in Italy, such as Valerio Faenzi in Verona, who wrote a scientific disputation about the formation of mountains, and Francesco Calzolari, who pointed out their importance for botany.[17]

During the seventeenth century scholars' interest in mountains seems to weaken: a possible explanation for this phenomenon is the century's relatively long period of instability. Thomas Burnet's work is indicative of this phase, which Marjorie Nicholson has labeled "mountain gloom."[18]

On the other hand, Johann Jakob Scheuchzer and Élie Bertrand are two clear examples of a revitalized form of the Renaissance debate, in which mountains are not only seen as useful because of the particularities of their nature (flora, fauna, and minerals) but are also admired for their beauty. Scheuchzer and Bertrand, among other scholars including Haller and Rousseau, contributed in a very crucial way to the developing perception of a new dimension of beauty during the eighteenth century, which was defined as *delightful horror* (the aesthetics of the sublime). Bertrand, however, went a step further than Scheuchzer in his valorization of mountain landscape: "Mountains are beautiful independently of their usefulness," he wrote in his *Essai sur les usages des montagnes* (1754).[19] The importance of Scheuchzer's and Bertrand's views lies not only in the valorization of mountains as useful and beautiful landscapes. In addition, in their research on alpine territories, they collected much information and many specimens, and—in so doing—they showed the variety of the alpine natural resources and "objectified" these places as worthy of scientific interest.[20]

Mountains in Johann Jakob Scheuchzer's and Élie Bertrand's Works

In Switzerland, echoes of the debate about the origin of the earth and about mountains can be encountered in Scheuchzer's works and, some decades later, in Bertrand's. Scheuchzer studied medicine and mathematics in Germany and in Holland and in 1695 took up the position of chief medical officer of the Foundling Hospital in Zurich. Some years later he was appointed professor of mathematics and then of physics in the city's most prestigious college, the Collegium Carolinum, which trained theologians to enter the Reformed Church. He was also named curator of the public library, the Bürgerbibliothek, and of the Kunstkammer (a sort of public museum of the city). He was awarded with membership in the Royal Society, in the Academia Leopoldina, and in the Accademia degli Inquieti in Bologna. Bertrand was trained as a theologian and acted as a special counselor to the king of Poland, Stanislas II August Poniatowski (1732–98), from 1765 to 1767, when he returned to Switzerland. He settled in Yverdon, where he established a library and founded the Société économique of Yverdon. He also served the Ökonomische Gesellschaft of Bern during many years as its French-speaking secretary.

Scheuchzer and Bertrand shared an excellent knowledge of the British physico-theological production and were both in touch with several savants of the island. Bertrand translated into French William Derham's *Astro-Theology*, the first edition of which was published in London in 1715. The book appeared in Zurich in 1760.[21] Scheuchzer was in close contact with the diluvian theorist John Woodward (1675–1728) from 1694 onward. They exchanged letters regularly between 1701 and 1726, and during this period Scheuchzer was a very active supporter of Woodward's interpretation of the European continent. In 1704 he translated Woodward's essay into Latin, which considerably enhanced its dissemination. In his treatise *Essay toward a Natural History of the Earth* (1695) Woodward severely criticized Burnet's *Sacred Theory*, affirming the similarity between the pre- and post-diluvian world. For Woodward, that is, the mountains had already existed before the biblical Flood. One of the most important arguments used by Woodward was that fossils were organic remainders of prediluvian animals or plants. Scheuchzer also collaborated with William Derham in collecting barometrical measurements in Zurich (Scheuchzer) and in Upminster (Derham); the results of their measurements were published in the *Philosophical Transactions* in 1708–9.[22]

Scheuchzer's and Bertrand's activities allow us to observe a shift in the debate about the origin and usefulness of mountains, which forced British savants to move from general considerations to a more specific discussion focusing on the Alps. In his *Helvetiae stoicheiographia*, the first book of his *Natural history of Switzerland* (first published in 1716) Scheuchzer defines the Alps as the most important gift of God to his country. The Alps play a central role as water reservoir for the whole of Europe (Scheuchzer mentions France, Italy, Germany, and Holland), a topic that we have already found in Ray's and Derham's works: "I only want to mention some aspects of the treasure of water, that our mountains deliver not only to the inhabitants of our country, but also to those of France, Italy, Germany and the Low Countries; as a result, thanks to God's dispensation, our mountains can be considered rich sources of the precious water needed by the European countries and peoples."[23] As for Ray and Derham, mountains and hills serve important functions as boundaries, as bulwarks, and for protection against enemies: "Our fortresses that allow us to sleep peacefully are our high mountains created not by humans but by Almighty God's wisdom; these walls protect our intellectual and physical liberties against foreign powers."[24] Like Ray and Derham, Scheuchzer also highlights the usefulness and richness of mountains: "Foreigners who cross our country think that it is rough and wild. But we can easily demonstrate the contrary, because our country possesses so many and such wonderful marvels and gifts of nature, which are impossible to find elsewhere."[25]

Élie Bertrand placed himself in the same tradition. In his writings we come across several references not only to Scheuchzer but also to his British colleagues, especially to Ray and to Derham. In his *Mémoires sur la structure interieure de la terre*, published in 1752 in Zurich, Bertrand repeatedly quotes the latter. The conclusion of his third *mémoire*, in which he analyzes the origin of fossils and of mountains, is very revealing:

> The relation that mountains and the arrangement of their layers bear to their usefulness to animals and humans, to the growth of plants, the very conservation of the globe, and the universal circulation of all things, proves clearly that it is not the result of chance, the effect of confusion, or the product of some blind movement. The more we know nature, the more perceptible is this truth. One would have to be completely biased or inattentive not to see in the structure of our Globe the work of a wise and powerful Being who has bound and arranged all its parts in an admirable way.[26]

Bertrand thus underlines the perfection of God's Creation and indirectly attacks scholars such as Burnet who refuse to see the magnificence of God reflected by the perfection of the earth. Two years later, in his essay *Essai sur les usages des montagnes*, he once more insists on the usefulness and on the beauty of the mountains; they are a wonderful example of God's providential care for humanity: "Come . . . into the mountains that the divine power raised up with so much majesty and resourcefulness, . . . admire the masterpieces of the beneficent hand of the one who, arranging the earth for our residence, so generously provided for our conservation, for our maintenance and even for our enjoyment. The beauty, and the necessity, the usefulness and the purposes of the mountains . . . —those are objects worthy of occupying everyone's minds."[27] In his chapter "On the Beauty of the Mountains," Bertrand stresses the aesthetic aspects of the mountainous landscapes as a source of inspiration for poets and painters. The diversity of this landscape, a topic that we have also encountered in the works of Ray and Derham, inspires artists.[28] Bertrand believed that the Swiss Alps were the highest mountains of Europe, and he reiterates the reflection, familiar from reading Ray, Derham, and Scheuchzer, that the mountains were the most important water reservoir for the whole European continent: "It is generally agreed that the Swiss mountains are the highest in Europe, for there are various rivers that carry the waters stemming from them to diverse, very distant seas."[29] And the mountains, again as in the works of Ray, Derham, and Scheuchzer, are hailed as an important protection for a country: "You must consult the chronicles of our country, in particular. . . . There you will discover how mighty enemies have been overcome by a handful of men descending on them from the mountains. . . . The terrain, together with courage and the will to attack, have occasioned valuable prodigies that have brought liberty to these fortunate regions and have maintained and confirmed it to this day."[30]

Conclusion

The various passages I have quoted from the works of John Ray, William Derham, Johann Jakob Scheuchzer, and Élie Bertrand about the role of mountains as an important part of God's providence and their usefulness for humankind are a significant example of the circulation and reception of this particular discourse between Britain and Switzerland. However, the arguments used by the physico-theologians—for example, the idea that mountains were a water reservoir and a harbor for various animals and plants because of the great diversity of their soils, that they could serve as bulwarks against enemies, and that they were, last but not least, beautiful—are arguments that were developed by Swiss scholars of the Renaissance. That generation of Swiss men were among the most important protagonists of a new approach to the Alps: they climbed mountains and collected all possible information about their flora, fauna, and minerals. As a botanist, John Ray was familiar with the works of Conrad Gessner, and Ray repeatedly quoted the Swiss botanist in his *Historia plantarum generalis* (1693–1704).

The debate about the mountains shows, in my view, two important elements that help us to understand the development of physico-theology, as well as the transfer of knowledge between Britain and the Continent at this time. First, I think it is important to develop a diachronic analysis of the physico-theological works: the debate about the origin of the earth shows the reception, circulation, and reuse of arguments about mountains that derive from the earlier work of Renaissance scholars, especially from naturalists, although they were not openly quoted. Second, knowledge transfer is always complex: British physico-theological scholars reused the arguments about the utility of the mountains in the wider debate about the perfection of God's creation and about the possibility of proving God's existence and magnificence. Likewise, they used them to demonstrate the perfection of nature, the usefulness of every part of it, and the veracity of the biblical account contained in the Book of Genesis. Originally specific arguments assumed a universal value and became a fundamental pillar of the rhetorical strategy used for proving the perfection of the earth. At a later date, the same arguments were received by Swiss scholars who incorporated them in the long-standing tradition of a political interpretation of the Alps as the place where freedom and democracy were born. In advancing a new aesthetics of the sublime, Scheuchzer and Bertrand, among others, transformed the general debate about mountains by arguing for the perfection of their mountainous country. Their tendency to celebrate the mountains as an expression of divine sublimity began to lead away from physico-theologians' erstwhile appreciation of a mechanical view of nature and, thus, pointed forward to the coming of romanticism. At the same time, Scheuchzer's and Bertrand's interest in the study of the natural history of mountains also pointed forward to the Alps' transformation into an open-air laboratory for geological research.

NOTES

1. Boscani Leoni, "Conrad Gessner"; Maissen, "Die Bedeutung der Alpen"; Marchal, "Johann Jakob Scheuchzer."

2. Burnet, *Telluris theoria sacra*, 91, 102; Burnet, *Sacred Theory of the Earth* (1697), 74, 86–87. See Nicolson, *Mountain Gloom*, chap. 5, esp. 195–200. See also Poole, *World Makers*.

3. Burnet, *Sacred Theory of the Earth* (1691), 111; (1697), 97.

4. Nicolson, *Mountain Gloom*, 187.

5. Ray, *Physico-theological Discourses*, 35–43.

6. Ray, *Wisdom of God*, 2nd ed. (1692), first part, 199–206: "But because Mountains have been look'd upon by some as Warts, and superfluous Excrescencies, of no use or benefit, . . . rather as signs and proofs, that the present Earth is nothing else but a heap of Rubbish and Ruins, I shall deduce and demonstrate in Particulars, the great Use, Benefit, and Necessity of them. I. They are of eminent Use for the Production and Original of Springs and Rivers. . . . II. They are of great Use for the Generation, and convenient digging up of Metals and Minerals; which how necessary Instruments they are of Culture and Civility I have before shewn. . . . III. They are useful to Mankind in affording them convenient Places for Habitation, and Situations of Houses and Villages, serving as Skreens to keep off the cold and nipping Blasts of the *Northern* and *Easterly* Winds [italic by Ray]. . . . IV. They are very ornamental to the Earth,

affording pleasant and delightful Prospects. . . . V. They serve for the production of great variety of Herbs and Trees. . . . VI. They serve for the Harbour, Entertainment, and Maintenance of various Animals, Birds, Beasts, and Insects, that breed, feed and frequent there. . . . The highest Ridges of many of those Mountains [i.e., the Alps] serve for the Maintenance of Cattel for the Service of the Inhabitants of the Valleys." Ray mentions, among other authors' works, Keill, *Dr. Burnet's theory.*

7. Derham, *Physico-Theology*, 6th ed. (1723), 70–71.
8. Derham, *Physico-Theology*, 6th ed. (1723), esp. 70–83.
9. Derham, *Physico-Theology*, 6th ed. (1723), 75–76.
10. The bibliography on the discovery of the Alps and mountain aesthetics is very substantial. Here, I only mention Nicolson, *Mountain Gloom*; Dirlinger, *Bergbilder*; Mathieu and Boscani Leoni, *Die Alpen!*; Mathieu, "Globalization of Mountains Perception"; Korenjak, "Why Mountains Matter." About Gessner and the Alps: Zoller, "La découverte des Alpes de Pétrarque à Gessner."
11. Gessner took the booklet to his friend Avenius when he went to visit him in Glarus. See Gessner, *Libellus de lacte*; Steiger, "Gessner und die Berge," 206. A German translation of the letter was published by Weiss, *Die Entdeckung der Alpen*; for Gessner, see pp. 1–12. William A. B. Coolidge published a collection of Alpine texts in *Simmler et les origines de l'alpinisme*. On Gessner: Boscani Leoni, "Conrad Gessner."
12. "Descriptio Montis Fracti sive Montis Pilati ut vulgo nominant iuxta Lucernam in Helvetia," in Gessner, *De raris et admirandis herbis*, 43–67.
13. Benedikt Marti was rector of the Latin School in Bern and professor of Latin and Greek. As a reputed botanist, he is known for being the first to cultivate tobacco in Europe. Bratschi, *Aretius und Rhellicanus*, 32–70.
14. Mela, Vadianus, and Alantse, *Pomponii Melae De Orbis Situ Libri Tres, Accuratissime emendati, unà cum Commentariis Ioachimi Vadiani Heluetii castigatioribus, & multis in locis auctioribus factis.*
15. Rhellicanus was professor of Greek and philosophy in Bern and in 1541 was appointed pastor of Biel. Intending to conduct botanical research, he climbed Mount Stockhorn in 1536. The description he wrote of that experience was published in 1555. Bratschi, *Aretius, and Rhellicanus, Berg-Besteigungen im 16. Jahrhundert*, 8–31.
16. Collinus was a minister in Dietikon-Urdorf and after 1600 at St. James's Church (St. Jakob), Zurich. For information on him and Rhellicanus: Wiegand, *Hodoeporica*, 190–95.
17. Faenzi, *De montium origine*; Calzolari, *Il viaggio de monte Baldo della magnifica citta de Verona.*
18. Nicolson, *Mountain Gloom*. Considering his description of mountains as "beautiful horror," Thomas Burnet is very often seen as a precursor of the aesthetic of the sublime. Poole, *World Makers*. See also Mathieu, "Alpenwahrnehmung." In this article Mathieu convincingly demonstrates the role of different disciplines in the diverging histories of the perception of mountains.
19. Bertrand, *Essai sur les usages des montagnes*, chap. 2, p. 9. All translations are my own. The best seller of Albrecht von Haller, "Die Alpen" (1732), was published in the era between Scheuchzer and Bertrand. This poem gave an important input to developing a positive perception of wild alpine landscapes and of mountaineering. About Haller, for example, see Steinke and Stuber, "Hallers Alpen."
20. About Scheuchzer's and Bertrand's research on natural disasters (like earthquakes), with focus on mountains, see Gisler, *Göttliche Natur?*
21. Derham, *Theologie Astronomique*. An earlier translation had appeared at The Hague in 1729.

22. Derham, "Barometrical altitudes."

23. Scheuchzer, *Helvetiae stoicheiographia* (Vol. 1, *Natur-Historie des Schweitzerlandes*), 146.

24. Scheuchzer, *Helvetiae stoicheiographia*, 148.

25. Scheuchzer, *Einladungs-Brief*, 1–2.

26. Bertrand, *Memoires sur la structure interieure de la terre*, 96.

27. Bertrand, *Essai sur les usages des montagnes*, 8.

28. Bertrand, *Essai sur les usages des montagnes*, 10. In what follows, Bertrand mentions Albrecht von Haller's poem "Die Alpen."

29. Bertrand, *Essai sur les usages des montagnes*, 23.

30. Bertrand, *Essai sur les usages des montagnes*, 28.

Printed Sources

Abbadie, Jacques. *Traité de la vérité de la religion chrétienne.* Rotterdam: R. Leers, 1684.

Alsted, Johann Heinrich. *Theologia naturalis exhibens augustissimam naturae scholam, in qua creatura Dei communi sermone ad omnes pariter docendos utuntur adversus atheos, Epicureos et sophistas huius temporis.* Frankfurt am Main: Hummius, 1615.

Anon. *A Short Discourse concerning Miracles.* London: Printed for Matt. Wotton, 1702.

Aristotle. *Complete Works of Aristotle.* Edited by Jonathan Barnes. 2 vols. Princeton, NJ: Princeton University Press, 1984.

[Arnauld, Antoine]. *Nouveaux Elemens de geometrie; contenant, outre un ordre tout nouveau, & de nouvelles demonstrations des propositions les plus communes, de nouveaux moyens de faire voir quelles lignes sont incommensurables, de nouvelles mesures de l'angle, dont on ne s'estoit point encore avisé, et de nouvelles manieres de trouver & de demontrer la proportion des Lignes.* Paris: Chez Charles Savreux, 1667.

[Arnauld, Antoine, and Pierre Nicole]. *La Logique ou L'Art de penser: Contenant, outre les Regles communes, plusieurs observations nouvelles propres à former le jugement.* Paris: Chez Charles Savreux, 1662.

Arndt, Johann. *Vier Bücher, Vom wahrem Christhentumb.* 4 vols. Magdeburg, 1610. Edited by Johann Anselm Steiger. 3 vols. Hildesheim: G. Olms, 2007.

Astruc, Jean. Report on an account about fossils at Boutonnet by Jean Astruc, presented in Montpellier on December 17th, 1707. *Mémoires de Trévoux* 8 (1708): 512–25.

Augustinus, Aurelius. *Enarrationes in Psalmos 1–50.* Vienna: Verlag der Österreichischen Akademie der Wissenschaften, 2003.

Bacon, Francis. *The Twoo Bookes of Francis Bacon of the Proficience and Aduancement of Learning.* London: Printed . . . for Henrie Tomes, 1605.

———. *The Instauratio Magna Part II: Novum Organum.* Vol. 11 of *The Oxford Francis Bacon.* Edited by Graham Rees, Maria Wakely, Michael Kiernan, et al. Oxford: Oxford University Press, 2004.

Barlaeus, Caspar. *Oratio de animae humanae admirandis: habita in . . . Amstelodamensium Gymnasio cum libros Aristotelis De anima interpretatur.* Amsterdam: Guilielmus Blaeu, 1635.

Bärmann, Georg Friedrich. "Abhandlung von den Absichten des Schöpfers, bey Darstellung der Blumen. . . ." In *Der deutschen Gesellschaft in Leipzig eigene Schriften und Uebersetzungen,* 3:278–316. Leipzig: Breitkopf, 1739.

Barrow, Isaac. *Works of the Learned Isaac Barrow.* 4 vols. London: M. Flesher, 1683.

Benemann, Johann. *Gedancken über das Reich derer Blumen, bey müssigen Stunden gesammelt.* Dresden und Leipzig: Walther, 1740.

Bentley, Richard. *A Confutation of Atheism from the Structure and Origin of Humane Bodies.* London: Printed for Henry Mortlock, 1692.

———. *The Folly and Unreasonableness of Atheism.* London: Printed by J.H. for H. Mortlock, 1693.

Bertrand, Élie. *Mémoires sur la structure interieure de la terre.* Zurich: Heidegger et Compagnie, 1752.

———. *Essai sur les usages des montagnes. Avec une lettre sur le Nil.* Zurich: Heidegger et Compagnie, 1754.

Blumenbach, Johann Friedrich. *Über den Bildungstrieb.* Göttingen: Johann Christian Dieterich, 1789.

———. *Beyträge zur Naturgeschichte.* 2nd ed. Göttingen: Heinrich Dieterich, 1806.

Boyle, Robert. *New Experiments Physico-Mechanicall, touching the Spring of the Air.* Oxford: Printed by Hall . . . for Thomas Robinson, 1660.

———. *Some Considerations touching the Usefulnesse of Experimental Naturall Philosophy.* Oxford: Hen. Hall, 1663.

———. *The Excellency of Theology, compar'd with Natural Philosophy (as both are Objects of Men's Study) Discours'd of in a Letter to a* Friend. *To which are annex'd some Occasional Thoughts about the Excellency and Grounds of the Mechanical Hypothesis.* London: Printed by T.N. for Henry Herringman, 1674.

———. *The Excellency and Grounds of the Mechanical Hypothesis.* 1674. In *The Excellency of Theology,* by Boyle.

———. *Some Physico-theological Considerations about the Possibility of the Resurrection.* London: Printed by T.N. for H. Herringman, 1675.

———. *A Disquisition about the Final Causes of Natural Things.* London: Printed by H.C. for John Taylor, 1688.

———. *The Christian Virtuoso.* 1690. In *The Works,* edited by Thomas Birch. 6 vols. 5:508–40. London: J. and F. Rivington, 1772.

———. *The Correspondence of Robert Boyle.* Edited by Michael Hunter, Antonio Clericuzio, and Lawrence M. Principe. 6 vols. London: Pickering & Chatto, 2001.

Brockes, Barthold Heinrich. *Irdisches Vergnügen in Gott, bestehend in Physicalisch- und Moralischen Gedichten.* 8 vols. Hamburg: Kißner, 1721–48. 2nd ed. Vol. 5. Hamburg: König, 1740.

———. *Auszug der vornehmsten Gedichte aus dem von Herrn Barthold Heinrich Brockes in fünf Teilen herausgegebenen Irdischen Vergnügen in Gott.* Hamburg: Herold, 1738.

Buffon, Georges-Louis Leclerc, Comte de. *Histoire Naturelle, Générale et Particulière.* Vol. 1. Paris: Imprimerie Royale, 1749.

Burnet, Thomas. *Telluris theoria sacra. orbis nostri originem & mutationes generales, quas aut jam subiit, aut olim subiturus est, complectens: libri duo priores de diluvio & Paradiso.* Londini: Typis R. N. impensis Gualt. Kettilby, 1681, 1689.

———. *The Sacred Theory of the Earth. Containing an Account of the Original of the Earth, and of All the General Changes which it hath Already Undergone, or is to Undergo, till the Consummation of all Things.* Vol. 1. London: Printed by R. Norton, for Walter Kettilby, 1684. 2nd ed. 1691.

———. *An Answer to the Late Exceptions made by Mr. Erasmus Warren against the Theory of the Earth.* London: Printed by R. Norton for Walter Kettilby, 1690.

———. *The Sacred Theory of the Earth. Containing an Account of the Original of the Earth, and of the General Changes which it hath already undergone, or is to undergo till the Consummation of all Things.* London: printed by R.N. for Walter Kettilby, 1697.

Calzolari, Francesco. *Il viaggio di monte Baldo della magnifica citta di Verona . . . Nuovamente dato in luce dall' honorato . . . Francesco Calzolaris.* Venice: Vincenzo Valgrisio, 1566.

Charleton, Walter. *The Darknes of Atheism Dispelled by the Light of Nature. A Physico-Theologicall Treatise.* London: Printed by J.F. for William Lee, 1652.

———. *Exercitationes physico-anatomicae, de oeconomia animali, novis in medicina hypothesibus superstructa, & mechanice explicata.* 2nd ed. Amsterdam: Johannes Ravenstein, 1659.

———. *Oeconomia animalis.* London: R. Daniel and J. Redman, 1659.

———. *Onomasticon zoicon.* London: James Allestry, 1668.

———. *Enquiries into Human Nature.* London: Printed by M. White for Robert Boulter, 1680.

———. *The Harmony of Natural and Positive Divine Laws.* London: Printed for Walter Kettilby, 1682.

———. *Three Anatomic Lectures. Concerning 1. The Motion of the Blood . . . : 2. The Organic Structure of the Heart; 3. The Efficient Causes of the Hearts Pulsation.* London: Walter Kettilby, 1683.

Chemnitz, Johann Hieronymus. *Kleine Beyträge zur Testaceotheologie oder zur Erkäntniß Gottes aus den Conchylien in einigen Sendschreiben herausgegeben.* Frankfurt and Leipzig: Seligmann, 1760.

——. *Neues systematisches Conchylien-Cabinet fortgesetzet durch Johann Hieronymus Chemnitz.* 8 vols. Nürnberg: Raspe, 1779–96.

Cheyne, George. *Den Schepper en Zijn bestier te kennen in Zyne schepselen.* Translated by Lambert ten Kate Hermansz. Amsterdam: Pieter Visser, 1716.

Clarke, Samuel. *A Discourse Concerning the Being and Attributes of God, The Obligations of Natural Religion, and the Truth and Certainty of the Christian Revelation.* 1705. 7th ed. London: William Botham for James Knapton, 1728.

——. *The Leibniz-Clarke Correspondence.* See Leibniz.

Cowper, William. *The Letters and Prose Writings of William Cowper.* Edited by James King and Charles Ryskamp. 5 vols. Oxford: Clarendon Press, 1979–86.

Creech, Thomas, ed. and trans. *T. Lucretius Carus. The Epicurean Philosopher, His Six Books, De Natura Rerum, Done into English Verse.* 2nd ed. Oxford: L. Lichfield for Anthony Stephens, 1683.

Crooke, Helkiah. *Mikrokosmographia: A Description of the Body of Man.* London: William Iaggard, 1615.

Cudworth, Ralph. *The True Intellectual System of the Universe.* London: Richard Royston, 1678.

Darwin, Charles. *The Descent of Man.* London: Murray, 1871.

Derham, William. "Tables of the barometrical altitudes at Zurich in Switzerland in the Year 1708 observed by Dr. Joh. Ja. Scheuchzer, F.R.S. and at Upminster observed . . . by Mr. W. Derham, F.R.S. as also the rain at Pisa in Italy in 1707 and 1708, observed there by Dr. Michael Angelo Tilli, F.R.S. . . ." *Philosophical Transactions* 26 (1708–9): 332–66.

——. *Physico-Theology: Or, A Demonstration of the Being and Attributes of God, from His Works of Creation. Being the Substance of sixteen Sermons preached in St. Mary-le-Bow Church, London, . . . in the years 1711 and 1712. . . .* London: Printed for W. Innys, 1713. 2nd ed. London: Printed for W. Innys, 1714. 3rd ed. London: Printed for W. Innys, 1714. 6th ed. London: Printed for W. Innys, 1723.

——, ed. *Philosophical Letters between the late learned Mr. Ray and Several of his Ingenious Correspondents.* London: William and John Innys, 1718.

——. *Dimostrazione della Essenza, ed Attributi d'Iddio dall Opere della Sua Creazione,* di Guglielmo Derham. Tradotta dall'Idioma Inglese. Florence: Stamperia di S.A.R., 1719.

——. *Théologie physique, ou Démonstration de l'existence et des attributs de Dieu, tirée des oeuvres de la creation. . . .* Traduite de l'anglois [by H. Lufneu]. Rotterdam: J.D. Beman, 1726.

——. *God-leerende natuurkunde, of Eene betooging van Gods wezen en eigenschappen uit de beschouwing van de werken der scheppinge . . . ; Dienende tot opheldering van B. Nieuwentyt wereldbeschouwing.* Translated by Abraham van Loon. Leiden: Isaac Severinus, 1728.

——. *Physicotheologie oder Natur-Leitung zu Gott, Durch aufmerksame Betrachtung der Erdkugel und der darauf befindlichen Creaturen. Zum augenscheinlichen Beweiß, Daß ein Gott, und derselbige ein Allergütigstes, Allweises, Allmächtiges Wesen sey,* übersetzt von C.L.W . . . , zum Druck befördert von Jo. Alberto Fabricio. Hamburg: Brand, 1730.

——. *Physico-Theologie, eller: til Gud ledande Naturkunnighet . . . ifrån Tyskan och Fransöskan med hwarandra jämförde, öfværsatt af A.N.* [i.e., Anders Nicander]. Stockholm: Joh. Laur. Horrn, 1736.

——. *Astro-Theology: or, A Demonstration of the Being and Attributes of God, from a survey of the heavens.* London: W. Innys, 1715. 5th ed. London: William and John Innys, 1726.

——. *Astrotheologie, Oder Himmlisches Vergnügen in Gott, Bey aufmercksamen Anschauen des Himmels, und genauerer Betrachtung der Himmlischen Cörper, Zum augenscheinlichen Beweiß, Daß ein Gott, und derselbige ein Allergütigstes, Allweises, Allmächtigstes Wesen sey.*

Translated from the 5th ed. by Johann Albert Fabricius. Hamburg: Theodor Felginers Wittwe, 1728. 2nd ed. Hamburg: Felginer, 1732.

———. *Theologie Astronomique, Ou Demonstration De L'Existence Et Des Attributs De Dieu, Par L'Examen Et La Description Des Cieux. Enrichie De Figures.* Translated by Élie Bertrand. Zurich: Heidegguer, 1760.

Encyclopaedia Britannica; Or, a Dictionary of Arts and Sciences. 3 vols. Edinburgh: A. Bell and C. Macfarquhar, 1771.

Fabricius, Johann Albert. "De Deo ex oculi contemplatione demonstrando." In *Ad Cohonestandum Praesentia sua benevola Tirocinium Oratorium Quod posituri sunt publice Quinque Juvenes Optimae spei atque florentissimi in Classe prima Scholæ Johannæ ad D. XXI. Octobr. . . . invitat Jo. Albertus Fabricius, D. Prof. Publ. ac Scholae Rector,* 492–97. Hamburg: Neumann, 1710.

———. *Delectus Argumentorum et Syllabus Scriptorum qui veritatem religionis Christianae adversus Atheos, Epicureos, Deistas, Seu Naturalistas, Idololatras, Judeos et Muhammedanos Lucubrationibus suis asseruerunt.* Hamburg: Theodor Christoph Felginer, 1725.

———. "Verzeichnis derer alten und neuen Scribenten, die sich haben lassen angelegen seyn durch Betrachtung der Natur und der Geschöpfe die Menschen zu Gott zu führen." Foreword, addressed to Barthold Heinrich Brockes, to *Astrotheologie oder Himmlisches Vergnügen in Gott Bey aufmercksamem Anschauen des Himmels, und genauerer Betrachtung der Himmlischen Cörper,* by William Derham. Translated from the 5th ed. by Fabricius. Hamburg: Bey Theodor Christoph Felginers Wittwe, 1728. 2nd ed. 1732.

———. *Pyrotheologiae sciagraphia, oder Versuch durch nähere Betrachtung des Feuers die Menschen zur Liebe und Bewunderung ihres gütigsten, weisesten, mächtigsten Schöpfers anzuflammen.* Hamburg: Theodor Christoph Felginers Witwe, 1732.

———. *Hydrotheologie oder Versuch, durch aufmerksame Betrachtung der Eigenschaften, reichen Austheilung und Bewegung der Wasser, die Menschen zur Liebe und Bewunderung Ihres Gütigsten, Weisesten, Mächtigsten Schöpfers zu ermuntern.* Hamburg: König und Richter, 1734.

Faenzi, Valerio. *De montium origine: Testo a fronte* [1561]. Edited by Paolo Macini. Di monte in monte, 31. Verbania: Tararà, 2006.

Fénelon, François Salignac de la Mothe. *Démonstration de l'existence de Dieu, tirée de la connaissance de la nature, & proportionnée à la foible intelligence des plus simples.* Paris: J. Estienne, 1713.

———. *A Demonstration of the Existence, Wisdom and Omnipotence of God, drawn from the Knowledge of Nature, particularly of Man, and fitted to the meanest capacity,* translated . . . by the same Hand that English'd that excellent Piece [*Telemachus*]. London: Printed for W. Taylor, 1713.

———. *Augenscheinlicher Beweis, daß ein Gott sey: Hergenommen aus der Erkäntniß der Natur, und also eingerichtet, daß es auch die Einfältigen begreiffen können.* Translated by Johann Albert Fabricius. Hamburg: Benjamin Schillers Wittwe, 1714.

———. *Bewys dat God is, genoomen uit de kennisse der Natuure, en geschikt naa't geringe verstant van de allereenvoudigste.* Amsterdam: Gerard onder de Linden, 1715.

Filleau de la Chaise, Nicolas Jean. *Discours sur les Pensées de M. Pascal, où l'on essaye de faire voir quel estoit son dessein. Avec un autre Discours sur les Preuves des Livres de Moyse.* Paris: Guillaume Desprez, 1672. Republished together with *Traité où l'on montre qu'il y a des demonstrations d'une autre espece et aussi certaines que celles de geometrie, et qu'on en peut donner de telles pour la religion chrestienne.* N.p., n.d.

———. *Monsieur Pascall's Thoughts, Meditations and Prayers, touching matters moral and divine, as they were found in his papers after his death. Together with a Discourse upon Monsieur Pascall's Thoughts, wherein is shewn what was his Design. As also another Discourse* on the Proofs of the Truth of the Books of Moses. And a Treatise, wherein is made appear that there are Demonstrations of a different Nature, but as certain as those of Geometry, and that such may be given of the Christian religion.* Translated by Joseph Walker. London: Printed for Jacob Tonson, 1688.

Filmer, Robert. *An Advertisement to the Jurymen of England touching Witches.* London: Printed by I.G. for Richard Royston, 1653.

Francke, August Hermann. "Kurtzer und Einfältiger Unterricht / Wie Die Kinder zur wahren Gottseligkeit / und Christlichen Klugheit anzuführen sind. . . ." In *Oeffentliches Zeugniß Vom Werck / Wort und Dienst Gottes,* vol. 1, edited by Francke, 113–72. Halle: Verlag des Waisenhauses, 1702.

———. *Ordnung und Lehr-Art / Wie selbige in dem Paedagogio zu Glaucha an Halle eingeführet ist.* Halle: Verlag des Waisenhauses, 1702.

Gale, Theophilus. *The Court of the Gentiles. Part II: Of Philosophie.* Oxford: Printed by William Hall for Thomas Gilbert, 1670.

Galilei, Galileo. *The Assayer.* In *Discoveries and Opinions of Galileo,* selections translated by Stillman Drake, 231–80. New York: Doubleday, 1957.

———. *Lettere a Cristina di Lorena.* Edited by O. Besomi, D. Besomi, and G. Reggi. Rome: Antenore, 2012.

Gaule, John. *Select Cases of Conscience touching Witches.* London: W. Wilson, 1646.

Gessner, Conrad. *Libellus de lacte, et operibus lactariis, philologus pariter ac medicus. Cum epistola ad Jacobum Avienum de montium admiratione.* Tiguri: apud Christophorum Froschouerum, 1541.

———. *Conradi Gesneri medici de raris et admirandis herbis, quae sive quod noctu luceant, sive alias ob causas, lunariae nominantur, commentariolus. & obiter de aliis etiam rebus quae in tenebris lucent; eiusdem Descriptio montis fracti, sive Montis Pilati, iuxta Lucernam in Helvetia. his accedunt Io. Du Choul G. F. Lugdunensis, Pilati montis in Gallia descriptio. Io. Rhellicani Stockhornias, qua Stockhornus mons altissimus in Bernensium Helvetiorum agro, versibus heroicis describitur.* Tiguri: apud Andream Gesnerum F. & Iacobum Gesnerum fratres, 1555.

Glanvill, Joseph. *Essays on Several Important Subjects in Philosophy and Religion.* London: Printed by J.D. for John Baker . . . and Henry Mortlock, 1676.

[Goclenius, Rudolf]. *Disputatio Duplex Ordine VIII: I. Theologica contra errorem Monoletharum; II. Physico-Theologica de pane et vino / A Cuius Veritate Propugnanda, foeliciter Deo aspirante, Praeside Clarissimo & Auctissimo Philosopho. Dn. Rodolpho Goclenio Sen. in Celeberrima Hassiæ Academia Marpurgensi. . . . In Collegio Gocleniano M. Casparus Josephi Catto-Witzenhusanus.* Marburg: Hutwelckerus, 1610.

Goeree, Willem. *Voor-bereidselen tot de bybelsche wysheid, en gebruik der heilige en kerklijke historien: uit de alder-oudste gedenkkenissen der Hebreen, Chaldeen, Babyloniers, Egyptenaars, Syriers, Grieken en Romeinen; tot eene merkelijke verligting der Goddelike boeken, en veel andere voortreffelijke gweschriften, by-een vergadert, en door meer dan honderd naauwkeurige print-verbeeldingen opgehelderd.* Amsterdam: Willem Goeree, 1690.

———. *Mosaize historie der Hebreeuwse kerke, zoo als dezelve was in de stam-huyzen der h. vaderen des ouden verbands, voor en onder de belofte . . . : Uyt d'aller-oudste geheugnissen der Hebreen, Kaldeen, Zabeen, Egyptenaaren, Syriers, Feniciers, Grieken en Romeynen opgehelderd, en doorgans met veel naauwkeurige printverbeeldingen gestoffeerd, door en liefhebber der joodse oudheden. Vervat in vier delen.* Amsterdam: Willem and David Goeree, 1700.

Grimm, Jakob, and Wilhelm Grimm. *Deutsches Wörterbuch.* Vol. 12. Leipzig: Hirzel, 1885.

Hamberger, Georg Albert (presiding), and Georg Friedrich Beer (responding). *Deum ex inspectione cordis investigatum.* Jena: Gollner, 1692.

Herbst, Johann Friedrich Wilhelm. *Versuch einer Naturgeschichte der Krabben und Krebse. Nebst einer systematischen Beschreibung ihrer verschiedenen Arten.* 3 vols. Zurich: Fuessly (vol. 1); Berlin and Stralsund: Lange (vols. 2–3), 1782–1804.

———. *Natursystem aller bekannten in- und ausländischen Insekten, Nach dem System des Ritters Carl von Linné bearbeitet.* Vols. 3–11. Berlin: Pauli, 1788–1806.

———. *Betrachtungen zur Veredlung des menschlichen Herzens.* Berlin: Rottmann, 1792.

Hoefnagel, Joris, and Jacob Hoefnagel. *Archetypa studiaque patris Georgii Hoefnagelii.* Frankfurt: [Christoph Weigel excudit], 1592.

Hoffmann, Friedrich. *Vernünfftige Physikalische Theologie und gründlicher Beweis Des Göttlichen Wesens und dessen vollkommensten Eigenschafften Aus reifer Betrachtung Aller in der Natur befindlicher Wercke Besonders des Menschen.* Halle: Renger, 1742.

Hooke, Robert. *Micrographia.* London: Royal Society, 1665.

Horst, Jakob. *Occulta Naturae Miracula: von den wunderbarlichen Geheimnissen der Natur, und derselben fruchtbarlichen betrachtung, nicht allein nützlich, sondern auch lieblich zulesen.* Leipzig: Steinman, 1569.

———. *Occulta naturae miracula. Von den wunderbarlichen Geheimnissen der Natur und derselben fruchtbarlichen Betrachtung, nicht allein nützlich, sondern auch lieblich zulesen.* Leipzig: Hans Steinman, 1572.

———. *Levini Lemnii Occulta Naturae Miracula. Wunderbarliche Geheimnisse der Natur in des Menschen leib und Seel / auch in vielen andern natürlichen dingen / als Steinen / Ertzt / Gewechs / und Thieren / etc.* Leipzig: Hans Steinman, 1580.

———. *Occulta Naturae Miracula. Wunderbarliche Geheimnisse der Natur.* Leipzig: Steinmann, 1588.

———. *Occulta Naturae Miracula. Wunderbarliche Geheimnisse der Natur.* Leipzig: typis Voegelianis, 1605.

John, Christoph Samuel. "Auszug eines Schreibens von eben demselben an den Herausgeber, Trankenbar den 15ten Octobr. 1785." *Neue Hallesche Berichte* 32 (1790): 898–900.

Kant, Immanuel. *Kritik der reinen Vernunft.* 2nd, rev. ed. Riga: Johann Friedrich Hartnoch, 1787.

———. "The Only Possible Argument in Support of the Demonstration of the Existence of God." In *Theoretical Philosophy, 1755–1770,* edited and translated by David Walford, 111–201. Cambridge: Cambridge University Press, 1992.

———. *Critique of Pure Reason.* Translated and edited by Paul Guyer and Allen W. Wood. Cambridge: Cambridge University Press, 1998.

———. *Critique of the Power of Judgment.* Edited by Paul Guyer. Translated by Paul Guyer and Eric Matthews. Cambridge: Cambridge University Press, 2000.

Keill, John. *An examination of Dr. Burnet's theory of the earth. Together with some remarks on Mr. Whiston's New theory of the earth.* Oxford: Printed at the theater, 1698.

Kepler, Johannes. *Prodromus Dissertationum Cosmographicarum, continens Mysterium Cosmographicum.* Vol. 1. Frankfurt: Godefr. Tampach, 1621.

———. *Gesammelte Werke.* Munich: C. H. Beck, 1937–45.

———. *Selbstzeugnisse.* Edited by Franz Hammer. Translated by Esther Hammer. Stuttgart-Bad Canstatt: F. Frommann, 1971.

———. *Mysterium Cosmographicum.* Translated by A. M. Duncan. Norwalk, CT: Abarus, 1999.

Kinner, Cyprian. *Cogitationum didacticarum diatyposis summaria.* N. pl., n. pub., 1648.

———. *A Continuation of M. John-Amos-Comenius School-Endeavours.* [London]: Printed for R.L. in Monks-well Street, 1648.

Kirby, William, and William Spence. *An introduction to entomology: or, Elements of the natural history of insects.* 4 vols. London: Printed for Longman, Hurst, Rees, Orme, and Brown, 1815–26.

————. *Einleitung in die Entomologie*. Translated by Lorenz Oken. 4 vols. Stuttgart: Cotta, 1823–33.

Koerbagh, Adriaen. *Een Bloemhof van allerley Lieflikheid*. Amsterdam: gedrukt voor den Schrijver, 1668.

Krüger, Johann Gottlob. *Physicotheologische Betrachtungen einiger Thiere*. Halle: Hemmerde, 1741.

[La Peyrère, Isaac de]. *Praeadamitae, sive, Exercitatio super versibus duodecimo, decimotertio, & decimoquarto, capitis quinti Epistolae D. Pauli ad Romanos: Quibus inducuntur primi homines ante Adamum conditi*. [Amsterdam: Louis and Daniel Elzevier], 1655.

[Leibniz, Gottfried Wilhelm, and Samuel Clarke]. *The Leibniz-Clarke Correspondence*. Edited by Henry Gavin Alexander. Manchester: Manchester University Press, 1956. Reprint, 1998.

Lemnius, Levinus. *De occultis naturae miraculis ac variis rerum documentis, probabili, ratione atque artificiosa conjectura explicatis, libri IIII. Quorum duo priores nunc accuratius sunt recogniti, ac multis in locis aucti: posteriores vero duo, nunc recens adjecti, mira rerum ac sententiarum varietate, ad lectoris usum atque oblectationem exornati*. Cologne: Apud Joannem Birckmanum, 1573. [First published 1559.]

Lesser, Friedrich Christian. *Lithotheologie, das ist: Natürliche Historie und geistliche Betrachtung derer Steine, also abgefaßt, daß daraus die Allmacht, Weißheit, Güte und Gerechtigkeit des grossen Schöpffers gezeuget wird, anbey viel Sprüche der Heiligen Schrifft erklähret und die Menschen allesamt zur Bewunderung, Lobe und Dienste des grossen Gottes ermuntert werden*. Hamburg: Brandt, 1735.

————. *Insecto-Theologia, oder: Vernunfft- und Schriftmäßiger Versuch, wie ein Mensch durch aufmercksame Betrachtung derer sonst wenig geachteten Insecten zu lebendiger Erkänntniß und Bewundering der Allmacht, Weißheit, der Güte und Gerechtigkeit des grossen Gottes gelangen könne*. Frankfurt and Leipzig: Michael Blochberger, 1738.

————. *Insecto-Theologia, oder Vernunfft- und schrifftmäßiger Versuch, wie ein Mensch durch aufmercksame Betrachtung derer sonst wenig geachteten Insecten zu lebendiger Erkäntniß und Bewunderung der Allmacht, Weißheit, der Güte und Gerechtigkeit des grossen Gottes gelangen könne*. 2nd ed. Frankfurt and Leipzig: Michael Blochberger, 1740.

————. *Théologie des insectes, ou demonstration des perfections de Dieu dans tout ce qui concerne les insectes*. Edited by Pierre Lyonet. 2 vols. The Hague: Jean Swart, 1742.

————. *Testaceo-Theologia, oder gründlicher Beweiß des Daseyns und der vollkomnesten Eigenschaften eines göttlichen Wesens, aus der natürlichen und geistlichen Betrachtung der Schnecken und Muscheln, zur gebührenden Verherrlichung des grossen Gottes und Beförderung des Ihm schuldigen Dienstes*. Leipzig: Michael Blochberger, 1744.

————. "Nachricht von seinem Naturalien- und Kunstcabinet." *Hamburgisches Magazin* 3 (1748): 549–58.

————. *Teologia degl'insetti, ovvero Dimostrazione delle divine perfezioni in tutto ciò che riguarda gl'insetti del sig. Lesser, colle osservazioni del sig. Lyonnet, tradotta già dal tedesco nel francese, ed ora dal francese nell'italiano*. 2 vols. Venice: Nella stamperia Remondini, 1751.

————. *Einige kleine Schriften theils zur Geschichte der Natur, theils zur Physicotheologie gehörig*. Leipzig and Nordhausen: Johann August Cöler, 1754.

————. *Insecto-Theology: or, A demonstration of the being and perfections of God, from a consideration of the structure and economy of insects*. Edited by Pieter Lyonet. Edinburgh: Printed for William Creech; and T. Cadell, Jun. and W. Davies, London, 1799.

Lessius, Leonardus. *De providentia numinis et animi immortalitate libri duo adversus Atheos & Politicos*. Antwerp: Ex Officina Plantiniana, apud Viduam & Filios Io. Moreti, 1613.

————. *Rawleigh his ghost; Or, a feigned apparition of Syr Walter Rawleigh to a friend of his, for the translating into English, the booke of Leonard Lessius (that most learned man) entituled, De prouidentia numinis, & animi immortalitate: written against atheists, and polititians of these dayes.* Translated by A.B. [Saint-Omer?]: [G. Seutin?] Permissu superiorum. N. pl., n. pub., 1631.

Linnaeus, Carl. *Caroli Linnaei, Sueci, Doctoris Medicinae. Systema Naturae, Sive Regna Tria Naturae Systematice Proposita Per Classes, Ordines, Genera Et Species—Caroli Linnaei... Natur-Systema...* Translated into German by Johann Joachim Lange. Halle: Gebauer, 1740.

Linnaeus, Carl (presiding), and Isacus J. Biberg (responding). "Oeconomia Naturae" [1749]. In Linnaeus, *Amoenitates academicae seu dissertationes variae physicae, medicae, botanicae,* 2:1–58. 10 vols. Leiden, Stockholm, Erlangen: Cornelius Haak, 1749–90.

———. *Reflections on the Study of Nature.* 1754. Translated by J. E. Smith. 1786. In *Buffon's Natural History,* by David C. Goodman. Milton Keynes: Open University Press, 1980.

Lüthard, Christoph (presiding), and Samuel Haberüter (responding). *Disputatio physico-theologica de operibus του εξαμερου, quam in illustri Bernesium Gymnasio sub umbone viri plurimum Venerandi, Clariβimi, Doctiβimi Dn. Christophori Lüthardi, SS.Theol. Professoris... subjicit Samuel Haberüterus.* Bern: Georgius Sonnleitherus, 1660.

Luther, Martin. *Large Catechism.* 1530. Translated by F. Bente and W. H. T. Dau. In *Triglot Concordia: The Symbolic Books,* 565–773. St. Louis: Concordia Publishing House, 1921.

———. *Biblia, Das ist: Die gantze Schrifft Alten und Neuen Testaments verteutscht durch D. Martin Luther: Jetzundt Nach dem letzten in Anno 1545 bey des Authoris Lebzeiten aussgangenen Exemplar neben Unterschied der Versicul fleissig nachgedruckt...; insonderheit aber mit den schönen und kunstreichen Original-Kupfferstücken Matthaei Merians gezieret, darinnen die fürnembsten Historien artig für Augen gestellet werden, neben beygefügten nützlichen Registern.* Strasbourg: Lazari Zetzners Erben, 1630.

———. *Werke: kritische Gesamtausgabe.* 73 vols. Weimar: Böhlau, 1883–2009.

Malebranche, Nicolas. *The Search after Truth.* Edited and translated by Thomas M. Lennon and Paul J. Olscamp. Cambridge: Cambridge University Press, 1997.

Martin, Benjamin, ed. *The General Magazine of Arts and Sciences.* 14 vols. London, 1755–65.

Martinet, Johannes Florentius. *Katechismus der Natuur.* 4 vols. Amsterdam: Johannes Allart, 1777–79.

Mela, Pomponius, Joachimus Vadianus, and Lucas Alantse. *Pomponii Melae De Orbis Situ Libri Tres, Accuratissime emendati, unà cum Commentariis Ioachimi Vadiani Heluetici castigatioribus, & multis in locis auctioribus factis: id quod candidus lector obiter, & in transcursu facile deprehendet.* Basel: Cratander, 1522.

Menz, Johann Friedrich. *Generatio. ΠΑΡΑΔΟΞΟΣ in Rana. Conspicva.* Leipzig: Titius, 1724.

Merian, Matthaeus. *Icones Biblicae: praecipue Sacrae Scripturae historias eleganter & graphice repraesentantes—Biblische Figuren darinnen die fürnembsten Historien in heiliger göttlicher Schrift begriffen... an Tag gegeben durch Matthaeum Merian von Basel....* 4 pts. Strasbourg: Lazari Zetzners Erben, [1625]–30.

Mersenne, Marin. *L'impieté des deistes, athees, et libertins de ce temps.* Paris: P. Bilaine, 1624.

Meyer, Gerhard (presiding), and Balthasar Mentzer (responding). *Disputationum Hamburgensium Decima aranearum telas divinae existentiae teste in scenam producit.* Hamburg: Conrad Neumann, 1697.

Moeller, Johann. *Similitudines Physico-Theologicae: Das ist: Mencherley schöne nützliche und Geistliche Gleichnüsse.* Leipzig: Gleditsch, Weidmann, 1696.

Moffett, Thomas. *Insectorum sive minimorum animalium theatrum: Olim ab Edoardo Wottono, Conrado Gesnero, Thomaque Pennio inchoatum: tandem Tho. Moufeti Londinatis opera sumptibusque maximis concinnatum, auctum, perfectum.* London: Thomas Cotes, 1634.

Monti, Giuseppe. *De Monumento Diluviano nuper in Agro Bononiensi Detecto Dissertatio.* Bologna: Rossi & socios, 1719.

More, Henry. *Conjectura Cabbalistica. Or, a Conjectural Essay of Interpreting the Mind of Moses.* London: Flesher, 1653.

————. *An Antidote against Atheism, or, An appeal to the natural faculties of the mind of Man, whether there be not a God.* London: Roger Daniel, 1653. Reprinted in *A Collection.*

————. *An Explanation of the Grand Mystery of Godliness.* London: Flesher, 1660.

————. *A Collection of Several Philosophical Writings.* 2nd ed. London: Flesher, 1662.

————. *Divine Dialogues, containing Sundry Disquisitions and Instructions concerning the Attributes of God and his Providence in the World.* 2nd ed. London: J. Downing, 1713.

Müller, Theodor. *Meditationes Physico-Theologicae, Geistlicher Andacht-Wecker aus der Natur: Wie der Mensch durch die Betrachtung des Werckes der Erschöpffung/ und der Creaturen Gottes zur geistlichen Andacht/ Erkäntnis Gottes/ und Gottesfurcht kan gebracht werden.* Hamburg: Rebenlein, 1642.

Neickel, Caspar Friedrich. *Museographia Oder Anleitung Zum echten Begriff und nützlicher Anlegung der Museorum oder Raritäten-Kammern.* Breßlau and Leipzig: Hubert, 1727.

Neuere Geschichte der Evangelischen Missions-Anstalten zu Bekehrung der Heiden in Ostindien (= *Neue Hallesche Berichte*). Edited by Johann Ludwig Schulze et al. 95 vols. Halle: Verlag des Waisenhauses, 1770–1848.

Newton, Isaac. *Philosophiae Naturalis Principia Mathematica.* Bk. 1. 1687. Translated by A. Motte. 1729. Edited by F. Cajori. Berkeley: University of California Press, 1934.

————. "Letter to Richard Bentley." 1692. In *Newton's Philosophy of Nature: Selections from His Writings,* edited by H. S. Thayer, 46–50. New York: Hafner, 1953.

[Nicole, Pierre]. "Discours contenant en abregé les preuves naturelles de l'existence de Dieu & de l'immortalité de l'ame." In *De l'Education d'un prince. Divisée en trois Parties, dont la derniere contient divers Traittez utiles à tout le monde,* 119–42. Paris: Chez la veuve Charles Savreux, 1670.

————. *Traité de la foy humaine.* In *Les Imaginaires, et les Visionnaires. Traité de la foy humaine. Jugement equitable, tiré des oeuvres de S. Augustin. Lettre de Messire Nicolas Pavillon . . . ,* 485–647. Cologne: Chez Pierre Marteau, 1683. [First published 1664.]

Nieuwentijt [*also* Nieuwentyt], Bernard. *Het regt gebruik der werelt beschouwingen, ter overtuiginge van ongodisten en ongelovigen.* Amsterdam: J. Wolters and J. Pauli, 1715.

————. *Gronden van zekerheid, of de regte betoogwyse der wiskundigen.* Amsterdam: J. Pauli, 1720.

————. *The religious philosopher: or, the right use of contemplating the works of the creator.* Translated by John Chamberlayne. 3rd ed. London: Printed for J. Senex . . . , E. Taylor . . . , W. Innys . . . , and J. Osborne . . . , 1724.

————. *L'Existence de Dieu démontrée par les merveilles de la nature.* Paris: Vincent, 1725.

————. *Die Erkänntnüß Der Weisheit, Macht und Güte Des göttlichen Wesens aus dem rechten Gebrauch der Betrachtungen aller irdischen Dinge dieser Welt: Zur Überzeugung derer Atheisten und Unglaubigen . . . Samt einer Vorrede von Christian Wolffen. . . . Und mit nützlichen Registern vermehret von Wilhelm Baumann. . . .* Frankfurt: Pauli, 1732.

Paley, William. *Natural Theology.* 1802. Edited with an introduction and notes by M. Eddy and D. Knight. Oxford: Oxford University Press, 2006.

Parker, Samuel. *Tentamina physico-theologica de Deo, sive, Theologica scholastica ad normam novae & reformatae philosophiae concinnata.* London: Typis A. M., 1665.

————. *A Free and Impartial Censure of the Platonick Philosophie Being a Letter Written to his much Honoured Friend Mr N.B.* Oxford: W. Hall, 1666.

————. *Disputationes de Deo et providentia divina.* London: typis M. Clark, impensis Jo. Martyn, 1678.

Pascal, Blaise. *Pensées de M. Pascal sur la religion, et sur quelques autres sujets, qui ont esté trouvées après sa mort parmy ses papiers.* Paris: Guillaume Desprez, [preedition 1669], 1670. Facsimile with an introduction by Georges Couton and Jean Jehasse. Saint-Etienne: Centre interuniversitaire d'éditions et de rééditions, 1971.

———. *Monsieur Pascall's Thoughts, Meditations and Prayers, touching matters moral and divine, as they were found in his papers after his death. Together with a Discourse upon Monsieur Pascall's Thoughts, wherein is shewn what was his Design. As also another Discourse on the Proofs of the Truth of the Books of Moses. And a Treatise, wherein is made appear that there are Demonstrations of a different Nature, but as certain as those of Geometry, and that such may be given of the Christian religion.* Translated by Joseph Walker. London: Printed for Jacob Tonson, 1688.

———. *Thoughts*, translated by W. F. Trotter; *Letters*, translated by M. L. Booth; *Minor Works*, translated by O. W. Wight. With introductions and notes. Harvard Classics, vol. 48. New York: P. F. Collier & Son, 1910.

———. *Œuvres complètes*. Edited by Jean Mesnard. Vols. 1–2. Paris: Desclée de Brouwer, 1964, 1970.

———. *Pensées*. Edited by Philippe Sellier. Paris: Classiques Garnier, 1991.

———. *Pensées*. Translated by W. F. Trotter with an introduction by T. S. Eliot. New York: E. P. Dutton, 1958. Reprint, Mineola, NY: Dover, 2018.

Pierquin, Jean. *Dissertations physico-théologiques, Touchant la Conception de Jésus-Christ dans le sein de la Vierge Marie sa Mère.* Amsterdam: [Jacques de la Rue], 1742. Reprinted as *Dissertation physico-théologique. . . .* Grenoble: Jérôme Millon, 1996.

Pliny the Elder. *The historie of the world: Commonly called the natural historie of C. Plinius Secundus.* Translated by Philemon Holland. London: Printed by Adam Islip, 1601.

———. *Natural history.* Translated by John Bostock and H. T. Riley. 6 vols. London: H. G. Bohn, 1855–57.

Pluche, Noël-Antoine. *Le Spectacle de la nature, ou Entretiens sur les particularités de l'histoire naturelle. . . .* 8 vols. Vol. 1: *Ce qui regarde les animaux et les plantes.* Paris: Veuve Etienne, 1732. Vol. VIII, 1–2: *Ce qui regarde l'homme en société avec Dieu.* Paris: Estienne, 1750. New ed. Paris: Veuve Estienne et fils, 1749–52.

———. *Catalogue des livres de feu M. l'Abbé Pluche.* Paris: Estienne, 1763. Reedited in *Trois introductions à l'Abbé Pluche: sa vie, son monde, ses livres*, by Benoît De Baere, 117–75. Geneva: Droz, 2001.

———. *Concorde de la géographie des différens âges.* Paris: Frères Estienne, 1764.

———. *Spectacle de la Nature or Nature Display'd. Being Discourses on such Particulars of Natural History.* Vol. 3. London: Davis et al., 1766.

Rathlef, Ernst Ludewig. *Akridotheologie Oder Historische und Theologische Betrachtungen über die Heuschrekken, bei Gelegenheit der ietzigen Heuschrekken in Siebenbürgen, Ungern, Polen, Schlesien und Engelland, nebst einer Muthmassung, daß die Selaven, welche die Israeliten zweimal in der Wüsten gegessen, weder Wachteln, noch Heuschrekken, sondern die Vögel Seleuciden gewesen.* Pt. 1. Hannover: Ernst Ludwig, 1748.

———. *Akridotheologie Oder Historische Physikalische und Theologische Betrachtungen über die Morgenländischen Heuschrekken, bei Gelegenheit ihrer Züge in Europa in den Jahren 1747, 1748, 1749.* Pt. 2. Hannover: Ernst Ludwig, 1750.

Ray, John. *Catalogus plantarum circa Cantabrigiam nascentium.* Cambridge: John Field, 1660.

———. *Historia plantarum generalis species hactenus editas aliasque insuper multas noviter inventas & descriptas complectens.* 3 vols. London: Samuel Smith and Benjamin Walford, 1693–1704.

———. *The Wisdom of God Manifested in the Works of Creation. Being the Substance of Some Common Places Delivered in the Chappel of Trinity-College in Cambridge.* London: Printed for Samuel Smith, 1691. 2nd ed., very much enlarged. London: printed for Samuel Smith, 1692. 4th ed. corrected and very much enlarged. London: Printed by J.B. for Sam. Smith, 1704. 7th ed. London: R. Harbin, 1717. 11th ed. London: Innys, 1743.

———. *L'Existence et la sagesse de Dieu manifestées dans les oeuvres de la Création*, traduit de l'anglois. Utrecht: C. Broedelet, 1714.

———. *Gloria Dei oder Spiegel der Weissheit und Allmacht Gottes*. Translated by Caspar Calvör. Goslar: Johann Christoph König, 1717.

———. *Gods wysheid geopenbaard in de werken der schepping*. Amsterdam: Isaak Tirion, 1732.

———. *Miscellaneous Discourses concerning the Dissolution and Changes of the World*. London: Printed for Samuel Smith, 1692.

———. *Three physico-theological Discourses. Concerning I. The primitive chaos and Creation of the World. II. the general deluge, its Causes and Effects. III. the dissolution of the world, and Future Conflagration*. (First published 1692.) 2nd ed. corrected, very much enlarged, and illustrated with copper-plates. London: Printed for Sam. Smith, 1693. 3rd ed. London: Printed for William Innys, 1713.

———. *De werelt van haar begin tot haar einde of dry natuurkundige godgeleerde redeneringen*. Rotterdam: Barent Bos, 1694.

———. *Sonderbares Klee-Blätlein, der Welt Anfang, Veränderung und Untergang*. [German translation of *Three physico-theological Discourses*; translator: n.n.]. Hamburg: Thomas von Wiering, 1698.

———. *The Correspondence of John Ray*. Edited by Edwin Lankester. London: Ray Society, 1848.

———. *Further Correspondence of John Ray*. Edited by Robert W. T. Gunther. London: Ray Society, 1928.

Réaumur, René-Antoine Ferchault de. *Mémoires pour servir à l'histoire des insectes*. 6 vols. Paris: Imprimerie royale, 1734–42.

———. "Histoire des teignes ou des insectes qui rongent les laines et les pelleteries. 1ère partie." In *Histoire de l'Académie royale des sciences. Année 1728. Avec les Mémoires de Mathématique et de Physique pour la même année*. Paris: Durand, 1753.

Rohr, Julius Bernhard von. *Phyto-Theologia: Oder Vernunfft- und schriftmäßiger Versuch, Wie aus dem Reiche der Gewächse die Allmacht, Weisheit, Güte und Gerechtigkeit des grossen Schöpfers und Erhalters aller Dinge von den Menschen erkannt . . . werden möge*. Frankfurt and Leipzig: Michael Blochberger, 1740.

Rottler, Johann Peter. "Herrn Rottlers Reise nach Ceylon und der Malabarküste, vom 2ten Januar bis 4ten August 1788." *Neue Hallesche Berichte* 37 (1790): 27–94.

Rousseau, Jean-Jacques. *Collected Writings of Rousseau*. Edited by Roger D. Masters and Christopher Kelly. Vol. 8. Hanover, NH: University Press of New England, 2000.

Sabunde, Raimundus [*also* Raymund Sebond]. *Theologia naturalis seu liber creaturarum*. 1434–36. Deventer: Richardus Pafraet, [ca. 1484–85].

Sander, Heinrich. *Ueber Natur und Religion für die Liebhaber und Anbeter Gottes*. 2 vols. Frankfurt: n. pub., 1779–80.

Scheuchzer, Johann Jakob. *Einladungs-Brief, zu Erforschung natürlicher Wunderen, so sich im Schweitzer-Land befinden*. [Zurich]: n. pub., 1699. Reprinted in Simona Boscani Leoni (ed.), *"Unglaubliche Bergwunder." Johann Jakob Scheuchzer und Graubünden. Ausgewählte Briefe 1699–1707*. Chur: Bündner Monatsblatt / Cultura Alpina, 2019, 33–49.

———. *Beschreibung der Natur-Geschichten des Schweizerlands*. 3 pts. Zurich: Michael Schaufelb[erger] s.E. and Christoff Hardmeier, 1706–8.

———. *Piscium Querelae et Vindiciae*. Zurich: Sumtibus Authoris, Typis Gessnerianis, 1708. Reprinted with a French translation in *Johann Jakob Scheuchzer: Les fossiles témoins du déluge*, edited by Jean Gaudant, 26–107. Paris: Presses de l'école des mines, 2008.

———. *Physica oder Natur-Wissenschaft*. 2nd ed. 2 vols. Zurich: n. pub., 1711.

———. *Helvetiae stoicheiographia, orographia et oreographia, oder, Beschreibung der Elementen, Grenzen und Bergen des Schweitzerlands.* Zurich: In der Bodmerischen Truckerey (Helvetiae Historia Naturalis, oder, Natur-Historie des Schweitzerlandes), 1716.

———. *Geestelyke natuurkunde.* Translated by F. H. J. van Halen. 6 vols. Amsterdam: Schenk, 1728–38.

———. *Kupfer-Bibel, In welcher die Physica sacra, Oder Geheiligte Natur-Wissenschafft derer In Heil. Schrifft vorkommenden Natürlichen Sachen, deutlich erklärt und bewährt.* 4 vols. Augsburg: Johann Andreas Pfeffel [editor]; Ulm: Christian Ulrich Wagner [printer], 1731–35.

———. *Natur-Geschichte des Schweitzerlandes, samt seinen Reisen über die Schweitzerische Gebürge.* New edition by Johann Georg Sulzer. Zurich: David Geßner, 1746.

Schirach, Adam Gottlob. *Melitto-Theologia: Die Verherrlichung des glorwürdigen Schöpfers aus der wundervollen Biene nach der Anleitung der Naturlehre und Heiligen Gottesgelahrtheit.* Dresden: Walther, 1767.

Schmidt, Johann Andreas (presiding), and Ernst Heinrich Wedel (responding). *Auris theodeiktos.* Jena: Krebs, [1694].

Shaftesbury, Earl of (Anthony Cooper). *Characteristics of Men, Manners, Opinions, Times.* 1st ed.1711. Edited by Lawrence Klein. Cambridge: Cambridge University Press, 2000.

Sprat, Thomas. *History of the Royal Society.* London: J. Martyn, 1667.

Steno, Nicolaus. *The Prodromus of Nicolaus Steno's Dissertation concerning a solid body enclosed by process of nature within a solid. Edited and translated by John Garrett Winter.* New York: Macmillan, 1916.

Stillingfleet, Edward. *Origines Sacrae, or, A Rational Account of the Grounds of the Christian Faith.* London, 1662.

Stubbe, Henry, *A reply unto the letter written to Mr. Henry Stubbe.* Oxford: Printed for Richard Davis, 1671.

Sturm, Johann Christoph (presiding), and Johann Andreas Volland (responding). *Oculus theoscopos, hoc est De Visionis Organo et Ratione Genuina, Dissertatio Physica.* Altdorf: Meyer, 1678.

Swammerdam, Jan. *Historia insectorum generalis, ofte algemeene verhandeling van de bloedelose dierkens.* Utrecht: Meinardus van Dreunen, 1669.

———. *Ephemeri vita: of afbeeldingh van 's menschen leven, vertoont in de wonderbaarelijcke en nooyt gehoorde Historie van het vliegent ende een-dagh-levent Haft of Oever-Aas.* Amsterdam: Abraham Wolfgang, 1675.

———. *Ephemeri vita: Or the Natural history and anatomy of the ephemeron, a fly that lives but five hours.* London: Printed for Henry Faithorne, and John Kersey, 1681.

———. *Historia insectorum generalis, in qua vibrissae mutationem, seu lentae in membra epigeneseos rationes, duce experentia, redduntur, recepta vulgo insectorum metamorphosis solide refutatur.* Utrecht: Ex officina Otthonis de Vries, 1685.

———. *Biblia naturae, sive historia insectorum lingua Batava conscripta. Praefatio in qua vita auctoris descripsit Hermannus Boerhaave.* 2 vols. Leiden: Isaac Severinus et al., 1737–38.

———. *The Letters of Jan Swammerdam to Melchisedec Thévenot.* Translated by G. A. Lindeboom. Amsterdam: Swets & Zeitlinger, 1975.

Swift, Jonathan. *The Correspondence of Jonathan Swift.* Edited by Harold Williams. 5 vols. Oxford: Clarendon Press, 1963–65.

Topsell, Edward. *The history of four-footed beasts and serpents.; whereunto is now added, The theater of insects, or, Lesser living creatures by T. Muffet.* Edited by J. R. London: Printed by E. Cotes, for G. Sawbridge, T. Williams, and T. Johnson, 1658.

Triller, Daniel Wilhelm. "Zufällige Poetische Gedanken. . . ." In *Irdisches Vergnügen in Gott, bestehend in Physicalisch- und Moralischen Gedichten,* by Barthold Heinrich Brockes, vol. 2., unpaginated. 1st ed. Hamburg: Kißner, 1727.

Vallisneri, Antonio. *Nuove osservazioni, ed esperienze intorno all'ovaja scoperta ne' vermi tondi dell'uomo, e de' vitelli, con varie lettere spettanti alla storia medica, e naturale.* Padua: Seminario, 1713.

———. *Lezione Accademica intorno all'ordine della progressione e della connessione che hanno insieme tutte le cose create.* Published with Vallisneri, *Istoria della generazone dell'uomo e degli animali se sia da' vermicelli spermatici, o dalle uova.* Venice: Gio. Gabbriele Hertz, 1721.

———. *De' corpi marini que su monti si trovano.* . . . Venice: D. Lovisa, 1721.

———. *Opere fisico-mediche stampate e manoscritte di Antonio Vallisneri.* 3 vols. Venice: Coleti, 1733.

———. *Notize della vita e degli studi del Kavalier Antonio Vallisneri.* Edited and published by Giovanni Artico di Porcia, 1733. Edited by Dario Generali. Bologna: Pátron, 1986.

———. *Epistolario (1714–1729).* Edited by Dario Generali. Milan: Franco Angeli, 2006.

———. *Istoria della generazione dell'uomo e degli animali.* Edited by Maria Teresa Monti. Edizione nazionale delle opere di A. V. Florence: Olschki, 2009.

Voltaire [François-Marie Arouet]. *Facéties.* [Kehl]: Société littéraire-typographique, 1785. In *Œuvres complètes,* vol. 61. Paris: Baudouin, 1825.

———. *Remerciement sincère à un homme charitable.* In *Voltaire, The Complete Works of Voltaire,* 32a:175–208. Edited by Mark Waddicor. Oxford: Voltaire Foundation, 2006.

Walpurger, Johann Gottlieb. *Cosmotheologische Betrachtungen deren wichtigsten Wunder und Wahrheiten im Reiche der Natur und Gnaden zur Verherrlichung ihres glorwürdigen Urhebers, zur Beschämung des Unglaubens und zur allgemeinen Erbauung; schrift- und vernunftmässig ausgefertigt von Johann Gottlieb Walpurgern, Pastore Primario und Inspectore zu Waldheim,* 4 pts. Chemnitz: Stößel, 1748–54.

Webster, John. *Metallographia, or, An History of Metals.* London: Kettilby, 1671.

Whiston, William. *A New Theory of the Earth.* London: R. Roberts, 1696.

———. *Astronomical Principles of Religion, natural and reveal'd.* London: Printed for J. Senex and W. Taylor, 1717.

———. *Memoirs of the Life and Writings of Mr. William Whiston.* 2nd ed. London: Printed for J. Whiston and B. Whit., 1753.

Wilkins, John. *Ecclesiastes; or, A Discourse concerning The Gift of Preaching.* 5th ed. London: A. Maxwell, 1669.

———. *Of the Principles and Duties of Natural Religion.* London: A. Maxwell, 1675.

Willich, A. F. M., ed. and trans. *Elements of the Critical Philosophy.* London: Printed for T. N. Longman, 1798.

Wolff, Christian. *Philosophia rationalis sive Logica.* Frankfurt and Leipzig: n. pub., 1728.

———. Preface to *Die Erkänntnüß Der Weisheit, Macht und Güte Des göttlichen Wesens aus dem rechten Gebrauch der Betrachtungen aller irdischen Dinge dieser Welt: Zur Überzeugung derer Atheisten und Ungauged . . . , Same diner Varied von Christian Wolffen. . . . Und mit nützlichen Registern vermehret von Wilhelm Conrad Baumann,* by Bernard Nieuwentyt. Frankfurt and Leipzig: Johannes Pauli, 1732.

———. *Vernünfftige Gedancken von dem Gebrauche der Theile in Menschen, Thieren, und Pflanzen.* 4th ed. Halle: Renger, 1743.

Woodward, John. *An Essay toward a Natural History of the Earth and Terrestrial Bodies, especially Minerals: as also of the Sea, Rivers, and Springs: with an Account of the Universal Deluge: and of the Effects that it had upon the Earth.* London: printed for Ric. Wilkin, 1695.

———. *The Natural History of the Earth, Illustrated, Inlarged* [sic], *and Defended.* Translation of Latin original by Benjamin Holloway. London: Thomas Edlin, 1726.

Zedler, Johann Heinrich, ed. *Großes Vollständiges Universal-Lexicon.* 64 vols. Halle and Leipzig: Zedler, 1732–54.

Secondary Literature

Albrecht-Birkner, Veronika. "'Ich verspreche Ihnen nochmals feyerlich, das Wort Pietist nie wieder im übelen Sinne zu gebrauchen . . .' Fromme Identitätsfindung im späten 18. Jahrhundert." *Pietismus und Neuzeit* 42 (2016): 183–204.

Allen, Don Cameron. *The Legend of Noah: Renaissance Rationalism in Art, Science and Letters.* 1949. Urbana: University of Illinois Press, 1963.

Armogathe, Jean-Robert. "Proofs of the Existence of God." In *The Cambridge History of Seventeenth-Century Philosophy*, edited by Daniel Garber and Michael Ayers, 1:305–30. Cambridge: Cambridge University Press, 2008.

Badaloni, Nicola. *Antonio Conti, un abate libero pensatore tra Newton e Voltaire.* Milan: Feltrinelli, 1968.

Bakhuizen van den Brink, Jan Nicolaus. *De Nederlandse belijdenisgeschriften in authentieke teksten met inleidig.* 2nd ed. Amsterdam: Tom Bolland, 1976.

Barker, Peter. "The Role of Religion in the Lutheran Response to Copernicus." In *Rethinking the Scientific Revolution*, edited by Margaret J. Osler, 59–88. Cambridge: Cambridge University Press, 2000.

———. "The Lutheran Contribution to the Astronomical Revolution." In *Religious Values and the Rise of Science in Europe*, edited by J. Brooke and E. Ihsanoglu, 31–62. Istanbul: Research Centre for Islamic Art History and Culture, 2005.

Barnes, Robin B. *Astrology and Reformation.* Oxford: Oxford University Press, 2015.

Barth, Hans-Martin. *Atheismus und Orthodoxie: Analysen und Modelle christlicher Apologetik im 17. Jahrhundert.* Göttingen: Vandenhoeck & Ruprecht, 1971.

Barton, William M. *Mountain Aesthetics in Early Modern Latin Literature.* London: Routledge, Taylor & Francis Group, 2017.

Baur, Jörg. *Luther und seine klassischen Erben.* Tübingen: Mohr, 1993.

———. "Ubiquität." In *Theologische Realenzyklopädie*, 34:224–41. Berlin: De Gruyter, 2002.

———. "Ubiquität." In *Creator est creatura: Luthers Christologie als Lehre von der Idiomenkommunikation*, edited by Oswald Bayer and Benjamin Gleede, 186–302. Berlin: De Gruyter, 2007.

Bayer, Oswald. *Schöpfung als Anrede: Zu einer Hermeneutik der Schöpfung.* Tübingen: Mohr Siebeck, 1986.

———. *Martin Luthers Theologie: Eine Vergegenwärtigung.* Tübingen: Mohr Siebeck, 2003.

———, ed. *Creator est creatura: Luthers Christologie als Lehre von der Idiomenkommunikation.* Berlin: De Gruyter, 2007.

Bellucci, Dino. *Science de la nature et Réformation: La physique au service de la Réforme dans l'enseignement de Philippe Mélanchton.* Paris: Vivere, 1998.

Berg, Hein van den. "The Wolffian Roots of Kant's Teleology." *Studies in History and Philosophy of Biological and Biomedical Sciences* 44 (2013): 724–34.

Berkel, Klaas van, and Arjo Vanderjagt, eds. *The Book of Nature in Antiquity and the Middle Ages.* Leuven: Peeters, 2005.

———, eds. *The Book of Nature in Early Modern and Modern History.* Leuven: Peeters, 2006.

Bernardi, Walter. *Le metafisiche dell'embrione: Scienze della vita e filosofia da Malpighi a Spallanzani, 1672–1793.* Florence: Olschki, 1986.

Bernhardt, Reinhold. *Was heißt "Handeln Gottes"? Eine Rekonstruktion der Lehre von der Vorsehung.* Gütersloh: Christian Kaiser / Gütersloher Verlagshaus, 1999.

Beutel, Albrecht. *In dem Anfang war das Wort: Studien zu Luthers Sprachverständnis.* Tübingen: Mohr Siebeck, 1991.

———. "Wort Gottes." In *Luther Handbuch*, edited by Beutel, 362–71. 3rd ed. Tübingen: Mohr Siebeck, 2010.

Biagioli, Mario. *Galileo Courtier*. Chicago: University of Chicago Press, 1994.

Blair, Ann. "Natural Philosophy and the 'New Science.'" In *The Cambridge History of Literary Criticism*, vol. 3, *The Renaissance*, ed. Glyn Norton, 449–57. Cambridge: Cambridge University Press, 1999.

———. "Mosaic Physics and the Search for a Pious Natural Philosophy in the Late Renaissance." *Isis* 91 (2000): 32–58.

———. *Too Much to Know: Managing Scholarly Information before the Modern Age*. New Haven, CT: Yale University Press, 2010.

———. "Noël-Antoine Pluche as a Jansenist Natural Theologian." *Intellectual History Review* 26, no. 1 (2016): 91–99.

Bloomsbury Dictionary of Eighteenth-Century German Philosophers. Edited by Heiner F. Klemme and Manfred Kuehn. London: Bloomsbury Academic, 2016.

Blumenberg, Hans. *Die Lesbarkeit der Welt*. Frankfurt am Main: Suhrkamp, 1979.

Bono, James. *The Word of God and the Languages of Man: Interpreting Nature in Early Modern Science and Medicine*. Vol. 1, *Ficino to Descartes*. Madison: University of Wisconsin Press, 1995.

Boscani Leoni, Simona, ed. *Wissenschaft—Berge—Ideologien: Johann Jakob Scheuchzer (1672–1733) und die frühneuzeitliche Naturforschung / Scienza—montagna—ideologie: Johann Jakob Scheuchzer (1672–1733) e la ricercha naturalistica in epoca moderna*. Basel: Schwabe Verlag, 2010.

———. "Conrad Gessner and a Newly Discovered Enthusiasm for Mountains in the Renaissance." In *Conrad Gessner. Die Renaissance der Wissenschaften / The Renaissance of Learning*, edited by Urs Leu and Peter Opitz, 119–26. Berlin: De Gruyter, 2019.

Bots, Hans. *De Republiek der Letteren: De Europese intellectuele wereld, 1500–1760*. Nijmegen: Uitgeverij Vantilt Fragma, 2018.

Bots, Jan. *Tussen Descartes en Darwin: Geloof en natuurwetenschap in de achttiende eeuw in Nederland*. Assen: van Gorcum, 1972.

Bratschi, Max A. *Benedictus Aretius und Joannes Rhellicanus: Berg-Besteigungen im 16. Jahrhundert. Niesen und Stockhorn: zwei Lateintexte von Berner Humanisten*. Thun: Ott, 1992.

Brecht, Martin. "Das Aufkommen der neuen Frömmigkeitsbewegung in Deutschland." In *Geschichte des Pietismus*. Vol. 2, *Der Pietismus im achtzehnten Jahrhundert*, edited by Martin Brecht, 113–204. Göttingen: Vandenhoeck & Ruprecht, 1995.

———. "Die deutschen Spiritualisten des 17. Jahrhunderts." In *Geschichte des Pietismus*. Vol. 2, *Der Pietismus im achtzehnten Jahrhundert*, edited by Martin Brecht, 205–240. Göttingen: Vandenhoeck & Ruprecht, 1995.

Breitschuh, Wilhelm. *Die Feoptija V. K. Trediakovskijs: Ein physikotheologisches Lehrgedicht im Russland des 18. Jahrhunderts*, Munich: Otto Sagner, 2012.

Brockliss, L. W. B. *French Higher Education in the Seventeenth and Eighteenth Centuries*. Oxford: Clarendon Press, 1987.

Brooke, John Hedley. "Science and the Fortunes of Natural Theology: Some Historical Perspectives." *Zygon* 24 (1989): 3–22.

———. "Scientific Thought and Its Meaning for Religion: The Impact of French Science on British Natural Theology, 1827–1859." *Revue de Synthèse* 4 (1989): 33–59.

———. *Science and Religion: Some Historical Perspectives*. Cambridge: Cambridge University Press, 1991, 2014.

———. "The Natural Theology of the Geologists: Some Theological Strata." In *Images of the Earth*, edited by Ludmilla Jordanova and Roy Porter, 53–74. British Society for the History of Science monograph no. 1. 2nd rev. ed. Chalfont St. Giles: BSHS, 1997.

———. "'Wise Men Nowadays Think Otherwise': John Ray, Natural Theology and the Meanings of Anthopocentrism." *Notes and Records of the Royal Society* 54, no. 2 (2000): 199–213.

Brooke, John Hedley, and Geoffrey Cantor. *Reconstructing Nature: The Engagement of Science and Religion*. Oxford: Oxford University Press, 1998.

Brucker, Nicolas. "Noël-Antoine Pluche, entre sciences de la nature et apologétique." In *Apologétique 1650–1802: La nature et la grâce*, edited by Nicolas Brucker, 325–41. Recherches en littérature et spiritualité, no. 18. Bern: Peter Lang, 2010.

———. "Preuves physico-théologiques." In *Dictionnaire des anti-Lumières et des antiphilosophes: France, 1715–1815*, edited by Didier Masseau, 1244–52. Paris: Honoré Champion, 2017.

Buch, Hans-Christoph. *Ut Pictura Poesis: Die Beschreibungsliteratur und ihre Kritiker von Lessing bis Lukács*. Munich: Hanser, 1972.

Buchenau, Stefanie. "Die Teleologie zwischen Physik und Theologie." In "Die natürliche Theologie bei Christian Wolff," edited by Michael Albrecht. *Aufklärung* 23 (2011): 163–74.

Buckley, Michael J. *At the Origins of Modern Atheism*. New Haven, CT: Yale University Press, 1987.

———. *Denying and Disclosing God: The Ambiguous Progress of Modern Atheism*. New Haven, CT: Yale University Press, 2004.

Bunge, Wiep van. Introduction to *Adriaan Koerbagh: A Light Shining in Dark Places, to Illuminate the Main Questions of Theology and Religion*, edited by Michiel R. Wielema, 1–37. Leiden: Brill, 2011.

Burke, Peter. *A Social History of Knowledge II: From the Encyclopédie to Wikipedia*. Cambridge: Polity Press, 2012.

Bütikofer, Kaspar. *Der Frühe Zürcher Pietismus (1689–1721): Der soziale Hintergrund und die Denk- und Lebenswelten im Spiegel der Bibliothek Johann Heinrich Lochers (1648–1718)*. Göttingen: Vandenhoeck & Ruprecht, 2009.

Büttner, Manfred. "Theologie und Klimatologie im 18. Jahrhundert." *Neue Zeitschrift für Systematische Theologie und Religionsphilosophie* 6, no. 2 (1964): 154–91. Reprinted as "Protestantische Theologie und Klimatologie im 18. Jahrhundert." In *Zur Entwicklung der Geographie vom Mittelalter bis zu Carl Ritter*, edited by Büttner, 183–217. Abhandlungen und Quellen zur Geschichte der Geographie und Kosmologie, 3. Paderborn: Schöningh, 1982.

———. "Zum Übergang von der teleologischen zur kausalmechanischen Betrachtung der geographisch-kosmologischen Fakten: Ein Beitrag zur Geschichte der Geographie von Wolff bis Kant." *Studia Leibnitiana* 5, no. 2 (1973): 177–95.

———. "Kant und die Überwindung der physikotheologischen Betrachtung der geographisch-kosmologischen Fakten." *Erdkunde* 29, no. 3 (1975): 162–66.

———. "Kant and the Physico-theological Consideration of the Geographical Facts." In *Science and Religion / Wissenschaft und Religion*, edited by Änne Bäumer and Manfred Büttner, 82–92. Abhandlungen zur Geschichte der Geowissenschaften und Religion/ Umwelt-Forschung 3. Bochum: Universitätsverlag Dr. N. Brockmeyer, 1989.

———. "Zur neuen Epochen-Gliederung der Geschichte der Physikotheologie auf Grund neuerer Forschungergebnisse." In *Forschungen zur Physikotheologie im Aufbruch* III, edited by Manfred Büttner and Frank Richter, 9–22. Physikotheologie im historischen Kontext, 4. Münster: LIT Verlag, 1997.

Bynum, William F. "The Anatomical Method, Natural Theology, and the Functions of the Brain." *Isis* 64 (1973): 445–68.

Calloway, Katherine. *Natural Theology in the Scientific Revolution: God's Scientists.* London: Pickering and Chatto, 2014.

Carraud, Vincent. *Pascal et la philosophie.* Paris: Presses universitaires de France, 1992.

Chassot, Fabrice. *Le dialogue scientifique au XVIIIe siècle: Postérité de Fontenelle et vulgarisation des sciences.* Paris: Classiques Garnier, 2011.

Chraplak, Marc. *B. H. Brockes' fröhliche Physikotheologie: Poetische Strategien gegen Weltverachtung und religiösen Fanatismus in der Frühaufklärung.* Bielefeld: Aistesis Verlag, 2015.

Chrisman, Miriam U. *Lay Culture, Learned Culture: Books and Social Change in Strasbourg, 1480–1599.* New Haven CT: Yale University Press, 1982.

Clark, J. F. M. "History from the Ground Up: Bugs, Political Economy, and God in Kirby and Spence's *Introduction to Entomology* (1815–1856)." *Isis* 97, no. 1 (2006): 28–55.

Cobb, Matthew. "Malpighi, Swammerdam and the Colourful Silkworm: Replication and Visual Representation in Early Modern Science." *Annals of Science* 59 (2002): 111–47.

Coolidge, W. A. B. *Josias Simmler et les origines de l'alpinisme jusqu'en 1600.* 1st ed. 1904. 2nd ed. Grenoble: Glénat (Archives des Alpes), 1989.

Crawford, Robert. *The Bard: Robert Burns, A Biography.* Princeton, NJ: Princeton University Press, 2009.

Crowther, Kathleen M. "Sacred Philosophy, Secular Theology: The Mosaic Physics of Levinus Lemnius (1505–1568) and Francisco Valles (1524–1592)." In *Nature and Scripture in the Abrahamic Religions: Up to 1700*, edited by Jitse M. van der Meer and Scott H. Mandelbrote, 2:397–428. Leiden: Brill, 2008.

———. *Adam and Eve in the Protestant Reformation.* Cambridge: Cambridge University Press, 2010.

———. "The Lutheran Book of Nature." In *The Book of Nature and Humanity in Medieval and Early Modern Europe*, edited by David Hawkes and Richard Newhauser, 19–39. Arizona Studies in the Middle Ages and the Renaissance 29. Turnhout: Brepols, 2013.

Crowther-Heyck, Kathleen. "Wonderful Secrets of Nature: Natural Knowledge and Religious Piety in Reformation Germany." *Isis* 94, no. 2 (2003): 253–73.

Cunningham, Andrew. "Getting the Game Right: Some Plain Words on the Identity and Invention of Science." *Studies in the History and Philosophy of Science* 19, no. 3 (1988): 365–89.

———. "How the Principia Got Its Name: Or, Taking Natural Philosophy Seriously." *History of Science* 29 (1991): 377–92.

Daston, Lorraine, and Katharine Park. *Wonder and the Order of Nature, 1150–1750.* New York and Cambridge, MA: Zone Books and MIT Press, 1998.

Daston, Lorraine, and Michael Stolleis, eds. *Natural Law and Laws of Nature in Early Modern Europe.* Aldershot: Ashgate, 2008.

Dear, Peter. *Discipline and Experience.* Chicago: University of Chicago Press, 1995.

———. "Mixed Mathematics." In *Wrestling with Nature: From Omens to Science*, edited by Peter Harrison, Ronald L. Numbers, and Michael H. Shank, 149–73. Chicago: University of Chicago Press, 2011.

Debus, Allen G. *The French Paracelsians: The Chemical Challenge to Medical and Scientific Tradition in Early Modern France.* Cambridge: Cambridge University Press, 1991.

Deutsche Biographie. https://www.deutsche-biographie.de/.

Dietz, Bettina. "Aufklärung als Praxis: Naturgeschichte im 18. Jahrhundert." *Zeitschrift für Historische Forschung* 36, no. 2 (2009): 235–57.

Dillenberger, John. *Protestant Thought and Natural Science: A Historical Interpretation.* New York: Doubleday, 1960. London: Collins, 1961.

Dirlinger, Helga. *Bergbilder: Die Wahrnehmung alpiner Wildnis am Beispiel der englischen Gesellschaft 1700–1850.* Historisch-anthropologische Studien 10. Frankfurt am Main: Peter Lang, 2000.

Dobbs, Betty J. T. *The Foundations of Newton's Alchemy*. Cambridge: Cambridge University Press, 1975.

Dooley, Brendan. *Science, Politics, and Society in Eighteenth-Century Italy: The Giornale de' letterati d'Italia and Its World*. New York: Garland, 1991.

———. "Talking Science at the University of Padua in the Age of Antonio Vallisneri." *History of Universities* 24 (2009): 117–38.

———. "Scienza parlata, scienza scritta: Il Giornale de' letterati nelle aule universitarie." In *Il "Giornale de' Letterati d'Italia" trecento anni dopo: Scienza, storia, arte, identità (1710–2010)*, edited by Enza del Tedesco, 78–89. Atti del convegno, Padova, Venezia, Verona, Nov. 17–19, 2010. Pisa: F. Serra, 2012.

———. "Learned Journals and Teaching at the University of Padua in the Early Eighteenth Century." In "L'Europe des journaux savants (XVIIe–XVIIIe siècles). Communication et construction des savoirs," edited by Jeanne Peiffer. Special issue, *Archives internationales d'histoire des sciences* 63, nos. 170–71 (2014): 343–57.

———. "The Experience of Time." In *Routledge Companion to Cultural History*, edited by Alessandro Arcangeli et al. Forthcoming.

Drake, Ellen Tan. *Restless Genius: Robert Hooke and His Earthly Thoughts*. Oxford: Oxford University Press, 1996.

Duchesneau, François. *Les modèles du vivant de Descartes à Leibniz*. Paris: Vrin, 1998.

Ducheyne, Steffen. "Curing Pansophia through Eruditum Nescire: Bernard Nieuwentijt's (1654–1718) Epistemology of Modesty." *HOPOS* 7 (2017): 272–301.

Eamon, William. *Science and the Secrets of Nature: Books of Secrets in Medieval and Early Modern Culture*. Princeton, NJ: Princeton University Press, 1994.

Ehrenpreis, Stefan, and Ute Lotz-Heumann. *Reformation und konfessionelles Zeitalter*. Darmstadt: Wissenschaftliche Buchgesellschaft, 2002.

Encarnación, Karen Rosoff. "The Proper Uses of Desire: Sex and Procreation in Reformation Anatomical Fugitive Sheets." In *The Material Culture of Sex, Procreation, and Marriage in Premodern Europe*, ed. Anne L. McClanan and Karen Rosoff Encarnación, 222–49. New York: Palgrave, 2002.

Engelsing, Rolf. *Analphabetentum und Lektüre: Zur Sozialgeschichte des Lesens in Deutschland zwischen feudaler und industrieller Gesellschaft*. Stuttgart: J. B. Metzler, 1973.

Feldhay, Rivka. "The Simulation of Nature and the Dissimulation of the Law on a Baroque Stage: Galileo and the Church Revisited." In *Science in the Age of the Baroque*, edited by Ofer Gal and Raz Chen-Morris, 285–303. Dordrecht: Springer, 2013.

Ferrone, Vincenzo. *Scienza, natura, religione: Mondo newtoniano e cultura italiana nel primo Settecento*. Naples: Jovene, 1982.

———. *The Intellectual Roots of the Italian Enlightenment: Newtonian Science, Religion and Politics in the Early Eighteenth Century*. Translated by Sue Brotherton. Atlantic Heights, NJ: Humanities Press, 1995.

Findlen, Paula. "Jokes of Nature and Jokes of Knowledge: The Playfulness of Scientific Discourse in Early Modern Europe." *Renaissance Quarterly* 43, no. 2 (1990): 292–331.

Flubacher, Silvia. "Wunderbare Wesen: Die Ordnung der Tierwelt und das Schreiben der Naturgeschichte um 1700." Unpublished PhD thesis, University of Basel, 2015.

Fournier, Marian. "The Book of Nature: Jan Swammerdam's Microscopical Investigations." *Tractrix* 2 (1990): 1–24.

Freedman, Joseph S. *Philosophy and the Arts in Central Europe, 1500–1700*. Aldershot: Ashgate, 1999.

Frettlöh, Magdalene L. "'Gott ist im Fleische. . . .' Die Inkarnation Gottes in ihrer leibeigenen Dimension beim Wort genommen." In *"Dies ist mein Leib": Leibliches, Leibeigenes und*

Leibhaftiges bei Gott und den Menschen, edited by Jürgen Ebach et al., 186–229. Gütersloh: Gütersloher Verlagshaus, 2006.

Friedrich, Markus. *Die Grenzen der Vernunft: Theologie, Philosophie und gelehrte Konflikte am Beispiel des Helmstedter Hofmannstreits und seiner Wirkungen auf das Luthertum um 1600.* Göttingen: Vandenhoeck & Ruprecht, 2004.

Funkenstein, Amos. *Theology and the Scientific Imagination from the Middle Ages to the Seventeenth Century.* Princeton, NJ: Princeton University Press, 1986.

Fyfe, Aileen. "The Reception of William Paley's Natural Theology in the University of Cambridge." *British Journal for the History of Science* 30 (1997): 321–36.

Gaab, Hans, Pierre Leich, and Günter Löfflandt, eds. *Johann Christian Sturm (1635–1703).* Frankfurt am Main: H. Deutsch, 2004.

Garber, Daniel. *Descartes' Metaphysical Physics.* Chicago: University of Chicago Press, 1992.

Gascoigne, John. "From Bentley to the Victorians: The Rise and Fall of British Newtonian Natural Theology." *Science in Context* 2, no. 2 (1988): 219–56.

Gaukroger, Stephen. *The Emergence of a Scientific Culture: Science and the Shaping of Modernity, 1210–1685.* Oxford: Oxford University Press, 2006.

Gawlick, Günter. "Reimarus und der englische Deismus." In *Religionskritik und Religiosität in der Deutschen Aufklärung,* edited by Karlfried Gründer and Karl Friedrich Rengstorf, 43–54. Wolfenbütteler Studien zur Aufklärung 11. Heidelberg: Lambert Schneider, 1989.

Generali, Dario. *Antonio Vallisneri: Gli anni della formazione e le prime ricerche.* Florence: L. S. Olschki, 2007.

———, ed. *Antonio Vallisneri: La figura, il contesto, le immagini storiografiche.* Milan: Franco Angeli, 2008.

Gestrich, Christof. "Deismus." In *Theologische Realenzyklopädie*, 8:392–406. Berlin: De Gruyter, 1981.

Geyer, Hermann. *Verborgene Weisheit: Johann Arndts "Vier Bücher vom Wahren Christentum" als Programm einer spiritualistisch-hermetischen Theologie.* Berlin: De Gruyter, 2001.

Geyer-Kordesch, Johanna. *Pietismus, Medizin und Aufklärung in Preußen im 18. Jahrhundert: Das Leben und Werk Georg Stahls.* Tübingen: Max Niemeyer, 2000.

Giacomotto-Charra, Violaine. "Entre traduction et vulgarisation: L'astronomie en français au XVIe siècle." In *Traduire la science hier et aujourd'hui,* ed. Pascal Duris, 45–67. Pessac: Maison des Sciences de l'Homme d'Aquitaine, 2008.

Gillespie, Neal C. *Charles Darwin and the Problem of Creation.* Chicago: University of Chicago Press, 1979.

———. "Natural History, Natural Theology, and Social Order: John Ray and the 'Newtonian Ideology.'" *Journal of the History of Biology* 20, no. 1 (1987): 1–49.

Gipper, Andreas. *Wunderbare Wissenschaft: Literarische Strategien naturwissenschaftlicher Vulgarisierung in Frankreich, von Cyrano de Bergerac bis zur Encyclopédie.* Munich: Fink, 2002.

Gisler, Monika. *Göttliche Natur? Formationen im Erdbebendiskurs der Schweiz des 18. Jahrhunderts.* Zurich: Chronos, 2007.

Gould, Stephen J. *Rocks of Ages: Science and Religion in the Fullness of Life.* New York: Ballantine, 1999.

Grafton, Anthony. *Joseph Scaliger: A Study in the History of Classical Scholarship.* 2 vols. Oxford: Clarendon Press, 1983–93.

———. *Defenders of the Text: The Traditions of Scholarship in an Age of Science, 1450–1800.* Cambridge, MA: Harvard University Press, 1991.

———. "Isaac Vossius, Chronologer." In *Isaac Vossius (1618–1689) between Science and Scholarship,* edited by Eric Jorink and Dirk van Miert, 43–84. Leiden: Brill, 2012.

Grand-Carteret, John. *La montagne à travers les âges; rôle joué par elle; façon dont elle a été vue.* 1st ed. 1903. 2nd rev. ed. 2 vols. Geneva: Slatkine, 1983.

Grant, Edward. *The Foundations of Modern Science in the Middle Ages.* Cambridge: Cambridge University Press, 1996.

———. *A History of Natural Philosophy: From the Ancient World to the Nineteenth Century.* Cambridge: Cambridge University Press, 2007.

———. *The Nature of Natural Philosophy in the Late Middle Ages.* Washington, DC: Catholic University Press, 2010.

Greyerz, Kaspar von. *Religion and Culture in Early Modern Europe, 1500–1800.* Translated by Thomas Dunlap. New York: Oxford University Press, 2008.

———. "Early Modern Protestant Scientists and Virtuosos: Some Comments." *Zygon* 51, no. 3 (2016): 698–717.

———. "Protestantism, Knowledge, and Science between 1650 and c. 1760." In *The Cultural History of the Reformation: Current Research and Future Perspectives,* edited by Susan Karant-Nunn and Ute Lotz-Heumann. Forthcoming.

Grimm, Jakob, and Wilhelm Grimm. *Deutsches Wörterbuch.* 16 vols. Leipzig: Hirzel, 1854–1960.

Grosse, Sven. "Abgründe der Physikotheologie: Fabricius—Brockes—Reimarus." In *Das Akademische Gymnasium zu Hamburg (gegr. 1613) im Kontext frühneuzeitlicher Wissenschafts- und Bildungsgeschichte,* edited by Johann Anselm Steiger in cooperation with Martin Mulsow and Axel E. Walter, 319–39. Berlin: De Gruyter, 2017.

Grote, Simon. "Review-Essay: Religion and Enlightenment." *Journal of the History of Ideas* 75, no. 1 (2014): 137–60.

———. *The Emergence of Modern Aesthetic Theory: Religion and Morality in Enlightenment Germany and Scotland.* Cambridge: Cambridge University Press, 2017.

Häfner, Ralph. "Literaturgeschichte und Physikotheologie: Johann Albert Fabricius." In *500 Jahre Theologie in Hamburg: Hamburg als Zentrum christlicher Theologie und Kultur zwischen Tradition und Zukunft,* edited by Johann Anselm Steiger, 35–57. Arbeiten zur Kirchengeschichte, vol. 95. Berlin: De Gruyter, 2005.

Hahn, Philipp. "Lutheran Sensory Culture in Context." *Past & Present* 234, suppl. 12 (2017): 90–113.

Hahn, Roger. "Laplace and the Mechanistic Universe." In *God and Nature: Historical Essays on the Encounter between Christianity and Science,* edited by David C. Lindberg and Ronald L. Numbers, 256–76. Berkeley: University of California Press, 1986.

Hall, A. Rupert. *Henry More and the Scientific Revolution.* Cambridge: Cambridge University Press, 2002.

Hamm, Berndt. "Wie mystisch war der Glaube Luthers?" In *Gottes Nähe unmittelbar erfahren: Mystik im Mittelalter und bei Martin Luther,* edited by Berndt Hamm and Volker Leppin, 237–87. Tübingen: Mohr Siebeck, 2007.

Hardin, Jeff, Ronald L. Numbers, and Ronald A. Binzley, eds. *The Warfare between Science and Religion: The Idea That Wouldn't Die.* Baltimore: Johns Hopkins University Press, 2018.

Harkness, Deborah E. *The Jewel House: Elizabethan London and the Scientific Revolution.* New Haven, CT: Yale University Press, 2007.

Harrison, Peter. *The Bible, Protestantism and the Rise of Natural Science.* Cambridge: Cambridge University Press, 1998.

———. "The Influence of Cartesian Cosmology in England." In *Descartes' Natural Philosophy,* edited by Stephen Gaukroger, John Schuster, and John Sutton, 168–92. London: Routledge, 2000.

———. "Physico-theology and the Mixed Sciences: The Role of Theology in Early Modern Natural Philosophy." In *The Science of Nature in the Seventeenth Century,* edited by Peter Anstey and John Schuster, 165–83. Dordrecht: Springer, 2005.

———. *The Fall of Man and the Foundations of Science*. Cambridge: Cambridge University Press, 2007.

———. *The Territories of Science and Religion*. Chicago: University of Chicago Press, 2015.

Harrison, Peter, Ronald L. Numbers, and Michael H. Shank, eds. *Wrestling with Nature: From Omens to Science*. Chicago: University of Chicago Press, 2011.

Hazard, Paul. *The Crisis of the European Mind, 1680–1715*. Translated by J. Lewis May. New York: New York Review of Books, 2013. French original: *La crise de la conscience européenne (1680–1715)*. 3 vols. Paris: Boivin 1939–40.

Helm, Jürgen. *Krankheit, Bekehrung und Reform: Medizin und Krankenfürsorge im Halleschen Pietismus*. Wiesbaden: Harrassowitz, 2006.

Henry, John. "Metaphysics and the Origins of Modern Science: Descartes and the Importance of the Laws of Nature." *Early Science and Medicine* 9 (2004): 73–114.

Heyberger, Bernard. *Les Chrétiens du Proche-Orient au temps de la Réforme catholique*. Rome: École française, 1994.

Hillgarth, J. N. *Ramon Lull and Lullism in Fourteenth-Century France*. Oxford: Clarendon Press, 1971.

Hirsch, Emanuel. *Geschichte der neuern evangelischen Theologie im Zusammenhang mit den allgemeinen Bewegungen des europäischen Denkens*. 5 vols. Gütersloh: Bertelsmann, 1949–54.

Hölscher, Lucian. *Geschichte der protestantischen Frömmigkeit*. Munich: C. H. Beck, 2011.

Hommel, Karsten. "Naturwissenschaftliche Forschungen." In *Geliebtes Europa—Ostindische Welt: 300 Jahre interkultureller Dialog im Spiegel der Dänisch-Halleschen Mission*, edited by Heike Liebau, 163–79. Halle: Frankeschen Stiftungen, 2006.

———. "Physico-theology as Mission Strategy: Missionary Christoph Samuel John's (1746–1813) Understanding of Nature." In *Halle and the Beginning of Protestant Christianity in India*, vol. 3, *Communication between India and Europe*, edited by Andreas Gross, Y. Vincent Kumaradoss, and Heike Liebau, 1115–33. Halle: Frankeschen Stiftungen, 2006.

Hoorn, Carel Maaijo van. *Levinus Lemnius, 1505–1568: Zestiende-eeuws Zeeuws geneesheer*. Kloosterzande: J. Duerinck-Krachten, 1978.

Hoppe, Brigitte. "Von der Naturgeschichte zu den Naturwissenschaften: Die Dänisch-Halleschen Missionare als Naturforscher in Indien vom 18. bis 19. Jahrhundert." In *Mission und Forschung: Translokale Wissensproduktion zwischen Indien und Europa im 18. und 19. Jahrhundert*, edited by Heike Liebau, Andreas Nehring, and Brigitte Klosterberg, 141–67. Halle: Verlag der Franckeschen Stiftungen, 2010.

Hoskin, Michael A. "Newton, Providence and the Universe of Stars." *Journal for the History of Astronomy* 8 (1977): 77–101.

Hotson, Howard. *Commonplace Learning: Ramism and Its German Manifestations, 1543–1630*. Oxford: Clarendon Press, 2007.

Hunfeld, Barbara. *Der Blick ins All: Reflexionen des Kosmos der Zeichen bei Brockes, Jean Paul, Goethe und Stifter*. Tübingen: Max Niemeyer, 2004.

Hunter, Michael. *The Royal Society and Its Fellows, 1600–1700: The Morphology of an Early Scientific Institution*. 2nd ed. Oxford: BSHS, 1994.

———. *Boyle: Between God and Science*. New Haven, CT: Yale University Press, 2009.

———. *Boyle Studies: Aspects of the Life and Thought of Robert Boyle (1627–91)*. Farnham: Ashgate, 2015.

Hyde, Elizabeth. *Cultivated Power: Flowers, Culture, and Politics in the Reign of Louis XIV*. Philadelphia: University of Pennsylvania Press, 2005.

Iliffe, Rob. "The Religion of Isaac Newton." In *The Cambridge Companion to Newton*, edited by Iliffe and George E. Smith, 485–523. Cambridge: Cambridge University Press, 2016.

———. *Priest of Nature: The Religious Worlds of Isaac Newton*. Oxford: Oxford University Press, 2017.

Israel, Jonathan. *Radical Enlightenment: Philosophy and the Making of Modernity, 1650–1750.* Oxford: Oxford University Press, 2001.

Jacob, Margaret C. *The Newtonians and the English Revolution, 1689–1720.* Ithaca, NY: Cornell University Press, 1976.

Jahn, Ilse, and Konrad Senglaub. *Carl von Linné.* Leipzig: Senglaub, 1978.

Jardine, Nick. *The Birth of History and Philosophy of Science: Kepler's "A Defence of Tycho against Ursus" with Essays on Its Provenance and Significance.* Cambridge: Cambridge University Press, 1988.

Jensen, Niklas Thode. "Making It in Tranquebar: Science, Medicine and the Circulation of Knowledge in the Danish-Halle Mission, c. 1732–1744." In *Beyond Tranquebar: Grappling across Cultural Borders in South India,* edited by Esther Fihl and A. R. Venkatachalapathy, 325–51. Delhi: Orient BlackSwan, 2014.

Johnson, Francis R. *Astronomical Thought in Renaissance England: A Study of English Scientific Writings from 1500 to 1645.* Baltimore: Johns Hopkins Press, 1937.

Jorink, Eric. "'Horrible and Blasphemous': Isaac la Peyrère, Isaac Vossius, and the Emergence of Radical Biblical Criticism in the Dutch Republic." In *Nature and Scripture in the Abrahamic Religions: Up to 1700,* edited by Jitse van der Meer and Scott Mandelbrote, 429–50. Leiden: Brill, 2009.

———. *Reading the Book of Nature in the Dutch Golden Age, 1575–1715.* Leiden: Brill, 2010.

———. "Noah's Ark Restored (and Wrecked): Dutch Collectors, Natural History and the Problem of Biblical Exegesis." In *Silent Messengers: The Circulation of Material Objects of Knowledge in the Early Modern Low Countries,* edited by Sven Dupré and Christoph Lüthy, 153–84. Berlin: LIT, 2011.

———. *De Ark, de Tempel, het Museum: Veranderende modellen van kennis in de Eeuw van de Verlichting.* Haarlem: Teylers Stichting, 2014.

Jorink, Eric, and Ad Maas, eds. *Newton and the Netherlands: How Isaac Newton Was Fashioned in the Dutch Republic.* Amsterdam: Leiden University Press, 2012.

Jorink, Eric, and H. Zuidervaart. "Newton's Reception in the Low Countries." In *The Reception of Newton in Europe,* edited by Scott Mandelbrote and Helmut Pulte, 1:59–99. 3 vols. London: Bloomsbury Academic, 2019.

Kaiser, Wolfram, and Werner Piechocki. "Medizinisch-zoologischer Unterricht im 18. Jahrhundert an der Universität Halle." *Hercynia* 6, no. 3 (1969): 258–84.

Karant-Nunn, Susan C. "'Gedanken, Herz und Sinn.' Die Unterdrückung der religiösen Emotionen." In *Kulturelle Reformation. Sinnformationen im Umbruch 1400–1600,* edited by Bernhard Jussen and Craig Koslofsky, 69–96. Göttingen: Vandenhoeck & Ruprecht, 1999.

Kardel, Troels, and Paul Maquet. *Nicolaus Steno: Biography and Original Papers of a 17th Century Scientist.* 2nd ed. Berlin: Springer Verlag, 2018.

Kargon, Robert. "Walter Charleton, Robert Boyle, and the Acceptance of Epicurean Atomism in England." *Isis* 55, no. 2 (1964): 184–92.

Kaufmann, Thomas. *Konfession und Kultur: Lutherischer Protestantismus in der zweiten Hälfte des Reformationsjahrhunderts.* Tübingen: Mohr Siebeck, 2006.

———. "Die Sinn- und Leiblichkeit der Heilsaneignung im späten Mittelalter und in der Reformation." In *Medialität, Unmittelbarkeit, Präsenz: Die Nähe des Heils im Verständnis der Reformation,* edited by Johanna Haberer and Berndt Hamm, 11–43. Tübingen: Mohr Siebeck, 2012.

Kempe, Michael. *Wissenschaft, Theologie, Aufklärung: Johann Jakob Scheuchzer (1672–1733) und die Sintfluttheorie.* Frühneuzeit-Forschungen 10. Epfendorf: bibliotheca academica Verlag, 2003.

Kemper, Hans-Georg. *Gottebenbildlichkeit und Naturnachahmung im Säkularisierungsprozeß: Problemgeschichtliche Studien zur deutschen Lyrik in Barock und Aufklärung.* 2 vols. Tübingen: Max Niemeyer, 1981.

———. "Brockes und das hermetische Schrifttum seiner Bibliothek." In *Barthold Heinrich Brockes (1680–1747) im Spiegel seiner Bibliothek und Bildergalerie* . . . , edited by Kemper, 223–71. Wolfenbütteler Forschungen 80. Wiesbaden: Harrassowitz, 1998.

Ketelsen, Uwe-Karsten. *Die Naturpoesie der norddeutschen Frühaufklärung.* Stuttgart: Metzler, 1974.

Keynes, Geoffrey. *John Ray, 1627–1705: A Bibliography 1660–1970.* Amsterdam: Van Heusden, 1976.

Kimber, Ida M. "Barthold Heinrich Brockes' *Irdisches Vergnügen in Gott* als zeitgeschichtliches Dokument." In *Barthold Heinrich Brockes (1680–1747), Dichter und Ratsherr in Hamburg: Neue Forschungen zu Persönlichkeit und Wirkung,* edited by Hans-Dieter Loose, 45–70. Hamburg: Hans Christians, 1980.

Klemme, Heiner F., and Manfred Kuehn, eds. *The Bloomsbury Dictionary of Eighteenth-Century German Philosophers.* London: Bloomsbury Academic, 2016.

Knuuttila, Simo, ed. *Reforging the Great Chain of Being: Studies of the History of Modal Theories.* Dordrecht: Springer Science & Business Media, 2013.

Korenjak, Martin. "Why Mountains Matter: Early Modern Roots of a Modern Notion." *Renaissance Quarterly* 70, no. 1 (Spring 2017): 179–219.

Krafft, Fritz. " . . . *Denn Gott schafft nichts umsonst!" Das Bild der Naturwissenschaft vom Kosmos im historischen Kontext des Spannungsfeldes Gott—Mensch—Natur.* Natur—Wissenschaft—Theologie 1. Berlin: LIT Verlag, 1999.

Krieger, Martin. *Patriotismus in Hamburg: Identitätsbildung im Zeitalter der Frühaufklärung.* Cologne: Böhlau, 2008.

Krolzik, Udo. *Säkularisierung der Natur: Providentia-Dei-Lehre und Naturverständnis der Frühaufklärung.* Neukirchen-Vluyn: Neukirchener Verlag, 1988.

———. "Das Wasser als theologisches Thema der deutschen Frühaufklärung." In *Kulturgeschichte des Wassers,* edited by Hartmut Böhme, 189–207. Frankfurt am Main: Suhrkamp, 1988.

———. "Physikotheologie." In *Theologische Realenzyklopädie,* 26:590–96. Berlin: De Gruyter, 1996.

Kuhn, Bernhard. "'A Chain of Marvels': Botany and Autobiography in Rousseau." *European Romantic Review* 17, no. 1 (Jan. 2006): 1–20.

Kusukawa, Sachiko. *The Transformation of Natural Philosophy: The Case of Philip Melanchthon.* Cambridge: Cambridge University Press, 1995.

Langen, August. *Anschauungsformen in der deutschen Dichtung des 18. Jahrhunderts: Rahmenschau und Rationalismus.* 1934. Darmstadt: Wissenschaftliche Buchgesellschaft, 1968.

Laube, Stefan. *Von der Reliquie zum Ding. Heiliger Ort—Wunderkammer—Museum.* Berlin: De Gruyter, 2011.

Lenoir, Timothy. "The Eternal Laws of Form: Morphotypes and the Conditions of Existence in Goethe's Biological Thought." In *Goethe and the Sciences: A Reappraisal,* edited by Frederick Amrine et al., 17–28. Boston: Dordrecht, 1987.

Leu, Urs, ed. *Natura Sacra: Der Frühaufklärer Johann Jakob Scheuchzer (1672–1733).* Zug: Achius, 2012.

Levitin, Dmitri. "Halley and the Eternity of the World Revisited." *Notes and Records of the Royal Society* 67, no. 4 (2013): 315–29.

———. "Rethinking English Physico-theology: Samuel Parker's Tentamina de Deo (1665)." *Early Science and Medicine* 19 (2014): 28–75.

———. *Ancient Wisdom in the Age of the New Science.* Cambridge: Cambridge University Press, 2015.

Liebau, Heike. "Über die Erziehung 'tüchtiger' Subjekte zur Verbreitung des Evangeliums: Das Schulwesen der Dänisch- Halleschen Mission." In *Weltmission und religiöse Organisationen: Protestantische Missionsgesellschaften im 19. und 20. Jahrhundert,* edited by Artur Bogner, Bernd Holtwick, and Hartmann Tyrell, 427–58. Würzburg: Ergon Verlag, 2004.

——. *Die indischen Mitarbeiter der Tranquebarmission (1706–1845): Katecheten, Schulmeister, Übersetzer.* Tübingen: Niemeyer, 2008.

——. *Cultural Encounters in India: The Local Co-workers of the Tranquebar Mission, 18th to 19th Centuries.* Translated from German by Rekha V. Rajan. New Delhi: Social Science Press, 2013.

Lindberg, David C., and Ronald L. Numbers, eds. *God and Nature: Historical Essays on the Encounter between Christianity and Science.* Berkeley: The University of California Press, 1986.

Lindeboom, G. A. *Het cabinet van Jan Swammerdam (1637–1680): Catalogus met ein inleiding.* Amsterdam: Rodopi, 1980.

Lindroth, Sten. "The Two Faces of Linnaeus." In *Linnaeus: The Man and His Work,* edited by Tore Frängsmyr, 1–62. Berkeley: University of California Press, 1983.

——. "Linnaeus (or von Linné) Carl." In *Complete Dictionary of Scientific Biography,* 8:374–81. Detroit: Charles Scribner's Sons, 2008.

Livesey, Steven J. "Divine Omnipotence and First Principles." In *Thinking Impossibilities: The Legacy of Amos Funkenstein,* edited by Robert S. Westman and David Biale, 13–33. Toronto: University of Toronto Press, 2008.

Lohr, Charles H. "Metaphysics and Natural Philosophy as Sciences: The Catholic and the Protestant Views in the Sixteenth and Seventeenth Centuries." In *Philosophy in the Sixteenth and Seventeenth Centuries,* edited by Constance Blackwell and Sachiko Kusukawa, 280–95. Aldershot: Ashgate, 1999.

Lombardi, Ivano. *Un nume del Settecento: Antonio Vallisneri.* Lucca: Titania Editore, 1998.

Lovejoy, Arthur O. *The Great Chain of Being: A Study of the History of an Idea.* Cambridge, MA: Harvard University Press, 1936.

Luzzini, Francesco. *Il miracolo inutile: Alessandro Vallisneri e le scienze della Terra in Europa tra XVII e XVIII secolo.* Florence: Olschki, 2003.

——. "Flood Conceptions in Vallisneri's Thought." In *Geology and Religion: A History of Harmony and Hostility,* edited by M. Kölbl-Ebert, 77–81. Geological Society Special Publications. Bath: Geological Society Publishing House, 2009.

Maehle, Andreas-Holger. "'Est Deus ossa probant': Human Anatomy and Physicotheology in 17th and 18th century Germany." In *Science and Religion / Wissenschaft und Religion,* edited by Änne Bäumer and Manfred Büttner, 60–66. Abhandlungen zur Geschichte der Geowissenschaften und Religion / Umwelt-Forschung 3. Bochum: Universitätsverlag Dr. N. Brockmeyer, 1989.

Magruder, Kerry V. "Thomas Burnet, Biblical Idiom, and Seventeenth-Century Theories of the Earth." In *Nature and Scripture in the Abrahamic Religions: Up to 1700,* edited by Scott Mandelbrote and Jitse van der Meer, 461–500. Leyden: Brill, 2008.

Maissen, Thomas. "Die Bedeutung der Alpen für die Schweizergeschichte von Albrecht von Bonstetten (ca. 1442/43–1504/05) bis Johann Jakob Scheuchzer (1672–1733)." In Boscani Leoni, *Wissenschaft* (2010), 161–78.

Mandelbrote, Scott. "Isaac Newton and Thomas Burnet: Biblical Criticism and the Crisis of Late Seventeenth-Century England." In *The Books of Nature and Scripture,* edited by J. E. Force and R. H. Popkin, 149–78. Dordrecht: Kluwer, 1994.

——. "The Uses of Natural Theology in Seventeenth-Century England" *Science in Context* 20, no. 3 (2007): 451–80.

——. "Early Modern Natural Theologies." In *The Oxford Handbook of Natural Theology,* edited by Russell Re Manning, 75–99. Oxford: Oxford University Press, 2013.

Marchal, Guy. "Johann Jakob Scheuchzer und der schweizerische 'Alpenstaatsmythos.'" In Boscani Leoni, *Wissenschaft* (2010), 179–94.

Margolin, Jean-Claude. "Vertus occultes et effets naturels d'après les *Occulta naturae miracula* de Levinus Lemnius." In *L'uomo e la natura nel Rinascimento*, edited by Luisa Rotondi Secchi Tarugi, 415–43. Milan: Nuovi orizzonti, 1996.

Martynov, Alexander V. "The Shell Collection of J. H. Chemnitz in the Zoological Institute, St. Petersburg." *Ruthenia* 12, no. 1 (2002), 1–18.

Mathieu, Jon. "Alpenwahrnehmung: Probleme der historischen Periodisierung." In *Die Alpen! Zur europäischen Wahrnehmungsgeschichte seit der Renaissance = Les alpes! Pour une histoire de la perception européenne depuis la Renaissance*, edited by Mathieu and Simona Boscani Leoni, 53–72. Studies on Alpine History 2. Bern: P. Lang, 2005.

———. "The Globalization of Mountain Perception: How Much of a Western Imposition?" *Summerhill: Indian Institute of Advanced Study Review* 20 (Summer 2016): 8–17.

Mathieu, Jon, and Simona Boscani Leoni, eds. *Die Alpen! Zur europäischen Wahrnehmungsgeschichte seit der Renaissance = Les alpes! Pour une histoire de la perception européenne depuis la Renaissance*. Studies on Alpine History 2. Bern: P. Lang, 2005.

McKenna, Antony. "Filleau de la Chaise et la réception des *Pensées*." *Cahiers de l'Association internationale des études françaises* 40, no. 1 (1988): 297–314.

———. *De Pascal à Voltaire: Le rôle des* Pensées *de Pascal dans l'histoire des idées entre 1670 et 1734*. Oxford: Voltaire Foundation, 1990.

———. *Entre Descartes et Gassendi: La première édition des* Pensées *de Pascal*. Oxford: Voltaire Foundation, 1993.

Meer, Jitse M. van der, and Scott H. Mandelbrote, eds. *Nature and Scripture in the Abrahamic Religions: Up to 1700*. 2 vols. Leiden: Brill, 2009.

Methuen, Charlotte. *Kepler's Tübingen: Stimulus to a Theological Mathematics*. St. Andrews Studies in Reformation History. Aldershot: Ashgate, 1998.

———. "Interpreting the Books of Nature and Scripture in Medieval and Early Modern Thought: An Introductory Essay." In *Nature and Scripture in the Abrahamic Religions: Up to 1700*, edited by Jitse M. van der Meer and Scott Mandelbrote, 1:179–218. Leiden: Brill, 2009.

Michel, Paul. *Physikotheologie: Ursprünge, Leistung und Niedergang einer Denkform*. Neujahrsblatt auf das Jahr 2008, edited by Gelehrte Gesellschaft in Zürich. Zurich: Editions à la carte, 2008.

Mojet, Emma. "Early Modern Mathematics in a Letter: Adriaen Verwer to David Gregory on Mathematics and Natural Philosophy." *LIAS: Journal of Early Modern Intellectual Culture and Its Sources* 44 (2017): 117–42.

Monk, James Henry. *The Life of Richard Bentley, D.D.* London: Rivington, 1830.

Müller-Bahlke, Thomas. *Die Wunderkammer—Die Kunst- und Naturalienkammer der Franckeschen Stiftungen zu Halle*. Halle: Verlag der Franckeschen Stiftungen, 1998.

———. "Naturwissenschaft und Technik: Der Hallesche Pietismus am Vorabend der Industrialisierung." In *Geschichte des Pietismus*, vol. 4, *Glaubenswelt und Lebenswelten*, edited by Hartmut Lehmann, 357–85. Göttingen: Vandenhoeck & Ruprecht, 2004.

———. *Die Wunderkammer der Franckeschen Stiftungen*. Wiesbaden: Harrassowitz, 2012.

Mulsow, Martin. "Johann Christoph Wolf (1683–1739) und die verbotenen Bücher in Hamburg." In *500 Jahre Theologie in Hamburg: Hamburg als Zentrum christlicher Theologie und Kultur zwischen Tradition und Zukunft*, edited by Johann Anselm Steiger, 81–111. Arbeiten zur Kirchengeschichte, vol. 95. Berlin: De Gruyter, 2005.

Mumme, Jonathan. "Der Geist, die Geister und der Buchstabe: Was Martin Luther vom Heiligen Geist und von der Heiligen Schrift lehrt." *Lutherische Beiträge* 17, no. 1 (2012): 13–22.

Müsch, Irmgard. *Geheiligte Naturwissenschaft: Die Kupfer-Bibel des Johann Jakob Scheuchzer*. Göttingen: Vandenhoeck & Ruprecht, 2000.

Nehring, Andreas. "Natur und Gnade: Zur Theologie und Kulturkritik in den Neuen Halleschen Berichten." In *Missionsberichte aus Indien im 18. Jahrhundert: Ihre Bedeutung für die europäische Geistesgeschichte und ihr wissenschaftlicher Quellenwert für die Indienkunde*, edited by Michael Bergunder, 220–45. Halle: Verlag der Franckeschen Stiftungen, 1999.

———. *Orientalismus und Mission: Die Repräsentation der tamilischen Gesellschaft und Religion durch Leipziger Missionare 1840–1940*. Wiesbaden: Harrassowitz, 2003.

Neumann, Hanns-Peter. *Natura Sagax—Die geistige Natur: Zum Zusammenhang von Naturphilosophie und Mystik in der frühen Neuzeit am Beispiel Johann Arndts*. Tübingen: Niemeyer, 2004.

Newman, William R., and Anthony Grafton, eds. *Secrets of Nature: Astrology and Alchemy in Early Modern Europe*. Cambridge, MA: MIT Press, 2001.

Nicolaidis, Efthymios. *Science and Eastern Orthodoxy: From the Greek Fathers to the Age of Globalization*. Baltimore: Johns Hopkins University Press, 2011.

Nicolson, Marjorie Hope. *Mountain Gloom and Mountain Glory: The Development of the Aesthetics of the Infinite*. 1st ed. 1959. Seattle: University of Washington Press, 1997.

Noak, Bettina. "Schule der Wahrnehmung: Johannes Florentinus Martinets 'Katechismus der natuur.'" In *Kulturen des Wissens im 18. Jahrhundert*, edited by Ulrich Johannes Schneider, 499–506. Berlin: De Gruyter, 2008.

Nørgaard, Anders. *Mission und Obrigkeit: Die Dänisch-Hallische Mission in Tranquebar 1706–1845*. Gütersloh: Gütersloher Verlagshaus, 1988.

Norris, John A. "The Mineral Exhalation Theory of Metallogenesis in Pre-Modern Mineral Science." *Ambix* 53, no. 1 (2006): 43–65.

Numbers, Ronald L., ed. *Galileo Goes to Jail and Other Myths about Science and Religion*. Cambridge, MA: Harvard University Press, 2009.

Ogilvie, Brian. "Natural History, Ethics, and Physico-Theology." In *Historia: Empiricism and Erudition in Early Modern Europe*, edited by Gianna Pomata and Nancy Siraisi, 75–103. Cambridge, MA: MIT Press, 2005.

———. "Insects in John Ray's Natural History and Natural Theology." In *Zoology in Early Modern Culture: Intersections of Science, Theology, Philology, and Political and Religious Education*, edited by Karl A. E. Enenkel and Paul J. Smith, 234–60. Leiden: Brill, 2014.

———. "Stoics, Neoplatonists, Atheists, Politicians: Sources and Uses of Early Modern Jesuit Natural Theology." In *For the Sake of Learning: Essays in Honor of Anthony Grafton*, edited by Ann Blair and Anja-Silvia Goeing, 2:761–79. Leiden: Brill, 2016.

Osler, Margaret J. *Divine Will and the Mechanical Philosophy*. Cambridge: Cambridge University Press, 1994.

———. "Whose Ends? Teleology in Early Modern Natural Philosophy." *Osiris* 16 (2001): 151–68.

Ozment, Steven. *When Fathers Ruled: Family Life in Reformation Europe*. Cambridge, MA: Harvard University Press, 1985.

Pantin, Isabelle. *La poésie du ciel en France dans la seconde moitié du seizième siècle*. Geneva: Droz, 1995.

Pasini, Mirella. *Thomas Burnet: Una storia del mondo tra ragione, mito e rivelazione*. Florence: La Nuova Italia, 1981.

Pécharman, Martine. "Filleau de la Chaise, Nicolas Jean, 1631–1688." In *Dictionnaire des philosophes français du XVIIe siècle: Acteurs et réseaux du savoir*, edited by Luc Foisneau, 714a–19b. Paris: Classiques Garnier, 2015.

Pérouse, Marie. *L'Invention des Pensées de Pascal: Les éditions de Port-Royal (1670–1678)*. Paris: Honoré Champion, 2009.

Peschke, Erhard. "Die Reformideen des Comenius und ihr Verhältnis zu A. H. Franckes Plan einer realen Verbesserung in der ganzen Welt." In *Der Pietismus in Gestalten und Wirkun-*

gen: *Martin Schmidt zum 65. Geburtstag*, edited by Heinrich Bornkamm, 368–82. Bielefeld: Luther-Verlag, 1975.

Peterfreund, Stuart. *Turning Points in Natural Theology from Bacon to Darwin: The Way of the Argument from Design*. New York: Palgrave Macmillan, 2012.

Peters, Günter. *Die Kunst der Natur: Ästhetische Reflexion in Blumengedichten von Brockes, Goethe und Gautier*. Munich: W. Fink, 1993.

Petersen, Henrik. "B. H. Brockes, J. A. Fabricius, H. S. Reimarus: Physikotheologien im Norddeutschland des 18. Jahrhunderts zwischen theologischer Erbauung und Wissensvermittlung." PhD thesis, University of Kiel, 2004. http://nbn-resolving.de/urn:nbn:de:gbv:8-diss-20408.

Philipp, Wolfgang. *Das Werden der Aufklärung in theologiegeschichtlicher Sicht*. Göttingen: Vandenhoeck & Ruprecht, 1957.

Pietsch, Andreas. *Isaac La Peyrère: Bibelkritik, Philosemitismus und Patronage in der Gelehrtenrepublik des 17. Jahrhunderts*. Berlin: De Gruyter, 2012.

Pinto-Correia, Clara. *The Ovary of Eve: Egg and Sperm and Preformation*. Chicago: University of Chicago Press, 1997.

Pippin, Robert. *Kant's Theory of Form: An Essay on the* Critique of Pure Reason. New Haven, CT: Yale University Press, 1982.

Plummer, Marjorie Elizabeth. *From Priest's Whore to Pastor's Wife: Clerical Marriage and the Process of Reform in the Early German Reformation*. Farnham: Ashgate, 2012.

Poole, William. *The World Makers: Scientists of the Restoration and the Search for the Origins of the Earth*. Past in the Present. Oxford: Peter Lang, 2010. Paperback ed. Oxford: Peter Lang, 2017.

Popkin, Richard. *Isaac La Peyrère (1596–1676): His Life, Work and Influence*. Leiden: Brill, 1987.

Preisendanz, Wolfgang. "Naturwissenschaft als Provokation der Poesie: Das Beispiel Brockes." In *Frühaufklärung*, edited by Sebastian Neumeister, 469–94. Munich: W. Fink, 1994.

Principe, Lawrence M. *The Aspiring Adept: Robert Boyle and His Chemical Quest*. Princeton, NJ: Princeton University Press, 1998.

Rappaport, Rhoda. "Vallisneri and Europe." *History of Science* 29 (1991): 73–98.

———. *When Geologists Were Historians, 1665–1750*. Ithaca NY: Cornell University Press, 1997.

Rattansi, Piyo M. "The Social Interpretation of Science in the Seventeenth Century." In *Science and Society, 1600–1900*, edited by P. Mathias, 1–32. Cambridge: Cambridge University Press, 1972.

Raupp, Werner. "Fabricius, Johann Albert." In *Biographisch-Bibliographisches Kirchenlexikon*, 25: cols. 393–408. 2005.

Raven, Charles. *John Ray, Naturalist: His Life and Works*. Cambridge: Cambridge University Press, 1950.

Reill, Peter. *Vitalizing Nature in the Enlightenment*. Berkeley: University of California Press, 2005.

Re Manning, Russell, ed. *The Oxford Handbook of Natural Theology*. Oxford: Oxford University Press, 2013.

Richards, Robert J. "Influence of Sensationalist Tradition on Early Theories of the Evolution of Behavior." *Journal of the History of Ideas* 40 (1979): 85–105.

———. *Darwin and the Emergence of Evolutionary Theories of Mind and Behavior*. Science and Its Conceptual Foundations. Chicago: University of Chicago Press, 1987.

———. *The Romantic Conception of Life: Science and Philosophy in the Age of Goethe*. Chicago: University of Chicago Press, 2002.

———. "Darwin's Theory of Natural Selection and Its Moral Purpose." In *The Cambridge Companion to the* Origin of Species, edited by M. Ruse and R. J. Richards, 47–66. Cambridge: Cambridge University Press, 2009.

Rieke-Müller, Annelore. "Die außereuropäische Welt und die Ordnung der Dinge in Kunst- und Naturalienkammern des 18. Jh.—das Beispiel der Naturalienkammer der Franckeschen Stiftungen in Halle." In *Das Europa der Aufklärung und die außereuropäische koloniale Welt*, edited by Hans J. Lüsebrink, 51–73. Göttingen: Wallstein, 2006.

Rivers, Isabel. "'Galen's Muscles': Wilkins, Hume, and the Educational Use of the Argument from Design." *Historical Journal* 36, no. 3 (1993): 577–97.

Roemer, B. van de. "Regulating the Arts: Samuel van Hoogstraten versus Willem Goeree." In "Art and Science in the Early Modern Low Countries," edited by Eric Jorink and Bart Ramakers. *Netherlands Yearbook for History of Art / Nederlands Kunsthistorisch Jaarboek* 61 (2011): 184–207.

Roger, Jacques. *Les sciences de la vie dans la pensée française du XVIIIe siècle: La génération des animaux de Descartes à l'Encyclopédie*. 2nd ed. Paris: Albin Michel, 1993.

Roling, Bernd. *Physica sacra: Wunder, Naturwissenschaft und historischer Schriftsinn zwischen Mittelalter und früher Neuzeit*. Leiden: Brill, 2013.

Roper, Lyndal. *Der feiste Doktor: Luther, sein Körper und seine Biographien*. Translated by Karin Wördemann. Göttingen: Wallstein, 2012.

Rossi, Paolo. *I segni del tempo: Storia della Terra e storia delle nazioni da Hooke a Vico*. Milan: Feltrinelli, 1979.

Rudwick, Martin J. S. "Biblical Flood and Geological Deluge: The Amicable Dissociation of Geology and Genesis." In *Geology and Religion: A History of Harmony and Hostility*, edited by Martina Kölbl-Ebert, 103–10. The Geological Society London, Special Publications 310. London: Geological Society, 2009.

Ruestow, Edward G. *The Microscope in the Dutch Republic: The Shaping of Discovery*. Cambridge: Cambridge University Press, 1996.

Ruhland, Thomas. "Pietistische Rivalität und Naturgeschichte in der Südasienmission des 18. Jahrhunderts." PhD thesis, Universität Kassel, 2014. Published as *Pietistische Konkurrenz und Naturgeschichte—Die Südasienmission der Herrnhuter Brüdergemeine und die Dänisch-Englisch-Hallesche Mission (1755–1802)*. Herrnhut: Herrnhuter Verlag, 2018.

———. "Objekt, Parergon, Paratext—Das Linnésche System in der Naturalia-Abteilung der Kunst- und Naturalienkammer der Franckeschen Stiftungen zu Halle." In *Parerga und Paratexte—Steine rahmen, Tiere taxieren, Dinge inszenieren: Sammlung und Beiwerk*, edited by Kristin Knebel, Cornelia Ortlieb, and Gudrun Püschel, 72–105. Dresden: Sandstein Kommunikation, 2018.

Ruppel, Sophie. "Von Pflanzen und Menschen. Botanophilie in der aufklärerisch-bürgerlichen Gesellschaft um 1800." Unpublished habilitation-thesis, University of Basel, 2018.

Sarti, Carlo. "Giuseppe Monti and Paleontology." *Nuntius* 8, no. 2 (1993), 443–55.

Savelli, Roberto. "L'opera biologica di A. V." *Physis* 3, no. 4 (1961): 269–308.

Scarry, Elaine. "Imagining Flowers: Perceptual Mimesis (Particularly Delphiniums)." *Representations* 57 (Winter 1997): 90–115.

Schierbeek, A. *Jan Swammerdam, 12 February 1637–17 February 1680: His Life and Works*. Amsterdam: Swets & Zeitlinger, 1967.

Schneider, Werner. *Hoffnung auf Vernunft: Aufklärungsphilosophie in Deutschland*. Hamburg: Meiner, 1990.

Schuster, John. *Descartes-Agonistes: Physico-mathematics, Method & Corpuscular-Mechanism, 1618–33*. Dordrecht: Kluwer, 2014.

Schwaiger, Clemens. "Philosophie und Glaube bei Christian Wolff und Alexander Gottlieb Baumgarten." In "Die natürliche Theologie bei Christian Wolff," edited by Michael Albrecht. *Aufklärung* 23 (2011): 213–27.

Scott, David. "Rousseau and Flowers: The Poetry of Botany." *Studies on Voltaire and the Eighteenth Century* 182 (1979): 73–86.

Seters, Wouter Hendrik van. *Pierre Lyonet, 1706–1789: Sa vie, ses collections de coquillages et de tableaux, ses recherches entomologiques.* La Haye: Martinus Nijhoff, 1962.

Shanahan, T. "Teleological Reasoning in Boyle's *Disquisition about Final Causes.*" In *Robert Boyle Reconsidered,* edited by Michael Hunter, 177–92. Cambridge: Cambridge University Press, 1994.

Shank, Michael H. "Natural Knowledge in the Latin Middle Ages." In *Wrestling with Nature: From Omens to Science,* edited by Peter Harrison, Ronald L. Numbers, and Shank, 83–115. Chicago: University of Chicago Press, 2011.

Shapin, Steven. *The Scientific Revolution.* Chicago: University of Chicago Press, 1996.

Shapin, Steven, and Barry Barnes. "Science, Nature and Control: Interpreting Mechanics' Institutes." *Social Studies of Science 7,* no. 1 (1977): 31–74.

Sheehan, Jonathan. *The Enlightenment Bible: Translation, Scholarship, Culture.* Princeton, NJ: Princeton University Press, 2005.

Sheehan, Jonathan, and Dror Wahrman. *Invisible Hands: Self-Organization and the Eighteenth Century.* Chicago: University of Chicago Press, 2015.

Sloan, Phillip R. "The Buffon-Linnaeus Controversy." *Isis* 67, no. 3 (Sept. 1976): 356–75.

Sparn, Walter. "Natürliche Theologie." In *Theologische Realenzyklopädie,* 24:85–98. Berlin: De Gruyter, 1994.

Starobinski, Jean. "Rousseau's Happy Days." *New Literary History* 11, no. 1 (Autumn 1979): 147–65.

Stebbins, Sara. *Maxima in minimis: Zum Empirie- und Autoritätsverständnis in der physikotheologischen Literatur der Frühaufklärung.* Microkosmos, vol. 8. Frankfurt am Main: Peter D. Lang, 1980.

Steiger, Johann Anselm. *Bibel-Sprache, Welt und jüngster Tag bei Johann Peter Hebel: Erziehung zum Glauben zwischen Überlieferung und Aufklärung.* Göttingen: Vandenhoeck & Ruprecht, 1994.

———. "Die *communicatio idiomatum* als Achse und Motor der Theologie Luthers: der 'fröhliche Wechsel' als hermeneutischer Schlüssel zu Abendmahlslehre, Anthropologie, Seelsorge, Naturtheologie, Rhetorik und Humor." *Neue Zeitschrift für Systematische Theologie* 38, no. 1 (1996): 1–28.

———. "Ästhetik der Realpräsenz: Abendmahl, Schöpfung, Emblematik und mystische Union bei Martin Luther, Philipp Nicolai, Valerius Herberger, Johann Saubert und Johann Michael Dilherr." In *Von Luther zu Bach: Bericht über die Tagung 22.–25. September 1996 in Eisenach,* edited by Renate Steiger, 21–41. Sinzig: Studio, 1999.

———. "Ist es denn ein Wunder? Die aufgeklärte Wunderkritik. Oder: Von Spinoza zu Reimarus." In *500 Jahre Theologie in Hamburg: Hamburg als Zentrum christlicher Theologie und Kultur zwischen Tradition und Zukunft,* edited by Steiger, 113–30. Berlin: De Gruyter, 2005.

Steiger, Rudolf. "Conrad Gessner und die Berge." In *Conrad Gessner, 1516–1565, Universalgelehrter, Naturforscher, Arzt,* edited by Hans Fischer, 204–11. Jubiläumspublikationen zur 450 jährigen Geschichte des Art. Instituts Orell Füssli AG und ihrer Vorfahren: 1519–1969. Zurich: Orell Fuessli, 1967.

Steigerwald, Jörn. "Das göttliche Vergnügen des Sehens: Barthold Hinrich Brockes' Techniken des Betrachters." *Convivium: Germanistisches Jahrbuch Polen* (2000), 9–41.

Steinke, Hubert, and Martin Stuber. "Hallers Alpen—Kontinuität und Abgrenzung." In *Boscani Leoni, Wissenschaft* (2010), 235–58.

Steinmann, Holger. *Absehen—Wissen—Glauben: Physikotheologie und Rhetorik 1665–1747.* Berlin: Kadmos, 2008.

Stelter, Marcus. "Möglichkeiten und Grenzen des Erwerbs und der Vermittlung von Wissen durch Schenkungen." In *Ordnen—Vernetzen—Vermitteln: Kunst- und Naturalienkammern der Frühen Neuzeit als Lehr- und Lernorte*, edited by Eva Dolezel and Rainer Godel, 179–204. Stuttgart: Wissenschaftliche Verlagsgesellschaft Stuttgart, 2018.

Stevens, P. F., and S. P. Cullen. "Linnaeus, the Cortex-Medulla Theory, and the Key to His Understanding of Plant Form and Natural Relationships." *Journal of the Arnold Arboretum* 71, no. 2 (Apr. 1990): 179–220.

Stolberg, Michael. "A Woman Down to Her Bones: The Anatomy of Sexual Difference in the Sixteenth and Early Seventeenth Centuries." *Isis* 94, no. 2 (2003): 274–99.

Strickland, Lloyd. "The Doctrine of 'The Resurrection of the Same Body' in Early Modern Thought." *Religious Studies* 46, no. 2 (2010): 163–83.

Sudduth, Michael. *The Reformed Objection to Natural Theology*. Farnham: Ashgate, 2009.

Sullivan, Louis H. "The Tall Office Building Artistically Considered." *Lippincott's Magazine*, Mar. 1896, 403–9.

Terrall, Mary. *Catching Nature in the Act: Natural History in the Eighteenth Century*. Chicago and London: University of Chicago Press, 2014.

Thomson, Keith. *Before Darwin: Reconciling God and Nature*. New Haven, CT: Yale University Press, 2005.

Toellner, Richard. "Die Bedeutung des Physico-theologischen Gottesbeweises für die nachcartesische Physiologie im 18. Jahrhundert." *Berichte zur Wissenschaftsgeschichte* 5 (1982): 75–82.

Topham, Jonathan R. "Beyond the 'Common Context': The Production and Reading of the Bridgewater Treatises." *Isis* 89 (1998): 233–62.

——. "Biology in the Service of Natural Theology: Paley, Darwin, and the *Bridgewater Treatises*." In *Biology and Ideology: From Descartes to Dawkins*, edited by D. R. Alexander and Ronald L. Numbers, 88–113. Chicago: University of Chicago Press, 2010.

——. "Natural Theology and the Sciences." In *Cambridge Companion to Science and Religion*, edited by Peter Harrison, 59–79. Cambridge: Cambridge University Press, 2010.

Touber, Jetze. "Finding the Right Measure: Architecture and Philology in Biblical Scholarship in the Dutch Early Enlightenment, 1670–1710." *Historical Journal* 58, no. 4 (2015): 959–85.

Trepp, Anne-Charlott. "'Nature' as Religious Practice in Seventeenth-Century Germany." In *Religious Values and the Rise of Science in Europe*, edited by John H. Brooke and E. Ihsanoglu, 81–110. Istanbul: Research Centre for Islamic History, Art and Culture, 2005.

——. "Zwischen Inspiration und Isolation: Naturerkundung als Frömmigkeitspraxis in der ersten Hälfte des 18. Jahrhunderts." *Zeitenblicke* 5, no. 1 (2006): unpaginated. http://www.zeitenblicke.de/2006/1/Trepp/index_html, URN: urn:nbn:de:0009-9-2811.

——. "Natural Order and Divine Salvation: Protestant Conceptions in Early Modern Germany (1550–1750)." In *Natural Law and Laws of Nature in Early Modern Europe,* edited by Lorraine Daston and Michael Stolleis, 123–42. Aldershot: Ashgate, 2008.

——. *Von der Glückseligkeit alles zu wissen: Die Erforschung der Natur als religiöse Praxis in der Frühen Neuzeit*. Frankfurt am Main: Campus, 2009.

——. "Von der Missionierung der Seelen zur Erforschung der Natur: Die Dänisch-Hallesche Mission im ausgehenden 18. Jahrhundert." *Geschichte und Gesellschaft* 36 (2010): 231–56.

——. "'Adam's Knowledge' as Promise of Salvation and Knowledge Dispositive in the Early Modern Era." In "Science and Religion," edited by Fernanda Alfieri and Kärin Nickelsen. Special issue, *Annali dell'Instituto storico italo-germanico in Trento/Jahrbuch des italienisch-deutschen historischen Instituts in Trient* 43, no. 1 (2017): 33–57.

——. "Adam benennt die Tiere: Zur Bedeutung der Namen für die Kenntnis der Dinge. Genesis 2, 19–20 als ein Erkenntnisdispositiv der Frühen Neuzeit." In *Religiöses Wissen im vormodernen Europa: Schöpfung—Mutterschaft—Passion*, edited by Renate Dürr, Annette Gerok-Reiter, Andreas Holzem, and Steffen Patzold. Paderborn: Schoeningh, 2018, 143–82.

Trinkle, Dennis. "Noël-Antoine Pluche's *Le Spectacle de la nature*: An Encyclopedic Best Seller." *Studies on Voltaire and the Eighteenth Century* 358 (1997): 93–134.

Tucci, Francesco Saverio. "'Il parlare della Santa Scrittura e l'operare della natura,' gli interrogative della geologia storica nella riflessione di Antonio Vallisneri." *Contributi (Biblioteca Municipale di Reggio Emilia)* 7, no. 14 (1983): 5–37.

Valter, Claudia. "Studien zu bürgerlichen Kunst- und Naturaliensammlungen des 17. und 18. Jahrhunderts in Deutschland." PhD thesis. Aachen: Self-published, 1995.

Van Hoof, Henri. *Histoire de la traduction en Occident*. Paris: Duculot, 1991.

Vermij, Rienk. "Religion and Mathematics in Seventeenth-Century Holland: The Case of Bernard Nieuwentijt." In *Science and Religion / Wissenschaft und Religion*, edited by Änne Bäumer and Manfred Büttner, 152–57. Abhandlungen zur Geschichte der Geowissenschaften und Religion / Umwelt-Forschung 3. Bochum: Universitätsverlag Dr. N. Brockmeyer, 1989.

———. *Secularisering en natuurwetenschap in de zeventiende en achttiende eeuw: Bernard Nieuwentijt*. Amsterdam: Rodopi, 1991.

———. "The Beginnings of Physico-Theology: England, Holland, Germany." In *Grenz-Überschreitung: . . . Festschrift zum 70. Geburtstag von Manfred Büttner*, edited by Heyno Kattenstedt, 173–84. Abhandlungen zur Geschichte der Geowissenschaften und Religion / Umwelt-Forschung 9. Bochum: Universitätsverlag Dr. N. Brockmeyer, 1993.

———. "The Flood and the Scientific Revolution: Thomas Burnet's System of Natural Providence." In *Interpretations of the Flood*, edited by Florentino Garcia Martinez and Gerard P. Luttikhuizen, 150–66. Leiden: Brill, 1999.

———. *The Calvinist Copernicans: The Reception of the New Astronomy in the Dutch Republic, 1575–1750*. Amsterdam: Koninklijke Nederlandse Akademie van Wetenschappen, 2002.

———. "The Formation of the Newtonian Philosophy: The Case of the Amsterdam Mathematical Amateurs." *British Journal for the History of Science* 36, no. 2 (2003): 183–200.

———. "Nature in Defense of Scripture: Physico-Theology and Experimental Philosophy in the Work of Bernard Nieuwentijt." In *The Book of Nature in Early Modern and Modern History*, edited by Klaas van Berkel and Arjo Vanderjagt, 83–96. Leuven: Peeters, 2006.

———. "Defining the Supernatural: The Dutch Newtonians, the Bible, and the Laws of Nature." In *Newton and the Netherlands: How Isaac Newton Was Fashioned in the Dutch Republic*, edited by Eric Jorink and Ad Maas, 185–206. Leiden: Leiden University Press, 2012.

———. "Translating, Adapting, Mutilating: Or, How to Make an Enlightenment Classic." *Isis* 109, no. 2 (2018): 333–38.

Verner, Mathilde. "Johann Albert Fabricius, Eighteenth-Century Scholar and Bibliographer." *Papers of the Bibliographical Society of America* 60, no. 3 (1966): 281–36.

Vidal, Fernando. "Extraordinary Bodies and the Physicotheological Imagination." In *The Faces of Natures in Enlightenment Europe*, edited by Lorraine Daston and Gianna Pomata, 61–96. Berlin: Berliner Wissenschaftsverlag, 2003.

Vignau-Wilberg, Thea. *Archetypa studiaque patris Georgii Hoefnagelii 1592: Natur, Dichtung und Wissenschaft in der Kunst um 1600; Nature, Poetry and Science in Art around 1600*. Munich: Staatliche Graphische Sammlung, 1994.

Wall, Ernestine G. E. van der. "Newtonianism and Religion in the Netherlands." *Studies in History and Philosophy of Science* 35 (2004): 493–514.

Wallmann, Johannes. *Der Pietismus*. Göttingen: Vandenhoeck & Ruprecht, 1990.

Weimar, Klaus. "Gottes und der Menschen Schrift." *Merkur* 513, no.12 (1991): 1089–95.

Weiss, Richard. *Die Entdeckung der Alpen: Eine Sammlung schweizerischer und deutscher Alpenliteratur bis zum Jahr 1800*. Frauenfeld: Huber, 1934.

Wels, Volkhard. *Manifestationen des Geistes*. Göttingen: V&R unipress, 2014.

Westfall, Richard S. *Science and Religion in Seventeenth Century England*. New Haven, CT: Yale University Press, 1958.

———. *Force in Newton's Physics: The Science of Dynamics in the 17th Century*. London: Elsevier, 1971.

Westman, Robert S. "The Astronomer's Role in the Sixteenth Century: A Preliminary Study." *History of Science* 18 (1980): 105–47.

Whitmer, Kelly Joan. *The Halle Orphanage as Scientific Community: Observation, Eclecticism, and Pietism in the Early Enlightenment*. Chicago: University of Chicago Press, 2015.

Wiegand, Hermann. *Hodoeporica: Studien zur neulateinischen Reisedichtung des deutschen Kulturraums im 16. Jahrhundert*. Baden-Baden: Verlag Valentin Koerner, 1984.

Wielema, Michiel R., ed. *Adriaan Koerbagh: A Light Shining in Dark Places, to Illuminate the Main Questions of Theology and Religion*. Leiden: Brill, 2011.

Wilson, Catherine. *The Invisible World: Early Modern Philosophy and the Invention of the Microscope*. Princeton, NJ: Princeton University Press, 1995.

Wulf, A. *The Invention of Nature*. London: Murray, 2015.

Zande, Johan van der. "Johann Georg Sulzer, Spaziergänge im Berliner Tusculum." In *Berliner Aufklärung*, edited by Ursula Goldenbaum and Alexander Kosenina, 41–68. Hannover: Wehrhahn, 1999.

Zeitz, Lisa. "Natural Theology, Rhetoric, and Revolution: John Ray's Wisdom of God, 1691–1704." *Eighteenth-Century Life* 18, no. 1 (1994): 120–33.

Zöckler, Otto. *Geschichte der Beziehungen zwischen Theologie und Naturwissenschaft mit besondrer Rücksicht auf Schöpfungsgeschichte*. Vol. 2 [Zweite Abtheilung], *Von Newton und Leibniz bis zur Gegenwart*. Gütersloh: C. Bertelsmann, 1879.

———. *Geschichte der Apologie des Christentums*. Gütersloh: Bertelsmann, 1907.

Zoller, Heinrich. "La découverte des Alpes de Pétrarque à Gessner." In *Une cordée originale: Histoire des relations entre science et montagne*, edited by Jean-Claude Pont and Jan Lacki, 417–28. Geneva: Georg, 2000.

Zuber, Mike A. "Copernican Cosmotheism: Johann Jacob Zimmermann and the Mystical Light." *Aries: Journal for the Study of Western Esotericism* 15, no. 2 (2015): 215–45.

Accademia del Cimento, 202
accommodation, 73, 79, 115, 201
Aldrovandi, Ulisse, 173, 176, 179
Alfonso the Wise of Aragon, 26–27
Alps, the, 222–32
Alsted, Johann Heinrich, 7, 70–71
anima mundi, 198
anthropocentricism, and physico-theology, 27–28
antlion, 28
Aquinas, Thomas, 9, 70
argument from design. *See* design, argument from
Aristotle: *entelechy* or *energeia* and, 57; existence
    from eternity and, 118; on form versus
    appearance, 52, 59–60; fossils and, 201;
    Heraclitus and, 171; metaphysics and, 70;
    natural science, mathematics and, 41–42;
    questions by, 59
Arnauld, Antoine, 145
Arndt, Johann, 15, 95, 130–31, 219
Artico di Porcia, Gian, 194
astrology, 106, 210–11
astronomy, 42
Astruc, Jean, 201
atheism: argument from design and, 78; Bentley
    and, 7–8, 14–15, 72, 96, 122; Boyle Lectures and,
    72, 188; Charleton and, 44; Descartes and, 75,
    141; Diderot and, 23; Horst and, 112; More and,
    124; Pascal and, 15, 141, 148–49; physico-
    theology and, 7, 23; Scheuchzer and, 166–67;
    Spinozism and, 16
Augustine, 80, 147
autonomy of natural things, 12, 26, 35, 56, 61

Bacon, Francis, 33, 70, 121, 123
Baldi, Bernardino, 41

Barlaeus, Caspar, 92, 95
Bärmann, Georg Friedrich, 55–56
Barnes, Robin, 106
Barrow, Isaac, 93
Bartas, Guillaume de Salluste du, 5
Basil of Caesarea, 95
Baur, Jörg, 129
beauty, 31, 54–56, 61, 81, 117–18, 130, 134, 137, 150,
    183, 197, 212, 224, 228, 230
Beeckman, Isaac, 41
bees, 127
Bekker, Balthasar, 166
Bellarmine, Roberto, 93–94
Benemann, Johann Christian, 55
benevolence (goodness), divine, 7, 72–73, 105,
    108–9, 130, 136, 164, 169, 179
Bennedetti, G. B., 41
Bentley, Richard, 7–8, 14–15, 27, 72, 96, 124
Bergius, Conrad, 108
Bernard of Cluny, 212
Bertrand, Élie, 18, 222–23, 228–32
Bildungstrieb, 58
biology, evolutionary, 24
Blackmore, Richard, 5
Blair, Ann, 28, 189, 202
Blumenbach, Johann Friedrich, 58–60
Blumenberg, Hans, 211
body-soul unity, 131
Boerhaave, Herman, 78, 81, 173, 176, 184
Bonet, Théophile, 196
Bonnet, Charles, 28, 49
book of nature / *liber naturae*: approaches to
    reading, 128, 130, 136–37; Arndt and, 219;
    Augustine and, 80; Bentley and, 122; Boyle and,
    119–20; Brockes and, 18, 213; Calvin and, 79;

book of nature / *liber naturae* (cont.)
　　Descartes and, 81; Dutch Reformed Church
　　and, 80–83; Galileo and, 43; Horst and, 111;
　　Neickel and, 133–34; Pascal and, 16; Pluche and,
　　16, 183; Ray and, 117; Swammerdam and, 81;
　　Vallisneri and, 17. *See also* accommodation
Boreel, Adam, 83
botany, 56–57, 226, 228
Bots, Jan, 157
Bourguet, Louis, 203
Boyle, Robert: causes and, 53–54, 119, 122;
　　moderate theism and, 30; organic science and,
　　123; rational God and, 7; reconciling natural
　　philosophy and revealed truth, 45–48;
　　resurrection and, 9; topics of interest to, 49
Boyle Lectures, 14–15, 72, 122, 175, 188
Bridgewater Treatises, 6, 30
Brockes, Barthold Heinrich, 4, 10, 17–18, 91–93,
　　96, 131, 209, 216–20
Brooke, John Hedley, 8, 11–12
Browne, Thomas, 174–75
Brucker, Nicolas, 16–17
Bruno, Giordano, 93
Bucer, Martin, 129
Buckley, Michael, 23, 175
Buffon, Georges-Louis Leclerc, Comte de, 58, 60
Burnet, Thomas: Derham and, 75; diluvialism
　　and, 18, 47, 160, 167; Flood and, 47; fossils and,
　　194; mountains and, 18; natural philosophy,
　　Scripture and, 45; origin of earth and, 223, 229;
　　Ray and, 73–74; reconciling natural philosophy
　　and revealed truth, 45
Burns, Robert, 116
Buteo, Johannes, 85
Büttner, Manfred, 4, 90

Caelius Secundus Curione, 97
Calloway, Katherine, 13–15, 31–32, 44, 69
Calvin, John, 79, 129
Calzolari, Francesco, 228
Cambridge Platonists, 25, 34, 45, 121, 198. *See also*
　　More, Henry
Cartesianism, 4, 6, 45, 75, 158
causes/causality/causation: efficient, 7; final, 7, 53,
　　67–68, 70, 122
censorship, 110–12
Cesalpino, Andrea, 201
chance, 53–56, 118, 164–65, 177–78, 225, 230
Charleton, Walter, 13, 23, 30–32, 36, 43–44, 67–69,
　　125n10

Chemnitz, Johann Hieronymus, 132, 134
Cheyne, George, 78
Chraplak, Marc, 10
Christ, Jesus, 9, 128, 142, 147–50
Christology, Luther's "ubiquitarian," 15, 127–29,
　　132–33, 136
Chytraeus, David, 108
Cicero, 81
Clark, J. F. M., 180
Clarke, Samuel, 27
Cluver, Dethlev, 167–68
Cockburn, John, 72
collecting, 130, 134, 136, 180, 194, 216. *See also*
　　libraries, private; *naturalia*
Collinus (Theodor Ambühl), 227
communicatio idiomatum, 128–29
conflagration, final (future), 47, 98, 160–61, 223
Conti Antonio, 198
Copernicanism, 80
Cowper, William, 75
Creation: Alfonso the Wise of Aragon and, 27;
　　Arndt and, 130; aspects of Lutheran view of, 14;
　　Bärmann and, 55; Bertrand and, 230; Burnet
　　and, 85; Calvin and, 79; Charleton and, 69;
　　Darwin and, 35–36; defects in, 34; Derham and,
　　8, 72, 74; Descartes and, 118; Dillenberger and,
　　23; Dutch Reformed Church and, 80–81;
　　Fabricius and, 92, 95; God and, 132–34, 136, 144;
　　as God's materialized word, 128–29; Goeree
　　and, 85; Horst and, 106, 110; Kepler and, 42;
　　Lemnius and, 106; Lesser and, 129, 177; Luther
　　and, 128–30, 136; More and, 46; mountains
　　and, 222–26, 230–32; natural theology and, 5;
　　physico-theology and, 209–10; physics and, 167;
　　Pluche and, 185, 189–90; Ray and, 47, 117; Rist
　　and, 28; special, 3, 56
criticism, radical biblical, 81–83
Crowther, Kathleen, 14
Cudworth, Ralph, 2, 34, 47, 115, 117–18
Cunaeus, Petrus, 84

Darwin, Charles, 28, 30
Dear, Peter, 41
de Geer, Charles, 179
deism, 7, 10–11, 14, 33–35, 112, 148
Delany, Mary, 75
Delessert, Madeleine-Catherine, 57
Derham, William: astronomy and, 210–11;
　　atheism and, 189; Bertrand and, 229; Boyle
　　and, 72; Brockes and, 219; Burnet and, 75;

cosmology and, 27–28; cosmology of, 27; defects in nature and, 34; deism and, 10; Descartes and, 27; Fabricius and, 13, 90–91, 95; Huyghens and, 27; influence of, 12–13, 23, 40; insects and, 175–76; language and, 211; mountains and, 225–26, 230–31; "physico-theology" and, 23; Pluche and, 184–85; publication history of *Astro-theology*, 74; purpose and, 75; Ray and, 5–6, 71–75, 225; Ray and Nieuwentijt and, 186–87; responses to nature and, 32; Scheuchzer and, 229–31; scholarly approach of, 186; special providence and, 72; translation into German, 129, 210–11

Descartes, René: Boyle and, 26, 119, 125n18; Burnet and, 195; challenges to, 24, 78–81, 118; Derham and, 27, 75; Genesis story and, 195; Keorbagh and, 82; laws of nature and, 158; matter in motion and, 6; mechanical philosophy of, 158; Swammerdam and, 81. *See also* mechanical philosophy

design, argument from, 1–3, 8–9, 12, 35–36, 78, 81, 90, 97–98, 119–20, 121, 141, 157–58, 159, 164, 166, 169, 172, 175, 197, 225; natural theology and, 1–3

Deutsche Gesellschaft (in Hamburg), 54–55

Diderot, Denis, 23

digestion, 110

Dillenberger, John, 23, 34–35

dissection, 29

Donat/Donati, Christian, 97, 99n24

Draper, John William, 3

Duplessis-Mornay, Philippe, 68

Eamon, William, 104

earth: age of, 199–200; origin of, 231–32

earthquake, 161

empiricism, 121, 195–96, 214–16

Enlightenment: *Aufklärung*, 6–7; Pietism and the, 131, 137; a right use of reason and the, 190, 213–14; *sodalitates*, 91; vernacular physico-theology and, 90

entelechy, 57

entomology, origins of, 16, 25, 172, 179–80. *See also* insectology; insects

Epicurus, 118–19

Estienne, Robert, 183

evidence: astronomical and geometric, 199–201; defined, 210; empirical, 196; evolution and, 180; from nature, 23–26, 44, 119–20, 124, 151–52, 179; physico-theology and, 209–20; "scientific," 45;

in/as text, 209–20; unfit or confusing phenomena and, 117. *See also* design, argument from

eye and eyesight, 52, 56, 58, 68, 120, 123, 134, 143, 174, 186, 209–20

Fabricius, Johann Albert: Arndt and, 131; Barrow and, 93; Bellarmine and, 93–94; Bentley and, 96; biography of, 90–91; Boyle and, 96–97; Brockes and, 219; Derham and, 27, 90, 129–30, 210; gratitude and, 129; Nieuwentijt and, 96; Parker and, 93; as physico-theology forerunner, 13–14; Ray and, 96; Register of, 90–98; Scheuchzer and, 98; sources, 13–14, 94, 96–97

Fabricius, Samuel, 93

factions, confessional, 104–5, 166

Faenzi, Valerio, 228

Falaguasta, Nani, 94

Fénelon, François de, 29, 72–73, 93

Feuerlein, Jakob Wilhelm, 95

Ficino, Marsilio, 93, 95

Filleau de la Chaise, Nicolas Jean, 142–46

Flood, biblical / diluvialism: Burnet and, 223–24; fossils and, 8–9, 98, 201; Leibniz and, 201; physico-theology and, 47; Scheuchzer and, 167–68; science and, 2; universality of, 82; Vallisneri and, 200, 202. *See also* Creation; fossils; Genesis; mountains

flowers, 54–62. *See also* seeds / seed theory

fly, 25, 118, 123, 172–74, 180, 186–87, 217–18

Fontana, M. Publius, 94, 97

Fontenelle, 189

form: as aesthetic judgment, 61; Aristotelian, 12, 57; contingent, 60; continuity of, 57–58; defined, 52; *entelechy* and *energeia*, 57; Kant and, 61; law of, 61; purposive, 59, 61; Rousseau and, 57; transcending material, 25. *See also* matter; purpose

Forster, John Reinhold, 135

fossils: Bertrand and, 230; Burnet and, 194; earth structure and, 194; Fabricius and, 98, 167; Vallisneri and, 199; Woodward and, 8, 98, 229. See also Flood, biblical / diluvialism

Fracastoro, Girolamo, 201

Francke, August Hermann, 131, 176

Franckesche Stiftungen (Francke's Foundations), 131, 134–35, 137

frogs, 127

Funkenstein, Amos, 43, 49, 104

Gale, Theophilus, 70
Galen, 44, 81, 118–19
Galileo, 12, 41–44, 93, 197, 203
Gassendi, Pierre, 31, 47
Gaukroger, Stephen, 29
generation of animals, 17, 49, 57, 59, 176, 194–96, 199
Genesis, 195, 199, 222–23, 232
geology, 201
Gerson, Jean, 93
Gessner, Conrad, 223, 226, 231
Giornale de' letterati d'Italia, 196
Glanvill, Joseph, 45–46
God: actively experiencing, 137; arguments for
    existence of, 31, 36, 44, 60, 69, 73, 79, 94, 112,
    117, 123, 128, 141–51, 189, 232; as artificer, 1, 81,
    85; *concursus* of, 132–33; as creator and
    moderator, 69; existence of, 60, 79, 94, 141–44,
    147; final causes and, 12; as a global creator, 189;
    naturalia and, 134; natural world and, 151;
    nature and, 112–13, 117; physico-theological
    attributes of, 7; special providence of, 10.
    See also benevolence (goodness), divine;
    providence, divine; wisdom, divine
Goeree, Willem, 13, 79, 83–89
Goethe, Johann Wolfgang von, 59
goodness, divine. *See* benevolence
Gould, Stephen J., 23
Gravesande, Willem Jacob 's, 78
gravity, 27, 36
Great Chain of Being, 198
Gregory, David, 123
Greyerz, Kaspar von, 13–14
Grimm brothers, 216
Grote, Simon, 55
Grotius, Hugo, 81
Gründler, Gottfried August, 134
Guidott, Thomas, 174–75
Guldin, Paul, 41

Haller, Albrecht von, 226, 228
Halley, Edmund, 199–200
Harrison, Peter, 4–5, 12–13, 33
Hartlib, Samuel, 67
Hartsoeker, Nicolas, 196
Harvey, William, 26, 198
Heister, Lorenz, 29
Heraclitus, 171
Herbst, Johann Friedrich Wilhelm, 132, 139n43
hexaemeron, 5, 95–96
historiography, 3–9, 157, 168, 191

Hobbes, Thomas, 70
Hoefnagel, Jacob, 172
Hoefnagel, Joris, 172
Hoffmann, Friedrich, 131
Hooke, Robert, 7, 196, 199
Horace, 217
Horst, Jakob, 14, 103–13
Hugh of Saint-Victor, 85
humanism, 44–45, 47, 119, 125n14, 226–27
Humboldt, Alexander von, 36
Hume, David, 7, 11, 36
Hunfeld, Barbara, 17–18
Hutchinson, John, 184
Huyghens, Christiaan, 27
hybridization, 12, 39–43, 59

incarnation, 129, 133, 136
India, 26–27, 147–49
individuation, 58
inquiry, experimental, 33
insectology, 16–17, 187. *See also* entomology,
    origins of
insects, 171–80; Derham and, 187; emblematic,
    173–74; Swammerdam and, 81
instinct, 177–79
"interior mold," 58
involucrism, 196
irenicism and consensus, 16, 169, 191. *See also*
    factions, confessional; polemics
Isidore of Seville, 92

Jablonski, Ernst, 96
Jansenism, 15, 191
Jesus Christ. *See* Christ, Jesus
John, Christoph Samuel, 135–37
Jorink, Eric, 13
Journal des sçavans, 196

Kant, Immanuel: argument from design and, 36;
    Hume and, 7, 11, 36, 49; natural theology and, 7,
    49; physico-theology and, 11–12, 30, 60–62,
    75–76, 128
Keckermann, Bartholomaeus, 70–71
Kemper, Hans-Georg, 10
Kepler, Johannes, 12, 41–42, 44
Kirby, William, 179–80
Kircher, Athanasius, 41, 85–86, 93, 95
knowledge: of God, 128–29; natural, 105, 109–13,
    131; Pluche and, 188, 191; popularization of
    scientific, 28–29

Koerbagh, Adriaan, 82, 85
Kölreuter, Joseph, 59
König, Johann Gerhard, 135–36
Kozak, Johann Sophron, 95
Kusukawa, Sachiko, 105–6

Lange, Johann Joachim, 134
Langen, August, 214
language: Brockes and, 211–20; Derham and, 211;
    of the heavens, 210–11
La Peyrère, Isaac, 82, 199
Laplace, Pierre-Simon, 11, 24
Le Clerc, Daniel, 198
Leeuwenhoek, Antoni van, 184–86, 196
Leibniz, Gottfried Wilhelm, 11, 27, 43, 58, 194
Lemarck, Jean-Baptiste, 24
Lemnius, Levinus, 14, 106–12
Leon, Jacob Jehuda, 83, 87
Leonardo da Vinci, 194
Leoni, Simona Boscani, 18
Lesser, Friedrich Christian, 16, 29, 127, 129, 131,
    176–79
Lessius, Leonard, 175
Leutwein, Christian Philipp, 95
Levitin, Dmitri, 45
Leopoldina (Deutsche Akademie der Natur-
    forscher Leopoldina), 131–32, 228
Lhwyd, Edward, 75
Libavius, Andreas, 95
libraries, private, 90–91
Linneaus, Carl (Carl von Linné), 8, 29, 57–58,
    134, 179
Locke, John, 9
Logos, 128, 130, 133, 136
Löscher, Valentin Ernst, 94–95
Lucretius, 31
Luiken, Jan, 84–85, 160
lusus naturae (jokes of nature), 9
Luther, Martin, 15, 108, 131
Lutheranism: characteristics of, 14; conflicts and,
    104; Horst and, 107–8, 111; natural philosophy
    and, 41–42, 70–71; natural theology and, 127–28;
    orthodoxy and physico-theology, 94–95, 99n24,
    104–6, 127–37; "sick," 28. *See also* Pietism
Luyken, Jan, 13
Luzzini, Francesco, 202
Lyonet, Pierre, 176–78

macrocosm/microcosm analogy, 198, 213
Magalotti, Lorenzo, 202

Malebranche, Nicolas, 58, 197
Malpighi, Marcello, 195
Mandelbrote, Scott, 13, 25, 32, 202
Manetho, 85
Martin, Benjamin, 75
Martinet, Johannes Florentinus, 136
materialism: Cartesianism and, 4; Epicurean,
    122; Mandelbrote and, 34; Newton and, 27
material word, 127
mathematics, 42
matter: adiaphorous, 48; Aristotle and, 70;
    corpuscularian theory of, 9, 26; Descartes and,
    141; gravity and, 27; inert, vitalized, 198–99;
    "meer," 75; More and, 120; in motion, 6; Pascal
    and, 148–49; "plastick power" of, 9, 47;
    potential state of, 57; purpose and, 60; Ray and,
    120. *See also* seeds / seed theory
maxima in minimis animalibus, 16, 171–72, 181n4,
    187
Mayer, Johann Friedrich, 91, 93, 95
mechanical philosophy, 24, 26, 36, 46, 48, 59, 69,
    75, 81, 119, 132–33, 141, 158, 198, 232. *See also*
    Descartes, René
Mela, Pomponius, 227
Melanchthon, Philip, 105, 108, 129
Menz, Friedrich, 127
Mersenne, Marin, 7, 41, 72, 95
metabasis, 41–42
metaphysics, 70–71
Meyer, Gerhard, 97
microscope, 16, 25, 85, 165, 173, 177
miracles, 103–13, 202–3
Moffett, Thomas, 173, 179
"molten sea" of King Solomon, 82, 88, 165, 168
More, Henry: Bentley and, 123–24; Descartes
    and, 46; empiricism and, 14–15; limits of
    human understanding and, 2; physico-
    theology and, 32–33, 45–48, 120–22; physico-
    theosophy and, 46; Ray and, 117, 120–22; topics
    of interest to, 49; wonders of the natural world
    and, 25, 74
Mosaic physics, 79–80, 83–85
mountains: Bertrand and, 228–31; Burnet and,
    224; Derham and, 224; Flood (Deluge), biblical
    and, 18, 222; Gessner and, 226–27; as an
    irregularity of the earth, 22; perception of,
    226–28; predating biblical Flood, 229; Ray and,
    224–26, 232n6; Scheuchzer and, 228–31;
    seventeenth century debate on, 223–26; Simler
    and, 227; as "warts" on earth's surface, 18.

mountains (cont.)
   See also Alps, the; Creation; Flood, biblical /
   diluvialism; Genesis
Müller, Johannes (Johannes Rhellicanus), 227
Müller, Theodor, 5
Mulsow, Martin, 92

natural history: eighteenth century, 171;
   physico-theology and, 179; piety and, 106; Ray
   and, 116–17; revealed truths and, 46;
   seventeenth century, 171; sixteenth century,
   171
naturalia, 133–37
natural knowledge, piety and, 103–6, 108
natural philosophy, experimental, 13, 16, 33, 42,
   59, 78, 83, 86, 118–19, 159, 162, 164, 166, 186,
   190–91
natural religion, 70
natural theology: eighteenth-century, 159;
   independent, 34–35; Pascal and, 15; physico-
   mathematics and, 42; physico-theology and,
   1–2, 24, 49, 71–72, 157, 168; pre-Darwinian, 3;
   rationality and, 146; seventeenth century, 159
nature: Darwinian, 35–36; God's providence and,
   108–9, 230; laws of, 158; use and design of, 188;
   utilitarian approach to, 7
nebular hypothesis, 24
necessity, 164
Neickel, Caspar Friedrich, 133
Newman, William R., 198
Newton, Isaac: Burnet and, 224; empiricism and,
   24; gravity and, 27, 36; laws of nature and,
   35–36, 123; Leibniz and, 11; Nieuwentijt and, 24;
   physico-theology and, 78; Pluche and, 184;
   Scheuchzer and, 167; "secular theology" and,
   43; time and, 200; Trinitarian creed and, 33;
   Voltaire and, 35
Nicole, Pierre, 15, 145, 151
Nicolson, Marjorie Hope, 224, 228
Nieuwentijt (also Nieuwentyt), Bernard, 162–66;
   argument from design and, 157; atheism and,
   78–79; Boyle and, 24; Cartesianism and, 6, 24;
   corpuscularian theory of, 99; Fabricius and,
   96; Newton and, 24; physico-theology and, 157;
   Pluche and, 184; Ray and, 16; Scheuchzer and,
   16, 167–68; scholarly approach of, 186;
   Spinozism and, 78–79, 163–64, 166
Nigrisoli, Francesco Maria, 198 nisus formati-
   vus, 58
Noah's Ark, 13, 82, 85–87

Ogilvie, Brian, 16, 25
Oken, Lorenz, 180
Oldenburg, Henry, 191
omnipotence, divine, 7, 72–73, 108, 129, 136, 142,
   147, 201–2
Origen, 85
ovism, 9, 17, 199

Paley, William, 3, 28–29, 32, 35, 176
Parker, Samuel, 12, 43–45, 93, 97, 117–19
parthenogenesis (virgin birth), 9
Pascal, Blaise, 15–16, 23, 141–52, 189
Patriotic Society (in Hamburg), 91
Paul (apostle), 31, 79, 130, 174
Pécharman, Martine, 15, 189
Pereyra, Benedict, 168
Périer, Étienne, 146
Périer, Gilberte, 151
Périer, Marguerite, 141
Pfäfferlin, Christoph, 227
Philipp, Wolfgang, 4, 10, 90, 94–95
philosopher's stone, 25–26
Philosophical Transactions, 184–85, 229
philosophy, 11, 24, 34, 45, 48, 81, 189; eclectic, 166;
   and physico-theology, 158–59. See also natural
   philosophy, experimental
physico-theology: basic pattern of, 209;
   characteristics of, 2, 4, 7–8; defined, 157;
   diversity of, 24–26; eighteenth-century, 6–7,
   24–25, 157; first use of term, 5; Galen and, 70;
   global, 135–36; impact of, 4; "parts" of, 75; root
   concerns of, 12; seventeenth-century, 5–6, 39–49
physico-theosophy, 46
physics, 71, 93, 122, 141, 167. See also natural
   philosophy, experimental
physiology, 69
Pico della Mirandola, 95
Pierquin, curé Jean, 9, 49
Pietism: continental, 45; Enlightenment and, 131;
   Francke and, 176; Halle and, 15, 131, 135–37;
   Lesser and, 176; Nieuwentijt and, 169;
   physico-theology and, 11, 128; radical versus
   mainstream, 20n32; Ray and, 169; Scheuchzer
   and, 169; sensory perception and, 134–35.
   See also Franckesche Stiftungen
"plastic" power of nature, 9, 34, 47, 118, 199
Platonism. See Cambridge Platonists
pleasure (Vergnügen), 56, 61, 133, 209–20, 226
Plinian maxim, 172–73. See also maxima in
   minimis animalibus

Pliny the Elder, 81, 171, 187

Pluche, abbé Noël-Antoine, 183–92; Boerhaave and, 184; Boyle and, 184; Derham and, 189–90; influence of, 4, 28; influences on, 16–17; Leeuwenhoek and, 184–85; Newton and, 184; purpose and, 55–56; Ray and, 184, 189–90; scholarly approach of, 185–86; Voltaire and, 183, 191

pneuma (vital heat), 198

polemics, 160, 165, 167, 190. *See also* irenicism and consensus

Preadamite theory, 199

preformationism/preformism, 17, 59, 196–97

Priestley, Joseph, 33

proof, 73, 117, 120, 142–50, 190

providence, divine, 10, 25, 28, 31, 35, 71–74, 105–9, 117, 119, 122–24, 129–30, 137, 147, 158, 161, 175, 179–80, 201, 222, 224, 230–31. *See also* God

purpose: Aristotle and, 52, 57; Harvey and, 26; internal, 57; natural, 54–59, 75, 81, 119–20; nature and divine, 7; as teleological judgment, 61; theological, and science, 29. *See also* form

Ramus, Petrus, 70

Raven, Charles, 120

Ray, John: anthropocentricism and, 28; Bentley and, 124; Boyle and, 43, 71–72, 74, 117–20, 122–23; Burnet and, 73–75, 224; calamity and, 159–63; Charleton and, 31–32, 36; Copernican astronomy and, 26; Cudworthy and, 34; Derham and, 13, 71–75, 186–87, 230–31; Descartes and, 26, 34, 47; extinct species and, 199; Fabricius and, 93, 96–98, 184, 186; forerunners to, 117–22; form and, 52–53; Galen and, 118; Gessner and, 231; God (*Maximus in minimis*) and, 187; heliocentric cosmos and, 26–27; inferences from nature and, 28–29; influence of, 115–16, 124n2, 124n7; influences on, 47; insects and, 175–76; More and, 15, 47, 120; mountains and, 224–26, 229–30, 232n6; nature, the divine artificer, and, 1; Nieuwentijt and, 36, 186–87; organic science and, 123; "physico-theology" and, 74, 159–62; Pliny and, 187; Pluche and, 189–90; Scheuchzer and, 167–68; scholarly approach of, 186; Scriptural authority and, 13–16; topics of interest to, 49

reading: the book of nature, 117, 131, 134, 136; Cartesian, 85; literal, of the Bible, 79, 83, 86; seeing and, 209–20

reason: aesthetic judgment and, 61; faith and, 146–47; judgment and, 177–78; "meer," 71; miracles and, 168, 202; natural theology and, 1–2; physics and, 167; proof and, 48; revelation and, 48; Scripture and, 150; theology and, 121; truth and, 32. *See also* Enlightenment; Pascal, Blaise

Réaumur, René-Antoine Ferchault de, 16, 176, 179, 184–87

redemption, 23, 33–34; of nature, 191, 211

Redi, Francesco, 85, 186

reproduction, 110–11, 196, 198

resurrection, 9, 47, 165

revelation, 1, 32–33, 45, 80

Richey, Michael, 91

Ries, Franz Ulrich, 92

Rist, Johann, 28

Rollins, Charles, 93

rosary, 108

Rottler, Johann Peter, 135–36

Rousseau, Jean-Jacques, 56–57, 60, 228

Roxburgh, William, 135

salvation, 32

Sbaraglia, Girolamo, 195

Scaliger, Joseph, 81, 85

Scarry, Elaine, 54

Scheuchzer, Johann Jakob, 166–68; argument from design and, 157; Bertrand and, 222, 228–32; Burnet and, 18; diluvialism and, 8, 203, 222; literal biblical interpretation and, 168; Lutheran orthodoxy, 95; Newton and, 167; Scriptural authority and, 16; Solomon's "molten sea" and, 168; Spinozism and, 167; Vallisneri and, 203; Woodward and, 98, 229

Schirach, Adam Gottlob, 127

science: legitimization of, 169; organic, 123, 126n37; physico-theology and, 26–30; Ray and, 116; and religion, 3–4, 23, 29, 104, 159, 183, 189–91

Scripture: authority of, 150–51, 165–67; Copernicus and, 168; literal reading of; science and, 158. *See also* Mosaic physics; Paul (apostle)

Sebond, Raymond, 68

"secular theology," 49, 159

seeds/seed theory, 54–55, 58–60, 110, 120, 181n4, 199

Seidel, Mathaeus, 96

semiotics, 210

Seneca, 81

Shaftesbury, Earl of, 55–56
Sheehan, Jonathan, 12
Simler (also Simmler), Josias, 227
skepticism, 68
Sloane, Hans, 75
solar system, 35. *See also* nebular hypothesis
Solomon's Temple, 80, 82–83, 85–86
Spectator, 184
Spence, William, 179–80
spiders, 97
Spinoza, Baruch: atheism and, 16; Clarke and, 27; Descartes and, 78–80, 158–59; Dutch concern with, 11, 16, 24–25, 78, 81–83; Goeree and, 13, 85, 88; miracles and, 82–83; Nieuwentijt and, 6, 78, 163–66; Scheuchzer and, 167–68
Sprat, Thomas, 46
stability, social and political, 30
Stanislas II August Poniatowski, 229
Starobinski, Jean, 57
Steiger, Johann Anselm, 10–11
Steinmann, Holger, 44
Stengel, Georg, 94
Steno, Nicholas (Niels Steensen), 173, 195
Stillingfleet, Edward, 117–18
Strabo, 194, 201
Straehler, Daniel, 93
Sturm, Johann Christoph, 92, 94–95, 97, 166
Sullivan, Louis, 52, 54, 61
Swammerdam, Jan, 25, 58, 81, 173–76, 186, 196

Tartaglia, Nicolò Fontana, 41
teleology, 54, 60
telescope, 74, 85, 186, 209
terra vergine (primigenia), 198
theism, 10
theology: division of natural, 125n34; "Newtonian" natural, 122–24; physico-, 1–2, 30–34; physico- versus natural, 1–2, 18n3
Theophilus of Antioch, 95
Theophrastus, 225
Thévenot, Melchisedec, 173
Thomson, James, 93
Tillotson, John, 33
Timpler, Clemens, 70–71
Topham, Jonathan, 30
Touber, Jetze, 88
Tranquebar (now Tharangambadi, India), 135
transcendence, 12, 25, 213
transubstantiation, 128
Treu, Catharina, 216–18

Triller, Daniel Wilhelm, 218–19
Trinitarian creed, 33
Trinkle, Dennis, 184
Tuscany, 195
Tyson, Edward, 174

ubiquitarian Christology. *See* Christology, Luther's "ubiquitarian"
Unitarianism, 33
usefulness 16, 29, 55, 68, 118, 134, 149, 183, 190, 202, 222–32; pleasantness and, 103–13, 133

Vadianus (Joachim von Watt), 227
Vallisneri, Antonio, 17, 184, 194–204
Venerable Bede, 92
verba creata, 128
Vermij, Rienk, 13, 16, 78
vernacular languages, 104, 111–12
Verwer, Adriaan, 6, 163–64
Vidal, Fernando, 49
vitalist hypothesis, 198–99
Vives, Juan Luis, 68
Voetius, Gisbertus, 80
Vogel, Jacob, 226
Voltaire (François-Marie Arouet), 11
von Schönau, Johann Heinrich, 95
Vossius, Isaac, 81, 199
"vulgar" opinion or theology, 27, 73, 174, 189, 197

Wallace, Alfred Russel, 35
Waller, Richard, 196
Walpurger, Johann Gottlieb, 8
Webster, John, 199
Wesley, John, 28
Whiston, William, 12, 36–37, 43, 45, 47, 95, 160, 167
White, John Dickson, 3
Wilkins, John, 25, 32, 117–19
Willis, Thomas, 75
Willughby, Francis, 179
wisdom, divine, 7, 23, 34, 53, 59–60, 74, 88, 97, 108, 117, 132, 137, 142, 164, 177–78, 230
Wolff, Christian, 11–12, 19n12, 53–55, 93, 99n15, 132
Woodward, John, 8, 93, 98, 160, 194, 203, 229

Zeuxis, legend of, 218
Zimmermann, Johann Jacob, 45–46, 49
Zöckler, Otto, 4, 18n3, 90
Zwingli, Huldrych, 129